Advances in Industrial Control

Springer
London
Berlin
Heidelberg
New York
Barcelona
Budapest
Hong Kong
Milan
Paris
Santa Clara
Singapore
Tokyo

Other titles published in this Series:

Intelligent Seam Tracking for Robotic Welding
Nitin Nayak and Asok Ray

Identification of Multivariable Industrial Process for Simulation, Diagnosis and Control
Yucai Zhu and Ton Backx

Nonlinear Process Control: Applications of Generic Model Control
Edited by Peter L. Lee

Microcomputer-Based Adaptive Control Applied to Thyristor-Driven D-C Motors
Ulrich Keuchel and Richard M. Stephan

Expert Aided Control System Design
Colin Tebbutt

Modeling and Advanced Control for Process Industries, Applications to Paper Making Processes
Ming Rao, Qijun Xia and Yiquan Ying

Robust Multivariable Flight Control
Richard J. Adams, James M. Buffington, Andrew G. Sparks and Siva S. Banda

Modelling and Simulation of Power Generation Plants
A.W. Ordys, A.W. Pike, M.A. Johnson, R.M. Katebi and M.J. Grimble

Model Predictive Control in the Process Industry
E.F. Camacho and C. Bordons

H_∞ Aerospace Control Design: A VSTOL Flight Application
R.A. Hyde

Neural Network Engineering in Dynamic Control Systems
Edited by Kenneth Hunt, George Irwin and Kevin Warwick

Neuro-Control and its Applications
Sigeru Omatu, Marzuki Khalid and Rubiyah Yusof

Energy Efficient Train Control
P.G. Howlett and P.J. Pudney

Hierarchical Power Systems Control: Its Value in a Changing Industry
Marija D. Ilić and Shell Liu

Steen Tøffner-Clausen

System Identification and Robust Control

A Case Study Approach

With 91 Figures

Springer

Steen Tøffner-Clausen
Department of Control Engineering, Institute of Electronic Systems,
Aalborg University, Fredrik Bajers Vej 7, DK-9220 Aalborg Ø, Denmark

Series Editors

Michael J. Grimble, Professor of Industrial Systems and Director
Michael A. Johnson, Reader in Control Systems and Deputy Director

Industrial Control Centre, Department of Electronic and Electrical Engineering,
Graham Hills Building, 60 George Street, Glasgow G1 1QE, UK

ISBN-13: 978-1-4471-1515-1 e-ISBN-13: 978-1-4471-1513-7
DOI: 10.1007/978-1-4471-1513-7

British Library Cataloguing in Publication Data
A catalogue record for this book is available from the British Library

Softcover reprint of the hardcover 1st edition 1996

Typesetting: Camera ready by author
69/3830-543210 Printed on acid-free paper

To Else and Nikolaj

SERIES EDITORS' FOREWORD

The series Advances in Industrial Control aims to report and encourage technology transfer in control engineering. The rapid development of control technology impacts all areas of the control discipline. New theory, new controllers, actuators, sensors, new industrial processes, computer methods, new applications, new philosophies, ..., new challenges. Much of this development work resides in industrial reports, feasibility study papers and the reports of advanced collaborative projects. The series offers an opportunity for researchers to present an extended exposition of such new work in all aspects of industrial control for wider and rapid dissemination.

The present text Steen Tøffner-Clausen deals with both system identification and robust control. It provides a very comprehensive tutorial introduction to some of the most difficult topics in robust control theory before considering applications problems. Traditional H_∞ robust control design concepts for multivariable systems are first considered and the problems of robust stability and performance are discussed.

The following chapter introduces the idea of the structured singular value and applies this to both analysis and synthesis problems. The author manages to provide a very straightforward introduction to this subject and also introduces some new ideas.

The application of μ-synthesis methods is illustrated in two chapters, the first concerned with the design of a compact disc servo drive and the second with an ASTOVL aircraft design problem. The key to the use of these techniques is to properly describe the uncertainty and to then utilise this in the standard system description to which the new analysis and synthesis tools are applied. These chapters provide a detailed but straightforward introduction to this rather lengthy process which is nevertheless an essential precursor to achieving robust solutions. The examples chosen were based on both real data and applications and they should provide some encouragement to those in industry who wish to apply these valuable and recent design tools.

Robust designs cannot proceed until system models are available and these may be obtained from physical system modelling or from plant identification. Several chapters deal with this latter issue and provide a state-of-the-art overview together with original contributions. For example, a new parameterization of the under modelling and the noise is introduced.

There is an attempt to combine the results of robust control and system identification theory so that an integrated philosophy may be developed. This is a subject which is attracting considerable attention from the research community and the approach described provides one route to obtaining a logical design philosophy in which the identification and robust design techniques are compatible.

An unusual feature of the text is the large number of useful engineering examples which are well chosen to illustrate the analysis and design procedures described. The text should therefore be valuable to the engineer in industry who wishes to explore the potential of these new theoretical control results for real applications.

M.J. Grimble and M.A. Johnson
Industrial Control Centre
Glasgow, Scotland, UK

PREFACE

The idea for this book was conceived when I wrote my PhD thesis [TC95b] concerning practical robust control design based on system identification model building. The last two decades have witnessed an intensive research in so-called ***robust control design*** where stability and performance have been explicitly considered under various perturbations, like noise and model errors, to the closed loop system. The literature on robust control design is extensive and several fundamentally different approaches to the robust control problem have been developed. Covering all aspects of these approaches would be a huge task. In this book, I have concentrated on the "American Approach" which is developed by Doyle, Stein, Morari and many others. When I surveyed the literature on this frequency domain based approach to robust control design during my PhD, I found that it suffered from the following weaknesses:

- Even though there exists many extremely well-written scientific papers on the subject, including some very nice tutorial papers, very few textbooks have been published covering the area. Some of exceptions are the fine books by Morari [MZ89] and Maciejowski [Mac89] and the more recent long announced book by Zhou, Doyle and Glover [ZDG96]. However, these books have a rather theoretical angle of approach and contain only few unrealistically simple examples.
- Most of the literature on the subject do not consider the choice of weighting matrices for the design in any great detail. This is true for both performance weights and, in particular, uncertainty weights. The uncertainty weights reflect the uncertainty in the nominal model. This uncertainty is almost exclusively assumed given a priori. However, it is clear that it is by no means a trivial task to obtain these bounds. Thus, I believe that any book on robust control theory should include the assessment of performance demands and model uncertainty in a more rigorous manner than it is done in most of the available robust control literature.

My main objective has been to fill these gaps by adopting a case-study oriented approach to the robust control problem and to include system identification as a tools for assessment of model quality. The main effort in this book is thus on bridging ***system identification*** and ***robust control.***

An Overview

In the early 1990'ies a growing awareness developed within the automatic control community that the necessary information on model quality needed in robust control could not be supplied by existing system identification methodologies. This spurred a renewed interest in methods for estimating models and model uncertainty from plant measurements. However, no concurrent treatment of the results of this research has been presented in connection with their application in robust control. Such a treatment is a main theme of this book.

The main purpose is thus to develop a coherent system identification based robust control design methodology by combining recent results from system identification and robust control. In order to accomplish this task, new theoretical results will be given in both fields.

Firstly, however, an introduction to modern robust control design analysis and synthesis will be given. It will be shown how the classical frequency domain techniques can be extended to multivariable systems using the singular value decomposition. An introduction to norms and spaces frequently used in modern control theory will be given. However, the main emphasis in this book will *not* be on mathematics. Proofs are given only when they are interesting in their own right and we will try to avoid messy mathematical derivations. Rather we will concentrate on interpretations and practical design issues.

A review of the classical $\mathcal{H}_\infty$ theory will be given. Most of the stability results from modern control theory can be traced back to the multivariable generalization of the famous Nyquist stability criterion. We will shown how the generalized Nyquist criterion is used to establish some of the classical singular value robust stability results. Furthermore, it will be shown how a performance specification may be cast into the same framework. The main limitation in the standard $\mathcal{H}_\infty$ theory is that it can only handle unstructured complex full block perturbations to the nominal plant. However, often much more detailed perturbation models are available, eg from physical modeling. Such structured perturbation models cannot be handled well in the $\mathcal{H}_\infty$ framework.

Fortunately theory exists which *can* do this. The structured singular value μ is an extension of the singular value which explicitly takes into account the structure of the perturbations. In this book, we will present a thorough introduction to the μ framework. A central result is that if performance is measured in terms of the ∞-norm and model uncertainty is bounded in the same manner, then, using μ, it is possible to pose one necessary and sufficient condition for both robust stability and robust performance. The uncertainty is restricted to block-structured norm-bounded perturbations which enter the nominal model in a linear fractional manner. This is, however, a very general perturbation set which includes a large variety of uncertainty such as unstructured and structured dynamic uncertainty (complex perturbations) and parameter variations (real perturbations). The uncertainty structures permitted by μ is definitely much more flexible than those used in $\mathcal{H}_\infty$.

Unfortunately, μ synthesis is a very difficult mathematical problem which is only well developed for purely complex perturbation sets. In order to develop our main result we will unfortunately need to synthesize μ controllers for mixed real and complex perturbation sets. A novel method, denoted *μ-K iteration*, has been developed to solve the mixed μ problem.

A general feature of all robust control design methods is the need for specifying not only a nominal model but also some kind of quantification of the quality of that model. In this book we will restrict ourselves to block-structured norm-bounded perturbations as described above. The specification of the uncertainty is, however, a non-trivial problem which to some extent has been neglected by the theoreticians of robust control. An uncertainty specification has simply been assumed given.

One way of obtaining a perturbation model is by physical modeling. Application of the fundamental laws of thermodynamics, mechanics, physics, etc will generally yield a set of coupled non-linear partial differential equations. These equations can then be linearized (in time and position) around a suitable working point and Laplace transformed for linear control design. The linearized differential equations will typically involve physical quantities like masses, inertias, etc which are only known with a certain degree of accuracy. This will give rise to real scalar perturbations to the nominal model. Furthermore working point deviations may also be addressed with real perturbations.

However, accurate physical modeling may be a complicated and time consuming task even for relatively simple systems. An appealing alternative to physical modeling for assessment of model uncertainty is system identification where input/output measurements are used to estimate a typically linear model of the true system. From the covariance of the parameter estimate, frequency domain uncertainty estimates may be obtained.

In classical (i.e Ljungian) system identification model quality has been assessed under the assumption that the only source of uncertainty is noisy measurements. Thus the structure of the model is assumed to be correct. This is, however, often an inadequate assumption in connection with control design.

Recently, system identification techniques for estimating model uncertainty have gained renewed interest in the automatic control community. In this book, a quick survey of these results will be given together with their Ljungian counterparts. Unlike the classical identification methods these new techniques enables the inclusion of both structural model errors (bias) and noise (variance) in the estimated uncertainty bounds. However, in order to accomplish this, some prior knowledge of the model error must be available. In general, this prior information is non-trivial to obtain. Fortunately, one of these new techniques, denoted *the stochastic embedding approach*, provides the possibility to estimate, given a parametric structure of certain covariance matrices, the required a priori information from the model residuals. Thus the necessary a priori knowledge is reduced from a quantitative measure to a qualitative measure. We believe that this makes the stochastic embedding approach superior to the other new techniques for estimating model uncertainty. In this work, new parameterizations of the undermodeling (bias) and the noise are investigated. Currently, the stochastic embedding approach is only well developed for scalar systems. Thus some work is needed to extend it to multivariable systems. This will, however, be beyond the scope of this book.

Using the stochastic embedding approach it is possible to estimate a nominal model and frequency domain uncertainty ellipses around this model. It will then be shown how these uncertainty ellipses may be represented by a mixed complex and real perturbation set. This is the link needed to combine the results in robust control and system identification into a step-by-step design philosophy for synthesis of robust control systems for scalar plant which is the main new theoretical result presented in this book.

Throughout the book, the presented results will be illustrated by practical design examples. Some of these examples are quite simple but a few are much more complex. The point of view taken is that the theories presented should be applicable to practical control systems design. The given examples thus represent a major part of the work behind this book. They consequently serve not just as illustrations but introduce many new ideas and should be interesting in their own right.

Readership

This book is aimed at post-graduate students and practicing feedback engineers who are well-acquainted with feedback and control systems. The book is *not* written as material for classroom courses. Its main objective is to illustrate how to combine system identification and robust control methods into a coherent design methodology and its application to real feedback systems.

Acknowledgments

Many people have contributed to the enjoyment and productivity of my research at the Department of Control Engineering. I would like to thank all my colleagues at the Department who has contributed or helped in various ways, in particular Palle Andersen for his enthusiastic support and guidance on robust control methodologies, Torben Knudsen and Morten Knudsen for discussions in connection with system identification, Carsten H. Kristensen for providing many hints and macros for LaTeX and last but not least our wonderful secretaries Karen Drescher and Jette Damkjær for invaluable general assistance.

Also I would like to thank Jakob Stoustrup and Hans H. Niemann from the Technical University of Denmark for their collaboration on mixed μ synthesis. They have provided much insight in μ methods.

A sincere thank goes to Professor Michael Grimble and the staff at the Industrial Control Center, University of Strathclyde in Glasgow for inviting me to stay with them for a three months period in the autumn 1994. My visit there was a very pleasant and rewarding one. Especially I would like to thank Stephen Breslin for being a good friend as well as a fine colleague. The flight control case study in Chapter 7 is the result of a joint effort by Stephen and myself.

Mostly, however, I am indebted to my wife Else and my son Nikolaj. Without their understanding, support and love this book would never have been written.

Aalborg, May 1996. Steen Tøffner-Clausen.

CONTENTS

Part II. System Identification and Estimation of Model Error Bounds

Part IV. Conclusions

Part V. Appendices

NOMENCLATURE

Part I, Robust Control – Theory and Design

Symbols

Symbol	Denotes
s/z	S/Z transform complex variable.
ω	Frequency [rad/sec].
$d(s), \delta(s)$	Disturbance.
$d'(s), \delta'(s)$	Normalized disturbance ($d(s) = W_{\mathrm{p1}}(s)d'(s)$).
$e(s)$	Control error ($e(s) = r(s) - y(s)$).
$e_0(s)$	Measured control error ($e_0(s) = r(s) - y_{\mathrm{m}}(s)$).
$e'(s)$	Weighted control error ($e'(s) = W_{\mathrm{p2}}(s)e(s)$).
$m(s)$	Model output ($m(s) = G(s)u(s)$).
$n(s)$	Sensor noise.
$r(s)$	Reference.
$u(s)$	Input.
$w(s)$	Perturbation.
$y(s)$	Output.
$y_{\mathrm{m}}(s)$	Measured output ($y_{\mathrm{m}}(s) = y(s) + n(s)$).
$z(s)$	Input to perturbation structure ($w(s) = \Delta(s)z(s)$).
$D(\omega), G(\omega)$	Scaling matrices for computation of μ upper bound.
$D(s)$	Scaling used in *D-K* iteration.
$G(s), G_{\mathrm{h}}(s)$	Additional scalings used in *D,G-K* iteration.
$\Gamma(s), \gamma(s)$	Additional scalings used in μ-*K* iteration.
$G(s)$	Plant transfer function matrix.
$G_{\mathrm{T}}(s)$	Perturbed plant transfer function matrix.
$J(s)$	DGKF parameterization.
$K(s)$	Controller transfer function matrix.
$M(s)$	Control sensitivity.
$N(s)$	Generalized plant.
$N_{\mathrm{D}}(s)$	Augmented generalized plant for *D-K* iteration. $N_{\mathrm{D}} = DND^{-1}$.
	continued on next page

continued from previous page	
Symbol	Denotes
$N_{\mathrm{DG}}(s)$	Augmented generalized plant for D,G-K iteration. $N_{\mathrm{DG}} = (DND^{-1} - \beta^{\star}G)G_{\mathrm{h}}$.
$N_{\mathrm{D\Gamma}}(s)$	Augmented generalized plant for μ-K iteration. $N_{\mathrm{D\Gamma}} = \Gamma DND^{-1}$.
$P(s)$	Generalized closed loop system.
$Q(s)$	Free stable transfer matrix used in DGKF parameterization.
$S_{\mathrm{o}}(s)/S_{\mathrm{i}}(s)$	Sensitivity evaluated at plant output and input.
$T_{\mathrm{o}}(s)/T_{\mathrm{i}}(s)$	Complementary sensitivity evaluated at plant output and input.
$W_{\mathrm{p1}}(s)$	Disturbance weight.
$W_{\mathrm{p2}}(s)$	Control error weight.
$W_{\mathrm{u1}}(s)$	Perturbation input weight.
$W_{\mathrm{u2}}(s)$	Perturbation output weight.
$\tilde{\Delta}(s)$	Perturbation, $\bar{\sigma}(\Delta) \leq \ell$.
$\Delta(s)$	Normalized perturbation, $\bar{\sigma}(\Delta) \leq 1$.
$\Delta_{\mathrm{p}}(s)$	Performance block.
$F_{\ell}(N(s), K(s))$	Lower LFT, $F_{\ell}(N(s), K(s)) = N_{11} + N_{12}K(I - N_{21}K)^{-1}N_{21}$.
$F_{\mathrm{u}}(P(s), \Delta(s))$	Upper LFT, $F_{\mathrm{u}}(P(s), \Delta(s)) = P_{22} + P_{21}\Delta(I - P_{11}\Delta)^{-1}P_{12}$.
A, B, C, D	State space matrices.
T_{s}	Sampling time [secs].
j	$\sqrt{-1}$, sometimes an index as in x_{ij}.
I_n	$n \times n$ identity matrix.
$\mathcal{J}_{\mathrm{np}}$	Nominal performance cost function.
$\mathcal{J}_{\mathrm{u}}$	Robust stability cost function.
$\mathcal{J}_{\mathrm{rp}}$	Robust performance cost function.
$\Delta\,\mathrm{arg}$	Change in argument as s traverses the Nyquist $\mathcal{D}$ contour.
$\det(A)$	Determinant of complex matrix A.
A^T	Transpose of A.
A^*	Complex conjugate transpose of A.
A_{ij}	The (i,j) element of A.
$\mathrm{tr}\,\{A\}$	Trace of A.
$\lambda_i(A)$	The i'th eigenvalue of A.
λ_{R_i}	The i'th real eigenvalue of A.
$\rho(A)$	The spectral radius of A.
$\rho_{\mathrm{R}}(A)$	The real spectral radius of A.
$\mu_{\mathbf{\Delta}}(A)$	The structured singular value of A.
$\bar{\mu}_{\mathbf{\Delta}}(A)$	Upper bound for $\mu_{\mathbf{\Delta}}(A)$.
	continued on next page

continued from previous page	
Symbol	Denotes
$\mathbf{Ric}(H)$	The Ricatti solution.

Sets, Norms and Spaces

Symbol	Denotes
$\|\cdot\|$	Norm.
$< \cdot >$	Scalar product.
$A = Y\Sigma U^*$	Singular value decomposition of A.
$\bar{\sigma}(A)$	Maximum singular value of A, $\bar{\sigma}(A) = \sigma_1(A) = \|A\|_2$.
$\underline{\sigma}(A)$	Minimum singular value of A, $\underline{\sigma}(A) = \sigma_k(A)$.
$\sigma_i(A)$	The i'th singular value of A.
$\kappa(A)$	Condition number of A, $\kappa(A) = \bar{\sigma}(A)/\underline{\sigma}(A)$.
$\|G(s)\|_{\mathcal{H}_2}$	Transfer function 2-norm, $\|G(s)\|_{\mathcal{H}_2} = \sqrt{\frac{1}{2\pi}\int_{-\infty}^{\infty} \mathrm{tr}\left\{G(j\omega)^H G(j\omega)\right\} d\omega}$.
$\|G(s)\|_{\mathcal{H}_\infty}$	Transfer function ∞-norm, $\|G(s)\|_{\mathcal{H}_\infty} = \sup_\omega \bar{\sigma}(G(j\omega))$.
$\mathbf{K}$	Field of real or complex numbers.
H	A linear space.
$\mathbf{C}^{n\times m}$	Set of complex $n \times m$ matrices.
$\mathbf{R}^{n\times m}$	Set of real $n \times m$ matrices.
$\mathcal{L}$	Lebesgue spaces.
$\mathcal{H}$	Hardy spaces.
$\mathcal{K}_{\mathrm{S}}$	Set of all stabilizing controllers.
$\mathcal{D}'$	Set of normalized generic disturbances, $\|\delta'\|_2 \le 1$.
$\mathcal{D}$	Set of generic disturbances, $\delta(s) = W_{\mathrm{p}1}(s)\delta'(s)$.
$\mathbf{\Delta}$	Block diagonal perturbation structure used with μ.
$\mathbf{\Delta}_{\mathrm{c}}$	Corresponding complex perturbation set.
$\mathbf{B\Delta}$	Bounded subset of perturbations, $\mathbf{B\Delta} = \{\Delta(s) \in \mathbf{\Delta} \| \bar{\sigma}(\Delta(j\omega)) < 1\}$.
$\mathbf{Q}, \mathbf{D}, \mathbf{G}, \hat{\mathbf{D}}, \hat{\mathbf{G}}$	Sets of scaling matrices used for μ upper and lower bounds.

Abbreviations

Abbreviation	Denotes
ASTOVL	Advanced short take-off and vertical landing.
CD	Compact disc.
DGKF	Doyle, Glover, Khargonekhar and Francis $\mathcal{H}_\infty$ parameterization.
	continued on next page

continued from previous page	
Symbol	Denotes
FDLTI	Finite dimensional linear time invariant.
LFT	Linear fractional transformation.
LHP/RHP	Left/right half plane.
LMI	Linear matrix inequality.
LQ	Linear quadratic.
LQG	Linear quadratic gaussian.
LTR	Loop transfer recovery.
MIMO	Multiple input multiple output.
SISO	Single input single output.
SSV	Structured singular value.

Part II, System Identification and Estimation of Model Error Bounds

Symbols

Symbol	Denotes
$\mathcal{S}$	Observed system.
$E\{\cdot\}$	Expectation.
$\bar{E}\{\cdot\}$	Statistical expectation.
$\mathcal{N}$	The normal distribution.
χ^2	The chi-square distribution.
Cov	Covariance.
q	Shift operator.
s, z	S/Z transform complex variable.
ω	Frequency [rad/sec].
μ	Mean value.
σ	Standard deviation.
H	Hessian matrix.
M	Fisher Information matrix.
T_s	Sampling time [sec].
j	$\sqrt{-1}$, sometimes an index like in x_{ij}.
I_N	The N dimensional identity matrix.
p	Number of model parameters.
θ	Parameter vector.
$\hat{\theta}_N$	N measurements parameter estimate.
$\hat{\theta}^*$	∞ measurements parameter estimate.
θ_0	True parameter vector.
N	Number of measurements.
L	Length of FIR model for the undermodeling.

continued on next page

continued from previous page

Symbol	Denotes
$G_T(z)$	True system discrete-time transfer function.
$G_\Delta(z)$	Undermodeling discrete-time transfer function.
$G(z, \hat{\theta}_N)$	N measurements model estimate.
$H(z)$	Noise discrete-time transfer function filter.
A, B, F, C, D	Black box model polynominals.
$y(k)$	Output.
$\hat{y}(k\|\theta)$	Predicted output.
$u(k)$	Input.
$e(k)$	Zero mean iid stochastic process.
λ	Variance of $e(k)$.
$\nu(k)$	Process noise.
$\delta(k)$	Undermodeling output.
$\eta(k)$	Undermodeling impulse response.
η	Undermodeling impulse response vector.
$h_\nu(k)$	Process noise impulse response.
f_Δ	Probability density for the undermodeling $\delta(k)$.
f_ν	Probability density for the process noise $\nu(k)$.
β	Parameter vector for the undermodeling impulse response covariance matrix C_η.
γ	Parameter vector for the process noise covariance matrix C_ν.
ζ	Combined parameter vector $[\beta \;\; \gamma]^T$.
Π	Undermodeling FIR regressor matrix.
ξ	Undermodeling filter $\Pi(q^{-1})u(k)$.
X	Undermodeling filter matrix.
Λ_k	kth fixed denominator model regressor.
Λ	Fixed denominator model regressor matrix.
ϕ	Predictor state vector, $\hat{y}(k\|\theta) = \phi(k, \theta)\theta$.
Φ	Predictor filter matrix.
$\psi(k, \theta)$	Model gradient $\partial\hat{y}(k\|\theta)/\partial\theta$.
Ψ	Model gradient filter matrix.
Y	Output vector.
V	Process noise vector.
$V_N(\theta, Z^N)$	Performance function for estimation problem.
$\bar{V}(\theta)$	Limit function for $V_N(\theta, Z^N)$.
$\bar{V}'(\theta)$	Derivative of $V_N(\theta, Z^N)$ with respect to θ.
Q	Variance of $\sqrt{N}V_N'(\theta, Z^N)$.
Ω	$(\Phi^T\Phi)^{-1}\Phi$.
P_θ	Parameter vector covariance matrix.
$\hat{P}_N$	Data estimate of P_θ.

continued on next page

continued from previous page	
Symbol	Denotes
C_η	Covariance matrix for the undermodeling impulse response.
C_ν	Covariance matrix for the process noise.
a, c, σ_e^2	Noise covariance parameters.
α, λ	Undermodeling covariance parameters.
$\hat{\rho}_N$	$\begin{bmatrix} \hat{\theta}_N & 0 \end{bmatrix}$
ρ_0	$\begin{bmatrix} \theta_0 & \eta \end{bmatrix}$
$\tilde{g}(z)$	Frequency domain model uncertainty.
$\Gamma(z)$	Uncertainty filter, $\tilde{g}(z) = \Gamma(z)(\theta^* - \hat{\theta}_N)$.
$P_{\tilde{g}}$	Covariance matrix for the total model error.
Υ	$E\left\{(\rho_0 - \hat{\rho}_N)(\rho_0 - \hat{\rho}_N)^T\right\}$.
$\epsilon(k, \theta)$	Prediction errors.
$\hat{R}_N^\epsilon(\kappa)$	Data estimate of residual covariance.
$\hat{R}_N^{\epsilon,u}(\kappa)$	Data estimate of residual/input cross-covariance.
ϖ	Transformed residuals vector.
R	Transformation matrix.
P_Φ	$[I - \Phi(\Phi^T\Phi)^{-1}\Phi^T]$.
$\mathcal{L}(\varpi \| U, \zeta)$	The likelihood of ϖ given U and ζ.
$\ell(\varpi \| U, \zeta)$	The loglikelihood of ϖ given U and ζ.
$f(\varpi \| U, \zeta)$	The probability density function for ϖ given U and ζ.
Σ	The covariance for the residuals ϖ.
$d_n(\mathcal{S}, \mathcal{B})$	n-width.
$\delta_{\alpha\beta}$	Kronecker delta.
$L_i(z, a)$	Laguerre filters.
$\Psi_i(z, b, c)$	Kautz filters.

Abbreviations

Abbreviation	Denotes
AR	Auto-regressive.
ARX	Auto-regressive with exogenous input.
FDLTI	Finite dimensional linear time invariant.
FIR	Finite impulse response.
LHP/RHP	Left/right half plane.
MIMO	Multiple input multiple output.
OE	Output error.
PEM	Prediction error methods.
SISO	Single input single output.

CHAPTER 1
INTRODUCTION

Control theory within the 20th century can conveniently be divided into 3 main periods: classical control, optimal control and robust control. In Table 1.1 some of the key results for each period is given. The period 1930–1960 can be classified as the "Classical Control" period. Here famous pioneers, like Bode and Nyquist, developed control design tools which made it possible to meet standard requirements on stability, robustness and performance for scalar (single-input single-output) systems. These are eg Bode and Nyquist plots, Nichols charts and root-locus plots. The chief paradigm for this period was the frequency domain. The results from the classical control period make up the bulk of industrial control practice to this day.

Table 1.1. *Control theory in this century [Dai90].*

	Classical Control 1930–1960	Optimal Control 1960–1980	Robust Control 1980–present
Analysis	Nyquist plots Bode plots Root-locus Gain/phase margins	State space models Controllability Observability Random processes	Singular value Bode plots μ analysis Balanced realizations Spectral factorizations
Synthesis	PID controllers Lead-lag compensation	LQ state feedback Kalman-Bucy filters LQG control	$\mathcal{H}_\infty$ synthesis μ synthesis LQG/LTR control Youla parameterizations
Chief paradigm	Frequency domain	Time domain	Frequency domain

The period from 1960–1980 can be termed "Optimal Control". In the early 1960'ies Kalman and others introduced concepts like controllability, observability, optimal state estimation and optimal state feedback. Concepts which were based on state space matrix equations rather than frequency domain transfer functions. It was shown that using the state space approach one could formulate sensible performance measures and optimize the control scheme with respect to these measures not only for scalar systems but also for multivariable systems. When Kalman in a series of papers, see eg [Kal60, KHN62, Kal64], provided a state-space solution to the *Linear Quadratic Gaussian* (LQG) optimal control problem, it almost created euphoria within the automatic control community. A multivariable control design method was discovered which was *optimal* for the given design criteria. Furthermore, it was possible to show that the full state feedback Linear Quadratic (LQ) optimal control law possessed some very strong robustness properties, namely at least 60° phase margin and infinite gain margin *regardless of the choice of performance weights*. Because of duality, the optimal state estimator had similar fine properties.

These and related results shifted the emphasis in control engineering from the frequency domain to the time domain and for 20 years the frequency domain approach was more or less considered to be obsolete.

Unfortunately, it turned out that LQG controllers did not always work well in practice. In fact, poor stability and, in particular, very poor performance was not an uncommon phenomena quite opposed to the theoretical results. However, these practical problem were to a large extent ignored by the theoreticians of automatic control and this created a gap between practitioners and theoreticians in control engineering.

Then in 1979 Doyle & Stein showed in their famous paper [DS79] that even though the control law and the state estimation law are optimal, there is no guarantee that the *combined* LQG scheme possesses equivalent fine properties. Thus the problems encountered by control engineers with LQG in practice had a firm theoretical explanation. Doyle & Stein used standard frequency domain methods to prove their point. The chief paradigm for the automatic control community was beginning to shift once again.

Another key paper from that period is the frequently quoted paper [DS81] also by Doyle & Stein. In this paper singular value sensitivity plots and norm bounded uncertainty descriptions was introduced. The similarities between (multivariable) singular value analysis and classical (scalar) frequency domain analysis was emphasized. The singular value plots represented the extension of the classical scalar Bode plots to multivariable systems. They are thus also frequently termed *singular value Bode plots*.

In both papers [DS79, DS81] a method for regaining the fine loop properties of the LQ state feedback from the LQG implementation was presented. This method, known as *loop transfer recovery* (LTR), works by modifying the intensity of the process noise in the Kalman-Bucy filter. Thus, even though that the main emphasis was on the frequency domain, state space (or time domain) algorithms were used to compute the controller. The papers by Doyle & Stein amongst many others introduced a new period in control engineering. The mixture of frequency domain analysis and state space formulations became symptomatic for the period from 1980 until today. This period is usually termed "Robust Control".

The next decade witnessed an intensive research in robust control designs. The term robust stressed the importance of developing methods which maintained closed loop stability and performance not only for the nominal model of the control object (plant) but also for a set of plants including the invariable discrepancy between the nominal model and the true plant. The robust control paradigm in fact represented a return to a more physically motivated basis for control design. In order to pose a meaningful problem, the model uncertainty or perturbations are assumed to be norm bounded. For example, a description of the true multivariable plant $G_{\mathrm{T}}(s)$ could be

$$G_{\mathrm{T}}(s) = (I + \Delta(s))\, G(s) \tag{1.1}$$

where $G(s)$ is the nominal transfer function matrix and $\Delta(s)$ a multiplicative perturbation at the plant output. $\Delta(s)$ could then be bounded in magnitude as

$$\|\Delta(j\omega)\|_2 = \bar{\sigma}(\Delta(j\omega)) < \ell(\omega) \tag{1.2}$$

where $\bar{\sigma}$ denotes the *maximum singular value*, *induced 2-norm* or *spectral norm* and $\ell(\omega)$ is a frequency dependent scalar. The choice of bounding norm is a compromise between those that best describe the plant perturbations and those that lead to mathematically tractable problems. The spectral norm has been a popular compromise since it describes well the effects of high frequency unmodeled dynamics, non-linearities and time delays and a solution to the corresponding control problem has been found. Using the celebrated *Small Gain Theorem* it is e.g. possible to show that for the perturbation structure in (1.1) with $\Delta(s)$ bounded by (1.2), the stability of the closed loop system is maintained if and only if

$$\bar{\sigma}(T(j\omega)) = \bar{\sigma}\left(G(j\omega)K(j\omega)(I + G(j\omega)K(j\omega))^{-1}\right) \leq \ell(\omega)\,. \tag{1.3}$$

If the process noise is assumed to be bounded by the same induced norm a unifying framework can be used for considering both stability and disturbance rejection.

Much of the literature on robust control have focused strongly on the *computational engine* of robust control, namely the $\mathcal{H}_\infty$ optimization algorithm. This has, unfortunately, in some sense hampered the application of robust control in practice since many practical issues have been left unattended. In fact, it turns out that the standard state space $\mathcal{H}_\infty$ framework is not very suitable for formulating practical control problems since the perturbation structures permitted in $\mathcal{H}_\infty$ is much too restricted to deal with many practical specifications. A much more general but less well known design framework is based on the *structured singular value* (μ) theory. μ theory allows the inclusion of much more natural and specialized perturbation models and it furthermore nicely separates stability and performance issues, a result which can not be obtained in the $\mathcal{H}_\infty$ framework. Interestingly, the computational engine in μ control synthesis is still the $\mathcal{H}_\infty$ optimization algorithm.

Unfortunately, even though μ theory is comprehensible straightforward with simple and insightful results, it has not been as widely recognized as $\mathcal{H}_\infty$ theory. There are several reasons for this. First of all, the practical computation of the structured singular value μ is a very difficult and in fact, yet unsolved mathematical problem. Fortunately, tight upper and lower bounds on μ may be effectively computed. Furthermore, algorithms for computing these bounds have become commercially available with the release of the MATLAB *μ-Analysis and Synthesis Toolbox* [BDG+93]. Thus for practical controller design, the mathematical complexity of μ is no longer such a critical issue. Another reason for μ theory not being more widespread is the limited amount of literature on μ. Usually the simplicity of the theory is hidden in messy mathematical expressions connected with the computation of μ.

Since μ cannot be directly computed unless in some special cases, it is clear that controller synthesis using μ must be problematic as well. The approach usually taken is to pose an upper bound problem which is solved iteratively through a series of $\mathcal{H}_\infty$ optimizations and μ upper bound computations. This scheme is usually denoted *D-K iteration.* However, *D-K* iteration is only well developed for purely complex perturbations. Many practical control problems unfortunately call for the use of both real and complex perturbations. Thus a method for μ synthesis for mixed real and complex perturbation sets is needed.

One objective of this book is to provide a thorough introduction to robust control design with special attention to μ methods. The treatment will not focus excessively on finer mathematical points but rather it will address practical design issues. A main result is a new algorithm, denoted *μ-K iteration*, which solves the mixed μ problem.

Modern robust control synthesis techniques aim at providing robustness with respect to both plant perturbations (robust stability) and additive disturbances (robust performance). It is assumed that information of noise and perturbations is available. One way of acquiring such information is by physical modeling. However, this is usually a quite complex task unless extremely simple systems are considered. A different approach could be by system identification. Unfortunately, classical identification methods assume that all uncertainty is in the form of additive noise. There is thus a gap between robust control and system identification. This gap has hampered the application of robust control methods to practical problems.

The gap between robust control methods and system identification results was realized by a number of researchers around 1990, see eg [GS89a, WL90b, Gev91]. The development of formal techniques for estimation of model uncertainty has since been the focus of active research and numerous results have been published, see [Bai92] and references therein.

In this book a discussion of these recent results will be provided. Unfortunately the estimation of model error bounds from finite noisy data becomes a very difficult problem when the true system cannot be represented within the model set. In fact, it is well-known that a priori knowledge of the noise and unmodeled dynamics is necessary to compute such bounds. Most of the attempts on estimation of model error bounds have relied upon hard bound prior knowledge of the noise and unmodeled dynamics to compute hard uncertainty bounds on the nominal model. However, the necessary a priori knowledge of the unmodeled dynamics seems very difficult to obtain in practice. Here we will consider some recent results on estimation of frequency domain model error bounds where the model bias is embedded in a stochastic framework. This approach is thus denoted the ***stochastic embedding approach.*** A central feature of the stochastic embedding approach is that the necessary a priori knowledge can be estimated from the model residuals.

The main point of this book is then that the results on the stochastic embedding approach for uncertainty estimation can be combined with the derived method for mixed real and complex μ synthesis to constitute a coherent design approach for linear systems control design. A step-by-step design method is proposed. Currently only scalar control problems can be considered since the stochastic embedding approach has been developed for scalar systems only. Future research will have to reveal whether an extension to multivariable systems is possible.

1.1 The Organization of the Book

The material in this book is split into 4 parts. Part I considers robust control, Part II system identification and estimation of model uncertainty and Part III the combination of the results given in the previous parts into a coherent design philosophy. Finally in Part IV the major results are reviewed and discussed.

1.1.1 Part I, Robust Control – Theory and Design

The purpose of this part is to formulate the necessary framework for the control analysis and synthesis part of the proposed synergistic control design methodology. First, in Chapter 3, an introduction to norms and spaces used in robust control design is given. Then, in Chapter 4, we will review "classical" singular value $\mathcal{H}_\infty$ control design. In $\mathcal{H}_\infty$ control it is assumed that the model uncertainty can be described by a single complex perturbation block. However, usually much finer perturbation structures are available including both real and complex blocks. Such uncertainty structures can only be handled conservatively in an $\mathcal{H}_\infty$ framework. Furthermore we will show how the *robust performance* problem, i.e guaranteed performance in presence of uncertainty, also cannot be non-conservatively addressed using an $\mathcal{H}_\infty$ approach *even if the perturbation can be described by a single complex block.*

This motivates us then, in Chapter 5, to consider the structured singular value μ. Using μ we may develop a robust control framework where real and complex perturbation blocks which enter the nominal system in a linear fractional manner can be addressed without conservatism. A very large class of perturbation models may be cast into this structure. Furthermore, the robust performance problem can be handled non-conservatively. We will start by presenting the "classical" results on μ. As discussed earlier, only upper and lower bounds on μ can be computed unless for some special very simple cases where μ can be exactly computed. Even though these bounds are usually quite tight at least for purely complex perturbation sets, the control synthesis problem becomes very difficult. The usual approach for complex perturbation sets is the iterative scheme *D-K iteration* which generally seems to work well in practice although convergence cannot be guaranteed.

Unfortunately, in order to develop our central result we will need a control synthesis procedure which can be applied for mixed real and complex perturbation sets. Some of the first results on mixed μ synthesis was given in the thesis by Peter Young [You93]. Young proposes an iterative scheme denoted *D,G-K iteration* which solves the mixed μ problem. Unfortunately the approach is much more involved that the corresponding D-K iteration for purely complex blocks. We will present a different approach, denoted *μ-K iteration* which sacrifices some of the convergence properties that can be obtained with D,G-K iteration. In return, μ-K iteration can be performed more easily. Furthermore, the iteration seems to work quite well in practice.

The μ-K iteration procedure is the central result of Part I.

Finally, we will present two design examples. In Chapter 6 mixed μ control of a CD servo drive is considered. The second design in Chapter 7 is concerned with control of a advanced short take off and vertical landing aircraft.

1.1.2 Part II, System Identification and Estimation of Model Error Bounds

The main purpose of this part is to investigate methods for quantification of frequency domain model uncertainty. First, in Chapter 9, we will review the "classical" results. In Chapter 10 we will consider some special cases, in particular fixed denominator models with basis functions which are orthonormal in the space of strictly proper stable transfer functions. The classical results can be applied in the case where the true system can be represented within the model set. This is however often an inadequate assumption in connection with robust control. Thus we will next consider methodologies where this is not a requirement. Specifically, in Chapter 11, we will consider the stochastic embedding approach where the undermodeling is embedded in a stochastic framework. It is then possible to extend the classical results to the case where undermodeling is present. In order to evaluate the frequency domain uncertainty estimates, the covariance matrices for the noise and the undermodeling must be known. It can be shown that if it is assumed that both the noise and the undermodeling impulse response are Gaussian and the covariance matrices are parameterized, the parameters themselves can be estimated through maximum likelihood techniques. The main result of Part II is new parameterizations for these covariance matrices.

Finally, in Chapter 12 we will present an example where different parameterizations of the undermodeling have been compared.

1.1.3 Part III, A Synergistic Control Systems Design Philosophy

In this part a coherent identification based robust control design philosophy is developed for scalar systems. In Chapter 13 it is shown how estimated frequency domain uncertainty ellipses may be represented with a mixed real and complex perturbation set. Then the proposed μ-K iteration procedure can be applied to solve the corresponding robust control problem. A step-by-step design procedure for scalar systems is outlined.

To illustrate the design methodology, an example is given in Chapter 14. This example considers control of a small domestic water supply system where the control problem is to maintain water pressure for flow disturbances.

1.1.4 Part VI, Conclusions

In Chapter 15 the main results of the book are summarized and directions for further research are given.

PART I

ROBUST CONTROL - THEORY AND DESIGN

CHAPTER 2
INTRODUCTION TO ROBUST CONTROL

Design of controllers with guaranteed closed loop stability and performance for uncertain plants have been the focus of active research for almost 2 decades now. Most of the research on robust control has focused on $\mathcal{H}_\infty$ like problems. However, it turns out that many practical problems do not readily fit the standard $\mathcal{H}_\infty$ problem setup since the involved model uncertainty is structured rather than unstructured. An $\mathcal{H}_\infty$ control design will then be potentially conservative and thus will limit the obtainable performance of the closed loop system.

2.1 μ Theory

Fortunately theory exists that non-conservatively handles these problems, namely the ***structured singular value*** (SSV) or μ theory. In many practical applications μ theory is more appropriate for system analysis and controller synthesis. μ theory has not been as widely recognized as $\mathcal{H}_\infty$ theory probably due to the small amount of literature on μ and to the computational difficulties associated with μ. Recently, however, algorithms for computing μ[1] have become commercially available through the MATLAB *μ-Analysis and Synthesis Toolbox* [BDG+93]. Also the literature on μ is improving, see eg the excellent introduction to μ analysis by Holohan [Hol94].

2.1.1 μ Synthesis

An approach to controller synthesis using μ for complex perturbations, frequently denoted *D-K iteration*, has been known for some time now [DC85] and controller synthesis for structured complex perturbation sets can be accomplished with the aid of the MATLAB μ toolbox.

Unfortunately, many practical application problem calls for the use of mixed real and complex perturbation sets. For example, analysis of plant parameter variations which is an often encountered problem rely on the use of mixed or even purely real perturbation sets. Until recently controller synthesis under mixed perturbation sets was an unsolved problem. A solution to this

[1] More accurately: upper and lower bounds on μ.

problem was probably first given by Young [You93, You94]. Unfortunately, the synthesis procedure proposed by Young is quite involved. Even though it relies on the same principles it is certainly more mathematically complex than *D-K* iteration used for purely complex perturbation sets.

The main emphasis in this part of the book will be on μ analysis and synthesis. However, in order to put things into perspective a survey of robust control design methods using the $\mathcal{H}_\infty$ approach will be given initially. The role of singular values will be discussed as will some of the commonly used singular value bounded unstructured uncertainty descriptions. The potential conservatism inherent in an unstructured uncertainty assumption will be emphasized. It will be shown how the nominal performance problem and the robust stability problem can be addressed with $\mathcal{H}_\infty$ methods. However, it turns out that the robust performance problem cannot be non-conservatively formulated in an $\mathcal{H}_\infty$ framework. Then it will be shown how the structured singular value may be used to overcome the shortcomings of the $\mathcal{H}_\infty$ approach. Controller design using μ will then be presented both for purely complex and for mixed real and complex perturbation structures. A novel approach to mixed μ synthesis, denoted μ-K iteration, will be presented. The μ-K iteration procedure is computationally simpler than the procedure proposed by Young. Furthermore, as we shall show, μ-K iteration seems to work quite well in practice.

2.2 An Overview

The remainder of this part of the book is organized as follows. In Chapter 3 an introduction to spaces and norms frequently used in modern robust control is given. In Chapter 4 an introduction to "classical" singular value based $\mathcal{H}_\infty$ robust control analysis and synthesis is given. In Chapter 5 robust stability and performance will be addressed using the structured singular value μ. Various μ synthesis techniques will be presented and discussed. In Chapter 6 a mixed μ optimal controller will be designed for a simple model of a CD player servo using μ-K iteration. Finally, in Chapter 7, a major flight control case study will be given. The control of an **A**dvanced **S**hort **T**ake-**O**ff and **V**ertical **L**anding (ASTOVL) aircraft will be considered at low speeds in the transition zone between jet-borne and fully wing-borne flight. In this flight condition the aircraft is unstable in the longitudinal axis and the nominal model is very ill-conditioned especially at low frequencies.

CHAPTER 3

SPACES AND NORMS IN ROBUST CONTROL THEORY

In order to fully comprehend and appreciate modern robust control theory some mathematical prerequisites from functional analysis are necessary. In many fine textbooks on robust control this prior knowledge is either assumed or discussed only very briefly. However, we believe that an introduction to the relevant spaces and norms used in modern control theory is mandatory. Conversely, if the reader is completely familiar with functional analysis he may skip this chapter. As it turns out, the necessary mathematical tools are conceptually quite straightforward even though some are computationally involved. The results presented here are taken mainly from [Fra87, Dai90, DFT92] and are all given without proof.

3.1 Normed Spaces

Let H be a *linear space* over the *field* $\mathbf{K}$ where $\mathbf{K}$ is either the field $\mathbf{C}$ of complex numbers or the field $\mathbf{R}$ of real numbers. A *norm* on H is a function denoted $\|\cdot\|$ from H to $\mathbf{R}$ having the following properties of

$$\text{(i)} \qquad \|f\| \geq 0\,. \tag{3.1}$$

$$\text{(ii)} \qquad \|f\| = 0\,, \qquad \text{iff } f = 0. \tag{3.2}$$

$$\text{(iii)} \qquad \|\alpha f\| = |\alpha|\|f\|\,. \tag{3.3}$$

$$\text{(iv)} \qquad \|f+g\| \leq \|f\| + \|g\|\,, \qquad \text{(triangle inequality)}. \tag{3.4}$$

where $f, g \in \mathsf{H}$ and $\alpha \in \mathbf{K}$. Thus a norm is a single real number measuring the "size" of an element of H. Given a linear space H there may be many possible norms on H. Given a linear space H and a norm $\|\cdot\|$ on H, the pair $(\mathsf{H}, \|\cdot\|)$ is denoted a *normed space*.

3.2 Vector and Matrix Norms

Let H be the space $\mathbf{C}^n$ which is a linear space. Then $x \in \mathbf{C}^n$ means that $x = (x_1, x_2, \cdots, x_n)$ with $x_i \in \mathbf{C}, \forall i$. Clearly, $\mathbf{C}^n$ is the space of complex n-dimensional vectors. For $x \in \mathbf{C}^n$ the Hölder or p-norms are defined by:

$$\|x\|_p = \left(\sum_{i=1}^{n} |x_i|^p\right)^{1/p} . \tag{3.5}$$

In control theory the 1-, 2- and ∞-norm are most important since they have obvious physical interpretations:

$$\|x\|_1 = \sum_{i=1}^{n} |x_i| , \qquad \text{resource.} \tag{3.6}$$

$$\|x\|_2 = \sqrt{\sum_{i=1}^{n} |x_i|^2} = \sqrt{x^* x} , \qquad \text{energy} \tag{3.7}$$

$$\|x\|_\infty = \max_i |x_i| , \qquad \text{peak.} \tag{3.8}$$

In (3.7), x^* denotes the complex conjugate transpose of x. Notice that the 2-norm $\|x\|_2$ is the usual Euclidean length of the complex vector x. All norms on $\mathbf{C}^n$ are *equivalent norms* which means that if $\|\cdot\|_\alpha$ and $\|\cdot\|_\beta$ are norms on $\mathbf{C}^n$, then there exists a pair $c_1, c_2 > 0$ so that

$$c_1\|x\|_\alpha \leq \|x\|_\beta \leq c_2\|x\|_\alpha , \qquad \forall x \in \mathbf{C}^n . \tag{3.9}$$

In particular, $\forall x \in \mathbf{C}^n$:

$$\|x\|_2 \leq \|x\|_1 \leq \sqrt{n}\|x\|_2 \tag{3.10}$$

$$\|x\|_\infty \leq \|x\|_2 \leq \sqrt{n}\|x\|_\infty \tag{3.11}$$

$$\|x\|_\infty \leq \|x\|_1 \leq n\|x\|_\infty . \tag{3.12}$$

Now let us consider the space $\mathsf{H} = \mathbf{C}^{m\times n}$, namely the space of $m \times n$ complex matrices. $\mathbf{C}^{m\times n}$ is also a linear space. Matrix p-norms on $\mathbf{C}^{m\times n}$ are defined in terms of the p-norms for vectors on $\mathbf{C}^n$:

$$\|A\|_\mathrm{p} = \sup_{x\in\mathbf{C}^n, x\neq 0} \frac{\|Ax\|_\mathrm{p}}{\|x\|_\mathrm{p}} , \qquad \forall A \in \mathbf{C}^{m\times n} . \tag{3.13}$$

Notice that the matrix p-norms are *induced norms*. They are induced by the corresponding p-norms on vectors. One can think of $\|A\|_\mathrm{p}$ as the maximum *gain* of the matrix A measured by the p-norm ratio of vectors before and after multiplication by A. In general matrix p-norms are difficult to compute. However, for $p = 1, 2,$ or ∞ there exist simple algorithms to compute $\|A\|_\mathrm{p}$ exactly. If $A = [a_{ij}] \in \mathbf{C}^{m\times n}$ we have

$$\|A\|_1 = \max_j \sum_{i=1}^{m} |a_{ij}| , \qquad \text{maximum column sum.} \tag{3.14}$$

$$\|A\|_2 = \bar{\sigma}(A) , \qquad \text{maximum singular value.} \tag{3.15}$$

$$\|A\|_\infty = \max_i \sum_{j=1}^{n} |a_{ij}| , \qquad \text{maximum row sum.} \tag{3.16}$$

The maximum singular value $\bar{\sigma}(\cdot)$ is defined in Section 3.2.1 below. A fourth norm which is important in modern control theory is the F-norm or *Frobenius norm*. It is given simply as the root sum of squares of the magnitude of all the matrix elements:

$$\|A\|_{\mathrm{F}} = \sqrt{\sum_{i=1}^{m}\sum_{j=1}^{n} |a_{ij}|^2}\,, \qquad \forall A \in \mathbf{C}^{m\times n}\,. \tag{3.17}$$

Notice that the F-norm is not an induced norm.

The F- and p-norms on $\mathbf{C}^{m\times n}$ are also equivalent norms. Thus there are upper and lower bounds on the ratio between any two different norms applied to the same matrix. If one type of norm for a given matrix tends towards zero or infinity, so do all other norms. Let $A = [a_{ij}] \in \mathbf{C}^{m\times n}$. Then

$$\|A\|_2 \le \|A\|_{\mathrm{F}} \le \sqrt{n}\|A\|_2 \tag{3.18}$$

$$\max_{i,j}|a_{ij}| \le \|A\|_2 \le \sqrt{mn}\max_{i,j}|a_{ij}| \tag{3.19}$$

$$\|A\|_2 \le \sqrt{\|A\|_1\|A\|_\infty} \tag{3.20}$$

$$\frac{1}{\sqrt{n}}\|A\|_\infty \le \|A\|_2 \le \sqrt{m}\|A\|_\infty \tag{3.21}$$

$$\frac{1}{\sqrt{m}}\|A\|_1 \le \|A\|_2 \le \sqrt{n}\|A\|_1\,. \tag{3.22}$$

The matrix 2-norm and F-norm are invariant under multiplication by unitary or orthogonal matrices. Assume that $Q^*Q = I$ and $Z^*Z = I$ for $Q \in \mathbf{C}^{m\times m}$ and $Z \in \mathbf{C}^{n\times n}$. Then

$$\|QAZ\|_{\mathrm{F}} = \|A\|_{\mathrm{F}} \tag{3.23}$$

$$\|QAZ\|_2 = \|A\|_2\,. \tag{3.24}$$

This property is crucial to many proofs in robust control theory.

3.2.1 Singular Values

In modern control theory singular values have been used to extend the classical frequency response Bode plot to multivariable systems. Consider the input-output relation

$$y(s) = G(s)u(s) \tag{3.25}$$

where $G(s)$ is a transfer function matrix. How do we then evaluate the frequency response $G(j\omega)$? An obvious choice would be to pick one of the induced matrix norms introduced previously. The 1-, 2- and ∞-norm all have potential engineering applications. However, the control theory for using them in design or analysis is only well-developed for the 2-norm. Thus let us evaluate the frequency response of $G(s)$ at the frequency ω by

$$\|G(j\omega)\|_2 = \sup_{u \in \mathbf{C}^n, u \neq 0} \frac{\|G(j\omega)u(j\omega)\|_2}{\|u(j\omega)\|_2} = \bar{\sigma}(G(j\omega)) \ . \tag{3.26}$$

Letting $0 \leq \omega \leq \infty$ we may compute the matrix 2-norm for every ω to obtain a upper bound for the "gain" of the transfer matrix $G(s)$. However, we would like to have a lower bound on $G(j\omega)$ as well. This lower bound can be obtained with the *minimum singular value* given by

$$\underline{\sigma}(G(j\omega)) = \inf_{u \in \mathbf{C}^n, u \neq 0} \frac{\|G(j\omega)u(j\omega)\|_2}{\|u(j\omega)\|_2} \ . \tag{3.27}$$

Thus if we measure the "gain" of the system $G(s)$ as the 2-norm ratio of the input and output, then the maximum and minimum singular values of $G(j\omega)$ will constitute upper and lower bounds on this gain. In fact, we may assess the system "gain" even more detailed using the *singular value decomposition* given below. Let us, however, first introduce the following important lemma which establishes the relation between the singular values and the eigenvalues of a complex matrix.

Lemma 3.1 (Singular values and Eigenvalues). *The singular values of a complex matrix $A \in \mathbf{C}^{m \times n}$, denoted $\sigma_i(A)$, are the k largest nonnegative square roots of the eigenvalues of A^*A where $k = \min\{n, m\}$. Thus*

$$\sigma_i(A) = \sqrt{\lambda_i(A^*A)} \qquad i = 1, 2, \cdots, k. \tag{3.28}$$

It is usually assumed that the singular values are ordered such that $\sigma_i \geq \sigma_{i+1}$.

Thus

$$\bar{\sigma}(A) = \sigma_1(A) = \sup_{u \in \mathbf{C}^n, u \neq 0} \frac{\|Au\|_2}{\|u\|_2} = \|A\|_2 \tag{3.29}$$

$$\underline{\sigma}(A) = \sigma_k(A) = \inf_{u \in \mathbf{C}^n, u \neq 0} \frac{\|Au\|_2}{\|u\|_2} = \left\|A^{-1}\right\|_2^{-1} \ , \quad \text{if } A^{-1} \text{ exists.} \tag{3.30}$$

The ratio between the maximum and minimum singular value is denoted the condition number κ:

$$\kappa(A) = \frac{\bar{\sigma}(A)}{\underline{\sigma}(A)} \ . \tag{3.31}$$

Let us then introduce the singular value decomposition.

Lemma 3.2 (Singular Value Decomposition). *Let $G \in \mathbf{C}^{m \times n}$ be a complex matrix. Then there exist two unitary[1] matrices $Y \in \mathbf{C}^{m \times m}, U \in \mathbf{C}^{n \times n}$ and a diagonal matrix $\Sigma \in \mathbf{R}^{m \times n}$ such that*

[1] A unitary matrix U satisfies the equation $U^*U = I$.

$$G = Y\Sigma U^* \tag{3.32}$$

$$= \begin{bmatrix} y_1 & y_2 & \cdots & y_m \end{bmatrix} \begin{bmatrix} \Sigma_k & 0 \\ 0 & 0 \end{bmatrix} \begin{bmatrix} u_1^* \\ u_2^* \\ \vdots \\ u_n^* \end{bmatrix} \tag{3.33}$$

$$= \sum_{i=1}^{k} y_i \sigma_i u_i^* \tag{3.34}$$

where

$$\begin{aligned} \Sigma_k &: \text{diag}(\sigma_1, \sigma_2, \cdots, \sigma_k). \\ y_1 \rightarrow y_m &: \textit{The } m \textit{ columns of } Y. \\ u_1^* \rightarrow u_n^* &: \textit{The } n \textit{ rows of } U^*. \end{aligned}$$

This is known as the singular-value decomposition (SVD) of the matrix G.

An interpretation for the SVD of a *real* matrix A is as follows. Any real matrix A, looked at geometrically, maps a unit radius hyper-sphere into a hyper-ellipsoid. The singular values $\sigma_i(A)$ give the lengths of the principal axis of the ellipsoid. The *singular vectors* y_i give the mutually orthogonal directions of these major axes and the singular vectors u_i are mapped into the y_i vectors with gain σ_i, that is, $Au_i = \sigma_i y_i$.

Example 3.1 (SVD of a Real Matrix). This example is taken from [MZ89]. Let A be given by

$$A = \begin{bmatrix} 0.8712 & -1.3195 \\ 1.5783 & -0.0947 \end{bmatrix}. \tag{3.35}$$

The SVD of A is given by $A = Y\Sigma U^*$ with

$$Y = \frac{1}{\sqrt{2}} \begin{bmatrix} 1 & 1 \\ 1 & -1 \end{bmatrix}, \Sigma = \begin{bmatrix} 2 & 0 \\ 0 & 1 \end{bmatrix}, U = \frac{1}{2} \begin{bmatrix} \sqrt{3} & -1 \\ -1 & -\sqrt{3} \end{bmatrix}. \tag{3.36}$$

A geometrical interpretation is given in Figure 3.1 with $U = [u_1, u_2]$ and $Y = [y_1, y_2]$. ■

In the following some of the important properties of singular values are stated:properties of

$$\bar{\sigma}(G) = \sup_{u \in \mathbf{C}^n, u \neq 0} \frac{\|Gu\|_2}{\|u\|_2} \tag{3.37}$$

$$\underline{\sigma}(G) = \inf_{u \in \mathbf{C}^n, u \neq 0} \frac{\|Gu\|_2}{\|u\|_2} \tag{3.38}$$

$$\underline{\sigma}(G) \leq |\lambda_i(G)| \leq \bar{\sigma}(G) \tag{3.39}$$

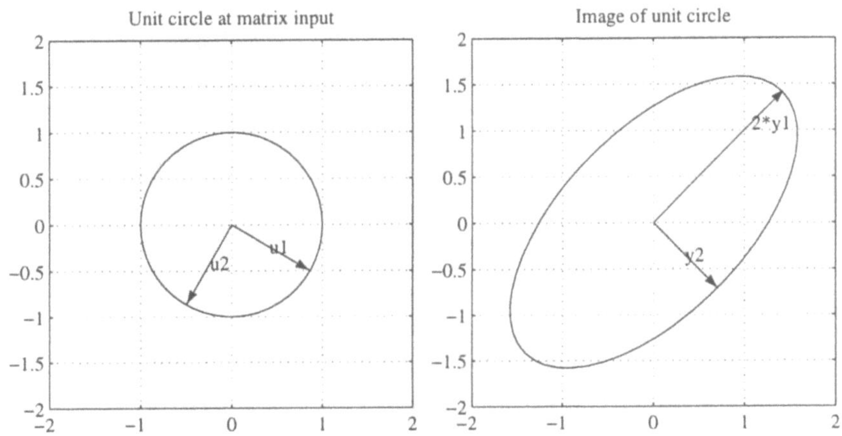

Fig. 3.1. Singular value decomposition of a real matrix.

$$\bar{\sigma}(G) = \frac{1}{\underline{\sigma}(G^{-1})} \tag{3.40}$$

$$\underline{\sigma}(G) = \frac{1}{\bar{\sigma}(G^{-1})} \tag{3.41}$$

$$\bar{\sigma}(\alpha G) = |\alpha|\bar{\sigma}(G) \tag{3.42}$$

$$\bar{\sigma}(G+H) \leq \bar{\sigma}(G) + \bar{\sigma}(H) \tag{3.43}$$

$$\bar{\sigma}(GH) \leq \bar{\sigma}(G)\bar{\sigma}(H) \tag{3.44}$$

$$\max\{\bar{\sigma}(G), \bar{\sigma}(H)\} \leq \bar{\sigma}([G\ H]) \leq \sqrt{2}\max\{\bar{\sigma}(G), \bar{\sigma}(H)\} \tag{3.45}$$

$$\sum_{i=1}^{n} \sigma_i^2 = \text{tr } \{G^*G\} \tag{3.46}$$

properties of where

$\lambda_i(G)$	: The i-th eigenvalue of G.
Property (3.40) and (3.41)	: Prerequisites the existence of G^{-1}.
α	: A constant (complex) scalar.
tr $\{G^*G\}$	: The trace of G^*G.

Consider the input-output matrix equation:

$$y(j\omega) = G(j\omega)u(j\omega) \; . \tag{3.47}$$

Using Equation (3.34) we may formulate this as:

$$y(j\omega) = \sum_{i=1}^{k} y_i \sigma_i u_i^* u(j\omega) \; . \tag{3.48}$$

Since U is unitary $u_i^* u_j$ will be orthogonal to each other so that $u_i^* u_j = 0$, for $i \neq j$ and $u_i^* u_i = 1$. Now assume that $u(j\omega) = \alpha u_j$. The input-output equation then becomes:

$$y(j\omega) = \sum_{i=1}^{k} y_i \sigma_i u_i^* u_j \alpha \tag{3.49}$$

$$= \sigma_j \alpha y_j \; . \tag{3.50}$$

This illustrates that if the input vector $u(j\omega)$ is in the direction of u_j then the gain of the system is precisely σ_j and the output vector $y(j\omega)$ is then precisely in the direction of y_j. The sets $\{u_1, u_2, \cdots, u_n\}$ and $\{y_1, y_2, \cdots, y_m\}$ are not surprisingly known as the *input* respectively *output principal directions.* The singular values σ_i are also known as the principal or directional gains of the system matrix G.

Thus when $G(s)$ is a transfer function matrix we can plot the singular values $\sigma_i(G(j\omega))$ for $i = 1, \cdots, k$ as functions of frequency ω. These curves are the multivariable generalization of the SISO amplitude-ratio Bode plot. For multivariable systems the amplification of the input vector sinusoid $ue^{j\omega t}$ depends on the direction of the complex vector u as illustrated above. The amplification is at least $\underline{\sigma}(G(j\omega))$ and at most $\bar{\sigma}(G(j\omega))$. The condition number $\kappa(G(j\omega))$, see (3.31), plotted versus frequency ω outlines the system gain sensitivity to the direction of the input vector. If $\kappa(G(j\omega)) \gg 1$ the gain of the transfer function matrix will vary considerably with the input direction and $G(s)$ is said to be *ill-conditioned.* Conversely, if $\kappa(G(j\omega)) \approx 1$, $\forall \omega \geq 0$, the gain of the transfer matrix will be insensitive to the input direction and the system is said to be *well-conditioned.* A well-conditioned multivariable system behaves much like a single-variable system and controller design for well-conditioned system are fairly straightforward. For ill-conditioned systems, however, much more care has to be taken in both design and analysis.

3.3 Operator Norms

The norm concept introduced in Section 3.2 for constant vectors and matrices can also be extended to time- and frequency domain functions. Such norms will be denoted *operator norms.*

3.3.1 Scalar Systems

Let us start by considering scalar functions $f(t)$ of a continuous-time variable t. For scalar systems the operator 1, 2 and ∞-norms are defined as:

$$\|f\|_1 = \int_{-\infty}^{\infty} |f(t)| dt\,, \qquad \text{resource.} \tag{3.51}$$

$$\|f\|_2 = \sqrt{\int_{-\infty}^{\infty} f^2(t) dt}\,, \qquad \text{energy.} \tag{3.52}$$

$$\|f\|_\infty = \sup_t |f(t)|\,, \qquad \text{peak.} \tag{3.53}$$

The reader may verify that the above operator norms fulfill the conditions given in Section 3.1. Notice that we use the notation $\|f\|$ for the operator norm and the notation $\|f(t)\|$ for the norm of the value of f at time t. Apart from these norms, the following measure is often used in control theory:

$$\|f\|_{\mathcal{P}} = \lim_{T\to\infty} \sqrt{\frac{1}{2T} \int_{-T}^{T} f^2(t) dt}\,. \tag{3.54}$$

$\|f\|_{\mathcal{P}}$ is known as the *power* or RMS value of f. However, $\|f\|_{\mathcal{P}}$ is *not* a true norm since property (ii) in Section 3.1 is not fulfilled. $\|f\|_{\mathcal{P}}$ is often denoted a semi-norm since it fulfill the remaining three properties. The power norm is useful since it is often defined where the 2-norm is not, eg for signals with limited but persistent power.

Equivalently, we may also define operator norms for scalar frequency domain functions $f(s)$. Here only the 2- and ∞ norm are of practical interest:

$$\|f\|_2 = \sqrt{\frac{1}{2\pi} \int_{-\infty}^{\infty} f^*(j\omega) f(j\omega) d\omega} \tag{3.55}$$

$$\|f\|_\infty = \sup_{\omega\in\mathbf{R}} |f(j\omega)|\,. \tag{3.56}$$

Note that by Parceval's theorem

$$\|f\|_2 = \sqrt{\int_{-\infty}^{\infty} f^2(t) dt} = \sqrt{\frac{1}{2\pi} \int_{-\infty}^{\infty} f^*(j\omega) f(j\omega) d\omega} = \|f\|_2 \tag{3.57}$$

such that the operator 2 norm of a time-domain function and its Laplace transform is the same. This fact is one of the main reasons for using the 2 norm in control theory. Note that a equivalent statement does *not* hold for the ∞ norm.

Finally, we can introduce the following 2 *transfer function* on a stable transfer functions $G(s)$:

$$\|G(s)\|_{\mathcal{H}_2} = \sqrt{\frac{1}{2\pi} \int_{-\infty}^{\infty} G^*(j\omega) G(j\omega) d\omega} \tag{3.58}$$

$$\|G(s)\|_{\mathcal{H}_\infty} = \sup_{\omega\in\mathbf{R}} |G(j\omega)|\,. \tag{3.59}$$

We use the subscripts $\mathcal{H}_2$ and $\mathcal{H}_\infty$ to emphasize that it is a transfer function norm. Here $\mathcal{H}$ stands for *Hardy* which is the name of a class of important spaces named after the mathematician G. H. Hardy. In Section 3.5 we will provide an introduction to some of these spaces.

Note how the definitions of transfer function norms and operator norms for frequency domain functions $f(s)$ are fully equivalent. Reliable methods exist for computing transfer function norms from state-space realizations of $G(s)$, see eg [DFT92].

Induced Norms for Scalar Systems. We will now define *induced operator norms* for scalar systems using similar techniques as when we defined induced matrix norms in Section 3.2. Let $g(t)$ be the impulse response associated with the scalar stable transfer function $G(s)$. If the input to the system is $u(t)$ and initial conditions are zero, then the output $y(t)$ is given by

$$y(t) = g(t) * u(t) = \int_{-\infty}^{\infty} g(t-\tau)u(\tau)d\tau \tag{3.60}$$

where $*$ denotes the convolution integral. We may then define an induced norm for $g(t)$ as

$$\|g\|_{\alpha\to\beta} = \sup_{u(t)\neq 0} \frac{\|y\|_\beta}{\|u\|_\alpha} = \sup_{u(t)\neq 0} \frac{\|g*u\|_\beta}{\|u\|_\alpha} \tag{3.61}$$

where α, β can be any combination of 1, 2, ∞ and $\mathcal{P}$. The induced norm of g is thus the maximum gain of the system if we measure the input u using the operator norm α and the output y using the operator norm β.

Using the 4 operator norms defined above we may thus construct 16 different induced norms for measuring the gain of the system with impulse response $g(t)$. However, some of these induced norms are trivially 0 or ∞. Let for example the input u is measured by the ∞ norm and the output y by the 2 norm. It is then clear that the induced norm $\|g\|_{\infty\to 2} = \infty$ since $u(t) \not\to 0$ for $t \to \infty$. Let us consider the case where both the input u and the output y are measured using the 2 norm[2]. Then by Parceval

$$\|g\|_{2\to 2} = \sup_{u(t)\neq 0} \frac{\|g*u\|_2}{\|u\|_2} = \sup_{u(t)\neq 0} \frac{\sqrt{\int_{-\infty}^{\infty} y^2(t)dt}}{\sqrt{\int_{-\infty}^{\infty} u^2(t)dt}} \tag{3.62}$$

$$= \sup_{u(j\omega)\neq 0} \frac{\sqrt{\int_{-\infty}^{\infty} y^*(j\omega)y(j\omega)d\omega}}{\sqrt{\int_{-\infty}^{\infty} u^*(j\omega)u(j\omega)d\omega}} \tag{3.63}$$

$$= \sup_{u(j\omega)\neq 0} \frac{\sqrt{\int_{-\infty}^{\infty} (G(j\omega)u(j\omega))^*(G(j\omega)u(j\omega))d\omega}}{\sqrt{\int_{-\infty}^{\infty} u^*(j\omega)u(j\omega)d\omega}} \tag{3.64}$$

[2] We will assume that it exists.

$$= \sup_{u(j\omega)\neq 0} \frac{\sqrt{\int_{-\infty}^{\infty} u(j\omega)^*|G(j\omega)|^2 u(j\omega))d\omega}}{\sqrt{\int_{-\infty}^{\infty} u^*(j\omega)u(j\omega)d\omega}} \tag{3.65}$$

$$= \sup_{\omega} |G(j\omega)| \tag{3.66}$$

$$= \|G(s)\|_{\mathcal{H}_\infty} \,. \tag{3.67}$$

The transfer function $\mathcal{H}_\infty$ norm thus measures the maximum possible increase in signal energy (as measured by the operator 2-norm) for all possible inputs for which the 2-norm exists. We may say that the transfer function $\mathcal{H}_\infty$ norm is induced by the operator 2-norm. The $\mathcal{H}_\infty$ norm measures the maximum gain of the system for a class of input signals characterized by their operator 2-norm.

Unlike the $\mathcal{H}_\infty$ norm, the $\mathcal{H}_2$ transfer function norm is *not* induced by any operator norm. In this sense, it corresponds to the Frobenius for matrices which is not induced by any vector norm. Assume that the input $u(t) = \delta(t)$, a unit impulse. Then the output is given by $y(t) = g(t)$ such that $\|y\|_2 = \|g\|_2$. But by Parceval we have that $\|g\|_2 = \|G(s)\|_{\mathcal{H}_2}$. Consequently, the transfer function $\mathcal{H}_2$ norm measures the operator 2-norm of the output given a specific input signal, namely a unit impulse.

Another interpretation of the $\mathcal{H}_2$ goes as follows. If $u(t)$ is unit variance white noise then the output power norm $\|y\|_{\mathcal{P}} = \|G(s)\|_{\mathcal{H}_2}$. Thus the transfer function $\mathcal{H}_2$ also measures the RMS value of the output $y(t)$ given unit variance white noise on the input $u(t)$. All quadratic norm control schemes like Linear Quadratic optimal control and Kalman-Bucy optimal filtering minimizes the $\mathcal{H}_2$ norm of a closed loop transfer function.

Remark 3.1. Throughout this section we have emphasized that the transfer function $G(s)$ must be stable. Does this then mean that these methods cannot be applied to unstable plants? The answer is no! In control design analysis the transfer function norm measures are applied to *closed loop* systems. Then it is clearly a reasonable assumption that the system is stable.

In Table 3.1 we have shown induced system gains for all combinations of the 2, ∞ and power.

Table 3.1. *Induced system gains for scalar functions [DFT92].*

	$\|u\|_2$	$\|u\|_\infty$	$\|u\|_{\mathcal{P}}$
$\|y\|_2$	$\|G(s)\|_{\mathcal{H}_\infty}$	∞	∞
$\|y\|_\infty$	$\|G(s)\|_{\mathcal{H}_2}$	$\|g\|_1$	∞
$\|y\|_{\mathcal{P}}$	0	$\leq \|G(s)\|_{\mathcal{H}_\infty}$	$\|G(s)\|_{\mathcal{H}_\infty}$

3.3.2 Multivariable Systems

Now let us extend the concept of operator norms to multivariable systems. Consider a vector-valued function $f(t)$ of a continuous-time variable t. We may then define operator norms for f by combining the vector norms in Equation (3.6)–(3.8) with the scalar operator norms in (3.51)–(3.53). This gives us 9 possible norms for f:

$$\|f\|_{1,1} = \int_{-\infty}^{\infty} \|f(t)\|_1 dt = \int_{-\infty}^{\infty} \sum_{i=1}^{n} |f_i(t)| dt \tag{3.68}$$

$$\|f\|_{1,2} = \int_{-\infty}^{\infty} \|f(t)\|_2 dt = \int_{-\infty}^{\infty} \sqrt{f^T(t) f(t)} dt \tag{3.69}$$

$$\|f\|_{1,\infty} = \int_{-\infty}^{\infty} \|f(t)\|_\infty dt = \int_{-\infty}^{\infty} \max_i |f_i(t)| dt \tag{3.70}$$

$$\|f\|_{2,1} = \sqrt{\int_{-\infty}^{\infty} \|f(t)\|_1^2 dt} = \sqrt{\int_{-\infty}^{\infty} \left(\sum_{i=1}^{n} |f_i(t)| \right)^2 dt} \tag{3.71}$$

$$\|f\|_{2,2} = \sqrt{\int_{-\infty}^{\infty} \|f(t)\|_2^2 dt} = \sqrt{\int_{-\infty}^{\infty} f^T(t) f(t) dt} \tag{3.72}$$

$$\|f\|_{2,\infty} = \sqrt{\int_{-\infty}^{\infty} \|f(t)\|_\infty^2 dt} = \sqrt{\int_{-\infty}^{\infty} \left(\max_i |f_i(t)| \right)^2 dt} \tag{3.73}$$

$$\|f\|_{\infty,1} = \sup_t \|f(t)\|_1 = \sup_t \sum_{i=1}^{n} |f_i(t)| \tag{3.74}$$

$$\|f\|_{\infty,2} = \sup_t \|f(t)\|_2 = \sup_t \sqrt{f^T(t) f(t)} \tag{3.75}$$

$$\|f\|_{\infty,\infty} = \sup_t \|f(t)\|_\infty = \sup_t \max_i |f_i(t)| \,. \tag{3.76}$$

The operator norm $\|f\|_{\alpha,\beta}$ thus denotes the norm we get by using the vector β norm at each time instant t and the scalar operator norm α over time.

The power $\|f\|_{\mathcal{P}}$ naturally extends to

$$\|f\|_{\mathcal{P}} = \lim_{T \to \infty} \sqrt{\frac{1}{2T} \int_{-T}^{T} \|f(t)\|_2^2 dt} \tag{3.77}$$

where $\|f(t)\|_2$ is the usual Euclidean length of the vector $f(t)$.

Furthermore, for transfer function matrices the norms (3.58) and (3.59) extend to

$$\|G(s)\|_{\mathcal{H}_2} = \sqrt{\frac{1}{2\pi}\int_{-\infty}^{\infty} \text{tr}\,\{G^*(j\omega)G(j\omega)\}d\omega} \tag{3.78}$$

$$\|G(s)\|_{\mathcal{H}_\infty} = \sup_{\omega\in\mathbf{R}} \bar{\sigma}(G(j\omega)) \,. \tag{3.79}$$

Induced Norms for Multivarable Systems. Let $g(t)$ be the impulse response matrix associated with the stable transfer function matrix $G(s)$. If the input vector to the system is $u(t)$ and initial conditions are zero, then the output vector $y(t)$ is given by the convolution integral

$$y(t) = g(t) * u(t) = \int_{-\infty}^{\infty} g(t-\tau)u(\tau)d\tau \tag{3.80}$$

as before. An induced norm for g can then be defined as:

$$\|g(t)\|_{(\alpha_1,\alpha_2)\to(\beta_1,\beta_2)} = \sup_{u\neq 0} \frac{\|y\|_{\beta_1,\beta_2}}{\|u\|_{\alpha_1,\alpha_2}} = \sup_{u\neq 0} \frac{\|g*u\|_{\beta_1,\beta_2}}{\|u\|_{\alpha_1,\alpha_2}} \tag{3.81}$$

where $\alpha_1, \alpha_2, \beta_1, \beta_2$ can be any combination of $1, 2, \infty, \mathcal{P}$. The induced operator norm $\|g(t)\|_{(\alpha_1,\alpha_2)\to(\beta_1,\beta_2)}$ thus measures the maximum possible "gain" of the system if the input is measured by $\|u\|_{\alpha_1,\alpha_2}$ and the output by $\|y\|_{\beta_1,\beta_2}$. The introduced concept of induced operator norms thus provides 100 ways of measuring the gain of a multivariable system. There are well motivated reasons for applying several of these induced norms in control theory, see Example 3.2. Others are trivially zero or ∞.

Example 3.2 (Induced Norms). Consider a multivariable control system with the purpose of minimizing the consumption of a given resource $e(t)$, eg fuel. It will then be reasonable to measure e using the norm

$$\|e\|_{2,1} = \sqrt{\int_{-\infty}^{\infty}\left(\sum_{i=1}^{n} |e_i(t)|\right)^2 dt} \,. \tag{3.82}$$

Furthermore, assume that the disturbances $d(t)$ on the system can be well described by their energy. Then the norm

$$\|d\|_{2,2} = \sqrt{\int_{-\infty}^{\infty} d^T(t)d(t)dt} \tag{3.83}$$

will be a sensible measure for d. A control design then should attempt to minimize the induced gain

$$\|s\|_{(2,1)\to(2,2)} = \sup_{d\neq 0}\frac{\|e\|_{2,1}}{\|d\|_{2,2}} = \sup_{d\neq 0}\frac{\sqrt{\int_{-\infty}^{\infty}\left(\sum_{i=1}^{n}|e_i(t)|\right)^2 dt}}{\sqrt{\int_{-\infty}^{\infty} d^T(t)d(t)dt}} \tag{3.84}$$

where s is the impulse response matrix of the transfer function $S(s)$ from $d(s)$ to $e(s)$. ∎

The application of induced operator norms in control theory is currently an area of intensive research and new tools like Linear Matrix Inequalities (LMI's) are being applied to solve the very difficult optimizations involved. However, these methodologies are not yet fully matured.

In both the "classical" $\mathcal{H}_\infty$ robust control theory (1980–1995) and the older LQG theory only a single induced norm is considered, namely the pure 2 norm:

$$\|g\|_{(2,2)\to(2,2)} = \sup_{u\neq 0} \frac{\|g * u\|_{2,2}}{\|u\|_{2,2}} = \sup_{u\neq 0} \frac{\sqrt{\int_{-\infty}^{\infty} y^T(t)y(t)dt}}{\sqrt{\int_{-\infty}^{\infty} u^T(t)u(t)dt}} \tag{3.85}$$

where both input and output are measured by $\|\cdot\|_{2,2}$. The $\|\cdot\|_{2,2}$ norm measures the energy of the signal and the induced operator 2 norm thus measures the maximum possible increase in signal energy for all possible inputs $u(t)$ for which the 2-norm is defined. As for scalar systems, we have the very important relationship:

$$\|g\|_{(2,2)\to(2,2)} = \|G(s)\|_{\mathcal{H}_\infty} = \sup_{\omega\in\mathbf{R}} \bar{\sigma}(G(j\omega)) \ . \tag{3.86}$$

The following lemma recapitulate our result.

Lemma 3.3 (Transfer Function $\mathcal{H}_\infty$ Norm). *Let $G(s)$ be a (stable) multivariable transfer function matrix. If both input $u(t)$ and output $y(t)$ are measured by the 2 norm $\|\cdot\|_{2,2}$, then the maximum gain of the system for all inputs $u(t)$ with $\|u\|_{2,2} \leq \infty$ is given by the transfer function $\mathcal{H}_\infty$ norm:*

$$\|G(s)\|_{\mathcal{H}_\infty} = \sup_{u\neq 0} \frac{\|y\|_{2,2}}{\|u\|_{2,2}} \ . \tag{3.87}$$

In the following we will define some spaces whose elements are time- or frequency domain functions which are useful in modern control theory. Let us first, however, introduce the concept of Banach and Hilbert spaces.

3.4 Banach and Hilbert Spaces

A *scalar product* (or inner product) on H is a function denoted $<\cdot,\cdot>$ from $\mathsf{H}\times\mathsf{H}$ to $\mathbf{K}$ having the following properties of

$$(i) \quad < f, g+h > = < f, g > + < f, h > \ . \tag{3.88}$$

$$(ii) \quad < f, \alpha g > = \alpha < f, g > \ . \tag{3.89}$$

$$(iii) \quad < f, g > = < g, f >^* \ , \quad \text{(Hermitian).} \tag{3.90}$$

$$(vi) \quad < f, f > \geq 0 \ . \tag{3.91}$$

$$(v) \quad < f, f > = 0 \ , \quad \text{iff } f = 0. \tag{3.92}$$

where $f, g, h \in \mathsf{H}$ and $\alpha \in \mathbf{K}$. A scalar product $<\cdot,\cdot>$ *induces* a norm, namely $\|f\| = < f, f >^{1/2}$. Given the linear space H and a scalar product $<\cdot,\cdot>$ on H, the pair $(\mathsf{H}, <\cdot,\cdot>)$ is denoted a *pre-Hilbert space*.

3.4.1 Convergence and Completeness

Let $(\mathsf{H}, \|\cdot\|)$ be a normed space. Having defined a norm $\|\cdot\|$ we can assess convergence in H. A sequence $\{f_n\}$, $n = 1, 2, \cdots$ in H converges to $f \in \mathsf{H}$ and f is the limit of the sequence if the sequence of real positive numbers $\|f - f_n\|$ converges to zero. If such f exists, then the sequence is *convergent.*

A sequence $\{f_n\}$, $n = 1, 2, \cdots$ in H is called a *Cauchy sequence* if for each $\epsilon > 0$ there exists a natural number $n_0 \in \mathbf{N}$ such that for $n, m \geq n_0$ we have $\|f_n - f_m\| \leq \epsilon$. Intuitively, the elements in a Cauchy sequence eventually cluster around each other. They are "trying to converge". Clearly, every convergent sequence is a Cauchy sequence. However, in an arbitrary normed space not every Cauchy sequence is convergent. A normed space $(H, \|\cdot\|)$ is said to be *complete* if every Cauchy sequence is convergent. A pre-Hilbert space $(\mathsf{H}, < \cdot, \cdot >)$ is said to be complete if it is complete with respect to the norm induced by the scalar product $< \cdot, \cdot >$.

A complete normed space is called a *Banach space* and a complete pre-Hilbert space is denoted a *Hilbert space.*

Example 3.3. $\mathbf{C}^n$ and $\mathbf{R}^n$ are Banach spaces under the norms $\|\cdot\|_1$, $\|\cdot\|_2$ and $\|\cdot\|_\infty$ defined by (3.6)-(3.8). ■

Example 3.4. $\mathbf{C}^n$ and $\mathbf{R}^n$ are Hilbert spaces under the scalar product

$$< x, y > = \sum_{i=1}^{m} x_i^* y_i = x^* y \ . \tag{3.93}$$

The corresponding induced norm is the 2-norm since

$$\|x\| = < x, x >^{1/2} = \sqrt{x^* x} = \|x\|_2 \ . \tag{3.94}$$

■

3.5 Lebesgue and Hardy Spaces

In Section 3.2 we only considered the space $\mathbf{C}^{m \times n}$, i.e spaces whose elements are constant matrices (or constant vectors as special cases). However, the concepts introduced above apply also to linear spaces whose elements are time domain or frequency domain functions (or operators). Two such spaces are important in robust control theory, namely *Lebesgue* and *Hardy spaces.* In $\mathcal{H}_\infty$ control theory, $\mathcal{H}$ stands for Hardy.

3.5.1 Time Domain Spaces

Let us consider vector-valued functions $f(t)$ of a continuous time variable t. Let $f(t)$ be defined for all time $t \in \mathbf{R}$ and taking values in $\mathbf{R}^n$. Restrict $f(t)$ to be *square-Lebesgue integrable*:

$$\int_{-\infty}^{\infty} \|f(t)\|_2^2 dt < \infty \tag{3.95}$$

such that the operator 2 norm $\|f\|_{2,2}$ exists, see Equation (3.72). The set of all such signals is a Banach space under the norm $\|f\|_{2,2}$. This space is called the *Lebesgue space* $\mathcal{L}_2(\mathbf{R}, \mathbf{R}^n)$. The first argument gives the *domain of* $(\mathcal{D}(f))$, or input space, and the second gives the *range of* $(\mathcal{R}(f))$, or output space. Quite often in the literature these arguments are dropped for brevity. Thus, which version of the Lebesgue space is meant must be determined form the context. To avoid any confusing we will keep the arguments here.

The subspace of $\mathcal{L}_2(\mathbf{R}, \mathbf{R}^n)$ for which $f(t) = 0, \forall t < 0$ (causal time functions) is the *Hardy space* $\mathcal{H}_2(\mathbf{R}, \mathbf{R}^n)$ under the norm $\|f\|_{2,2}$.

Similarly, we will consider spaces whose elements are time functions with finite operator 1- and ∞ norms:

- $\mathcal{L}_1(\mathbf{R}, \mathbf{R}^n)$ is the (Banach) space of *absolute-Lebesgue integrable* vector-valued functions of time:

$$\int_{-\infty}^{\infty} \|f(t)\|_1 dt = \int_{-\infty}^{\infty} \sum_{i=1}^{n} |f_i(t)| dt < \infty \tag{3.96}$$

  under the operator 1-norm $\|f\|_{1,1}$, see Equation (3.68). $\mathcal{H}_1(\mathbf{R}, \mathbf{R}^n)$ is the subspace of $\mathcal{L}_1(\mathbf{R}, \mathbf{R}^n)$ for which $f(t) = 0, \forall t < 0$.
- $\mathcal{L}_\infty(\mathbf{R}, \mathbf{R}^n)$ is the (Banach) space of bounded real vector-valued functions of time:

$$\sup_{t \in \mathbf{R}} \max_i |f_i(t)| < \infty \tag{3.97}$$

  under the ∞ norm $\|f\|_{\infty,\infty}$, see (3.76). $\mathcal{H}_\infty(\mathbf{R}, \mathbf{R}^n)$ is the subspace of $\mathcal{L}_\infty(\mathbf{R}, \mathbf{R}^n)$ for which $f(t) = 0, \forall t < 0$.

A large class of causal time domain signals have finite values for the above 3 norms and thus are members of all three spaces $\mathcal{H}_1(\mathbf{R}, \mathbf{R}^n)$, $\mathcal{H}_2(\mathbf{R}, \mathbf{R}^n)$ and $\mathcal{H}_\infty(\mathbf{R}, \mathbf{R}^n)$. The following example, taken from [Dai90], demonstrate some of the distinctions.

Example 3.5 (Hardy Spaces for Scalar Functions). Consider the two functions

$$f_1(t) = \begin{cases} 0 & \forall t < 0 \\ 2e^{-t/2}\sin(3t) & \forall t \geq 0 \end{cases} \tag{3.98}$$

$$f_2(t) = \begin{cases} 0 & \forall t < 0 \\ 1 & \forall t \geq 0 \end{cases} . \tag{3.99}$$

Notice that $f_2(t)$ is the ordinary step function. Now $f_1(t)$ belongs to all three spaces $\mathcal{H}_1(\mathbf{R},\mathbf{R})$, $\mathcal{H}_2(\mathbf{R},\mathbf{R})$ and $\mathcal{H}_\infty(\mathbf{R},\mathbf{R})$. Both its absolute and its square are integrable from 0 to ∞ and it has finite peak. In contrast, the step function $f_2(t)$ is a member of $\mathcal{H}_\infty(\mathbf{R},\mathbf{R})$ only, since neither of the 1- or 2-norms have finite values. $f_1(t)$ and $f_2(t)$ is shown in Figure 3.2. ■

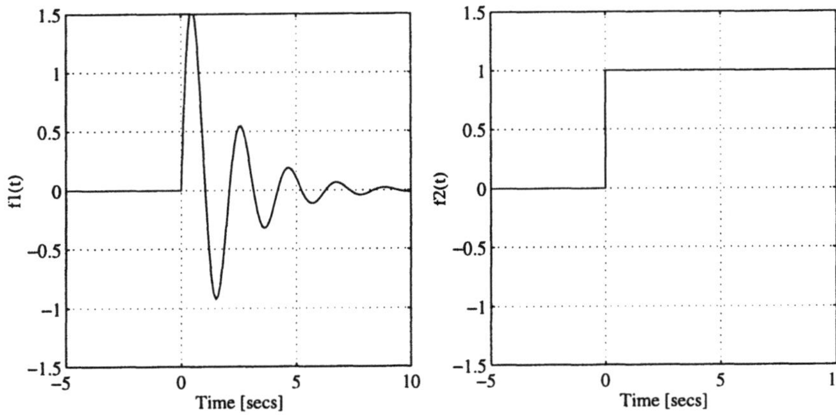

Fig. 3.2. The functions $f_1(t)$ (left) and $f_2(t)$ (right).

As mentioned above, in both robust control theory and the older LQG theory, only the operator 2 norm is used for time-domain signals. Only signals belonging to $\mathcal{H}_2(\mathbf{R},\mathbf{R}^n)$ is thus considered. The time domain spaces $\mathcal{H}_1(\mathbf{R},\mathbf{R}^n)$ and $\mathcal{H}_\infty(\mathbf{R},\mathbf{R}^n)$ do have potential engineering applications but the control theory for using them in design or analysis is not yet well-developed even though it is an area of intensive research. Since $\mathcal{H}_1(\mathbf{R},\mathbf{R}^n)$ can represent fuel usage, and $\mathcal{H}_\infty(\mathbf{R},\mathbf{R}^n)$ can treat the issue of actuator or sensor saturation, there are strong motivations for developing such theories.

Example 3.6. $\mathcal{L}_2(\mathbf{R},\mathbf{R}^n)$ is a Hilbert space under the scalar product

$$< f, g >= \int_{-\infty}^{\infty} f(t)^* g(t) dt \,. \tag{3.100}$$

The corresponding induced norm is the 2-norm since

$$\|f\| =< f, f >^{1/2}= \sqrt{\int_{-\infty}^{\infty} f(t)^T f(t) dt} = \|f\|_{2,2} \,. \tag{3.101}$$

■

3.5.2 Frequency Domain Spaces

Now consider the vector valued function $f(j\omega)$ which is defined for all frequencies $-\infty < \omega < \infty$ (that is, on the imaginary axis), takes values in $\mathbf{C}^n$ and is square-Lebesgue integrable on the imaginary axis. The space of all such functions is a Hilbert space under the scalar product

$$< f, g >= \frac{1}{2\pi} \int_{-\infty}^{\infty} f(j\omega)^* g(j\omega) d\omega \ . \tag{3.102}$$

It is denoted the Lebesgue space $\mathcal{L}_2(j\mathbf{R}, \mathbf{C}^n)$. The corresponding induced norm is

$$\|f\| =< f, f >^{1/2}= \sqrt{\frac{1}{2\pi} \int_{-\infty}^{\infty} f(j\omega)^* f(j\omega) d\omega} = \|f\|_{2,2} \ . \tag{3.103}$$

Next, $\mathcal{H}_2(\mathbf{C}, \mathbf{C}^n)$ is the (Hardy) space of all functions $f(s)$ which are analytic in $\Re e(s) > 0$ (i.e., no right half plane (RHP) poles), take values in $\mathbf{C}^n$ and satisfy the *uniform square-integrability condition*

$$\sup_{\xi} \frac{1}{2\pi} \int_{-\infty}^{\infty} f(\xi + j\omega)^* f(\xi + j\omega) d\omega < \infty \ . \tag{3.104}$$

$\mathcal{H}_2(\mathbf{C}, \mathbf{C}^n)$ is a Hilbert space under the scalar product (3.102) and with induced norm (3.103).

Because of Parceval's theorem $\mathcal{H}_2(\mathbf{C}, \mathbf{C}^n)$ is just the set of Laplace transforms of signals in $\mathcal{H}_2(\mathbf{R}, \mathbf{R}^n)$. Furthermore, if $f(t) \in \mathcal{H}_2(\mathbf{R}, \mathbf{R}^n)$ then its Laplace transform $f(s) \in \mathcal{H}_2(\mathbf{C}, \mathbf{C}^n)$.

Finally we will present 4 frequency domain spaces of matrix-valued functions.

- $\mathcal{L}_2(j\mathbf{R}, \mathbf{C}^{m \times n})$ is the space of all functions $F(j\omega)$, defined on the imaginary axis which take values in $\mathbf{C}^{m \times n}$ and are square-integrable on the imaginary axis. $\mathcal{L}_2(j\mathbf{R}, \mathbf{C}^{m \times n})$ is a Hilbert space under the scalar product

$$< F, G >= \frac{1}{2\pi} \int_{-\infty}^{\infty} \mathrm{tr} \ \{F(j\omega)^* G(j\omega)\} d\omega \ . \tag{3.105}$$

The corresponding induced norm is given by

$$\|F\|_{\mathcal{H}_2} =< F, F >^{1/2}= \sqrt{\frac{1}{2\pi} \int_{-\infty}^{\infty} \mathrm{tr} \ \{F(j\omega)^* F(j\omega)\} d\omega} \ . \tag{3.106}$$

Note that we have used subscript $\mathcal{H}_2$ in (3.106). However, since $F(j\omega)$ is only defined on the imaginary axis it is not a transfer function. Rather it is a frequency response. On the other hand, the norm (3.106) is identical to the transfer function 2-norm in (3.78). Thus the given notation.

- $\mathcal{H}_2(\mathbf{C}, \mathbf{C}^{m\times n})$ is the space of all functions $F(s)$ which are analytical in $\Re e(s) > 0$ (no RHP poles), take values in $\mathbf{C}^{m\times n}$ and satisfy the uniform square-integrability condition

$$\sup_{\xi>0} \frac{1}{2\pi} \int_{-\infty}^{\infty} \text{tr}\ \{F(\xi + j\omega)^* F(\xi + j\omega)\} d\omega < \infty\ . \tag{3.107}$$

  $\mathcal{H}_2(\mathbf{C}, \mathbf{C}^{m\times n})$ is a Hilbert space under the scalar product (3.105) and with induced norm:

$$\|F(s)\|_{\mathcal{H}_2} = \sqrt{\frac{1}{2\pi} \int_{-\infty}^{\infty} \text{tr}\ \{F(j\omega)^* F(j\omega)\} d\omega}\ . \tag{3.108}$$

- $\mathcal{L}_\infty(j\mathbf{R}, \mathbf{C}^{m\times n})$ is the space of all functions $F(j\omega)$, defined on the imaginary axis, which take values in $\mathbf{C}^{m\times n}$ and are bounded on the imaginary axis:

$$\sup_{\omega\in\mathbf{R}} \|F(j\omega)\|_2 = \sup_{\omega\in\mathbf{R}} \bar{\sigma}(F(j\omega)) < \infty \tag{3.109}$$

  $\mathcal{L}_\infty(j\mathbf{R}, \mathbf{C}^{m\times n})$ is a Banach space under the norm

$$\|F\|_{\mathcal{H}_\infty} = \sup_{\omega\in\mathbf{R}} \|F(j\omega)\|_2 = \sup_{\omega\in\mathbf{R}} \bar{\sigma}(F(j\omega))\ . \tag{3.110}$$

- $\mathcal{H}_\infty(\mathbf{C}, \mathbf{C}^{m\times n})$ is the space of all functions $F(s)$ which are analytic in $\Re e(s) > 0$ (no RHP poles), take values in $\mathbf{C}^{m\times n}$ and are bounded in the RHP:

$$\sup_{\Re e(s)>0} \|F(s)\|_2 = \sup_{\omega\in\mathbf{R}} \bar{\sigma}(F(j\omega)) < \infty \tag{3.111}$$

  $\mathcal{H}_\infty(\mathbf{C}, \mathbf{C}^{m\times n})$ is a Banach space under the transfer function ∞ norm:

$$\|F(s)\|_{\mathcal{H}_\infty} = \sup_{\omega\in\mathbf{R}} \bar{\sigma}(F(j\omega))\ . \tag{3.112}$$

As mentioned previously, (frequency domain) functions in $\mathcal{L}_2(j\mathbf{R}, \mathbf{C}^{m\times n})$ and $\mathcal{L}_\infty(j\mathbf{R}, \mathbf{C}^{m\times n})$ are defined only on the imaginary axis; their domain is $j\mathbf{R}$. It is thus not meaningful to talk about their poles and zeros or their stability. They are simply frequency responses, not transfer functions. Conversely, the domain of functions in $\mathcal{H}_2(\mathbf{C}, \mathbf{C}^{m\times n})$ and $\mathcal{H}_\infty(\mathbf{C}, \mathbf{C}^{m\times n})$ is the entire complex plane $\mathbf{C}$. They are (stable) transfer functions, not just frequency responses.

$\mathcal{H}_2(\mathbf{C}, \mathbf{C}^{m\times n})$ and $\mathcal{H}_\infty(\mathbf{C}, \mathbf{C}^{m\times n})$ are stable spaces since they do not allow poles in the RHP. Transfer functions in $\mathcal{H}_2(\mathbf{C}, \mathbf{C}^{m\times n})$, however, must roll off at high frequencies to fulfill (3.107). In contrast, transfer functions in $\mathcal{H}_\infty(\mathbf{C}, \mathbf{C}^{m\times n})$ may maintain non-zero gain as $\omega \to \infty$. In terms of state-space realizations (A, B, C, D), the D matrix must be zero for a transfer function in $\mathcal{H}_2(\mathbf{C}, \mathbf{C}^{m\times n})$.

Using the above spaces we may reformulate the result in Lemma 3.3:

Lemma 3.4 (Transfer Function $\mathcal{H}_\infty$ Norm II).
Let $G(s) \in \mathcal{H}_\infty(\mathbf{C}, \mathbf{C}^{m\times n})$ (a stable multivariable transfer function matrix) and $u(t) \in \mathcal{H}_2(\mathbf{R}, \mathbf{R}^n)$ (a causal time function with $\|\cdot\|_{2,2} < \infty$). Then $y(t) \in \mathcal{H}_2(\mathbf{R}, \mathbf{R}^m)$ and

$$\|G(s)\|_{\mathcal{H}_\infty} = \sup_{u\neq 0} \frac{\|y\|_{2,2}}{\|u\|_{2,2}} = \sup_{\omega\in\mathbf{R}} \bar{\sigma}(G(j\omega)) \,. \tag{3.113}$$

3.6 Summary

Sets, norms and spaces frequently used in robust control theory were discussed. It was illustrated how the familiar vector p-norms induce corresponding matrix norms. Of particular interest are the matrix 1, 2 and ∞ norms partly because they have obvious physical interpretations and partly because they can be effectively computed. In "classical" robust control theory the main emphasis has been put on the matrix 2-norm. However, recently also the 1- and ∞-norm have attracted considerable attention. It is well known that the matrix 2-norm equals the maximum singular value. In Section 3.2.1 a thorough introduction to singular values and the singular value decomposition were given. The singular value concept has played a key role in modern control theory for a number of reasons, in particular because it enables us to extend the classical Bode plot to multivariable systems. A well known problem with multivariable systems is that the "gain" of a transfer function matrix is not unique: it depends on the direction of the input vector. The maximum (minimum) singular value measures the maximum (minimum) possible gain of a transfer matrix frequency response $G(j\omega)$ in terms of the 2-norm of the input vector before and after multiplication by $G(j\omega)$. Thus the gain of $G(j\omega)$ is bounded by its maximum and minimum singular value as the input vector varies over all possible directions. By plotting the singular values $\bar{\sigma}(G(j\omega))$ and $\underline{\sigma}(G(j\omega))$ for each frequency ω we obtain a ***singular value Bode plot***; the multivariable generalization of the classical magnitude Bode plot.

In Section 3.3 we considered operator norms and induced operator norms. In particular, we showed that if the input and output of a stable multivariable system are measured by the operator 2-norm $\|\cdot\|_{2,2}$, then the induced norm is the transfer function $\mathcal{H}_\infty$ norm.

Next, in Section 3.4 the scalar product $<\cdot,\cdot>$ was used to define pre-Hilbert spaces. Completeness of normed spaces was discussed. A Banach space is a complete normed space and a Hilbert space is a complete pre-Hilbert space.

In Section 3.5 it was discussed how the concept of Banach and Hilbert spaces cover not only spaces of constant vectors and matrices like $\mathbf{C}^{m\times n}$, but also spaces whose elements are time domain or frequency domain functions. Two classes of Banach and Hilbert spaces are of particular interest in modern control theory. In fact, they have been given independent names, viz Lebesgue and Hardy spaces. In the famous $\mathcal{H}_\infty$ and $\mathcal{H}_2$ control theory, $\mathcal{H}$ stands for Hardy.

Using Lebesgue and Hardy spaces we can neatly classify some time domain and frequency domain functions. In Section'3.5 several different Lebesgue and Hardy spaces are introduced. For example, $\mathcal{L}_\infty(j\mathbf{R}, \mathbf{C}^{m\times n})$ is the Lebesgue space of all multivariable frequency responses $F(j\omega)$ for which the singular value upper bound across frequency exists $\sup_\omega \bar{\sigma}(F(j\omega)) < \infty$. The related Hardy space $\mathcal{H}_\infty(\mathbf{C}, \mathbf{C}^{m\times n})$ is the space of all multivariable stable transfer function matrices $F(s)$ for which the same condition $\sup_\omega \bar{\sigma}(F(j\omega)) < \infty$ holds. A central result in $\mathcal{H}_\infty$ control theory is that if $G(s) \in \mathcal{H}_\infty(\mathbf{C}, \mathbf{C}^{m\times n})$ and $u(t) \in \mathcal{H}_2(\mathbf{R}, \mathbf{R}^n)$, then $y(t) \in \mathcal{H}_2(\mathbf{R}, \mathbf{R}^m)$ and the induced operator 2-norm of $u(t)$ on $y(t)$ equals the $\mathcal{H}_\infty$ norm $\|G(s)\|_{\mathcal{H}_\infty}$.

CHAPTER 4
ROBUST CONTROL DESIGN USING SINGULAR VALUES

Consider the general multivariable feedback scheme shown in Figure 4.1. The plant and controller transfer function matrix are denoted $G(s)$ and $K(s)$ respectively.

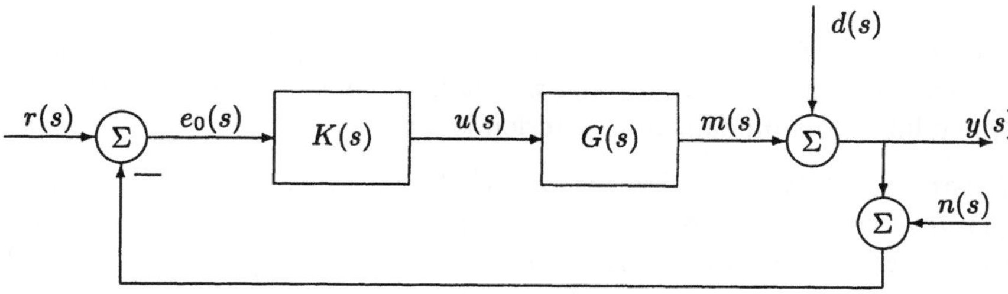

Fig. 4.1. *General feedback control configuration.*

Notice that the control configuration in Figure 4.1 is a *one-degree-of-freedom* control configuration. Disturbances $d(s)$ and reference signals $r(s)$ have, apart from the sign, the same effect on the control error $e(s) = r(s) - y(s)$. If r and d vary in a similar manner, then only a single controller is needed to compensate for r and d in e. However, if d and r behave differently then using a one-degree-of-freedom control scheme some compromise has to be found. If there are strict requirements on both set-point tracking and disturbance attenuation, an acceptable compromise might not exist. Then additional controller blocks have to be included into the control configuration. For example, a pre-compensator on $r(s)$ is often included to improve the transient response of the closed loop system. Using *two-degree-of-freedom* control configurations design of the feedback controller $K(s)$ and the pre-compensator can be handled sequentially and independently. First the feedback compensator $K(s)$ is designed to fulfill the following goals:

- Nominal closed loop stability.
- Rejection of disturbances and measurement noise for the nominal closed loop system (nominal performance).
- Robust closed loop stability.
- Robust performance.

Robust stability means that the closed loop system is stable under all possible perturbations to the plant $G(s)$. Robust performance will be used to indicate that the closed loop system is stable and that the performance requirements are met under all possible perturbations to the plant $G(s)$.

Having determined the feedback controller $K(s)$ it can secondly be judged whether a pre-compensator is needed to improve the transient response properties of the controlled system. The determination of such a compensator is then reduced to an open loop problem which can usually be solved through simple pole-zero cancelations.

In this book we will deal exclusively with the feedback problem.

4.1 Nominal Stability

The stability of a multivariable feedback control system is determined by the *extended* or *generalized Nyquist stability criterion*, see Theorem 4.1. In Appendix A this fundamental criterion is derived.

Theorem 4.1 (Generalized Nyquist Stability Criterion I). *If the open loop transfer function matrix $G(s)K(s)$ has p poles in the right-half s-plane, then the closed loop system is stable if and only if the map of* $\det(I + G(s)K(s))$, *as s traverses the Nyquist $\mathcal{D}$ contour once, encircles the origin p times anti-clockwise assuming no right-half s-plane zero-pole cancelations have occurred forming the product $G(s)K(s)$.*

Remember that the Nyquist $\mathcal{D}$ contour goes up the imaginary axis from the origin to infinity, then along a semicircular arc in the right half plane until it meets the negative imaginary axis and finally up to the origin. If any poles of $G(s)K(s)$ are encountered on the imaginary axis the contour is indented so as to exclude these poles.

An equivalent criterion can be established using *characteristic loci*. If $\lambda_i(\omega)$ denotes an eigenvalue of $G(j\omega)K(j\omega)$ the *characteristic loci* is defined as the graphs of $\lambda_i(\omega)$ for $1 \leq i \leq n$ where n is the size of the product $G(s)K(s)$ as $j\omega$ encircles the Nyquist $\mathcal{D}$ contour. Now let $\Delta \arg$ [rad] denote the change in the argument as s traverses the $\mathcal{D}$ contour so that $\Delta \arg/(2\pi)$ equals the number of origo encirclements. Since the determinant equals the product of the eigenvalues we then have that

$$\Delta \arg \{\det(I + G(s)K(s))\} = \Delta \arg \left\{ \prod_i \lambda_i(I + G(s)K(s)) \right\} \tag{4.1}$$

$$= \Delta \arg \left\{ \prod_i (1 + \lambda_i(G(s)K(s))) \right\} \tag{4.2}$$

$$= \sum_i \Delta \arg (1 + \lambda_i(G(s)K(s))) \ . \tag{4.3}$$

Since the number of encirclements of $1 + \lambda_i(G(j\omega)K(j\omega))$ around origo equals the number of encirclements of $\lambda_i(G(j\omega)K(j\omega))$ around the Nyquist point -1 we thus have the equivalent generalized Nyquist criterion in Theorem 4.2.

Theorem 4.2 (Generalized Nyquist Stability Criterion II). *If the open loop transfer function matrix $G(s)K(s)$ has p poles in the right-half s-plane, then the closed loop system is stable if and only if the characteristic loci of $G(s)K(s)$ encircle the point $(-1, 0j)$ p times anti-clockwise assuming no right-half s-plane zero-pole cancelations have occurred.*

The Generalized Nyquist Stability Criterion will be used in assessing not only nominal stability but also robust stability of an uncertain closed loop system, see Section 4.3.

4.2 Nominal Performance

From Figure 4.1 it is easily seen that

$$y(s) = T_o(s)\,(r(s) - n(s)) + S_o(s)d(s) \tag{4.4}$$
$$e_O(s) = S_o(s)\,(r(s) - d(s) - n(s)) \tag{4.5}$$
$$e(s) = r(s) - y(s) = S_o(s)\,(r(s) - d(s)) + T_o(s)n(s) \tag{4.6}$$
$$u(s) = M_o(s)\,(r(s) - n(s) - d(s)) \tag{4.7}$$

where

$$T_o(s) = (I + G(s)K(s))^{-1}\,G(s)K(s) = G(s)K(s)\,(I + G(s)K(s))^{-1} \tag{4.8}$$
$$S_o(s) = (I + G(s)K(s))^{-1} \tag{4.9}$$
$$M_o(s) = K(s)\,(I + G(s)K(s))^{-1} = (I + K(s)G(s))^{-1}\,K(s) \tag{4.10}$$

are the complementary sensitivity, sensitivity and control sensitivity functions respectively. The subscript $(\cdot)_o$ emphasizes that the sensitivity functions are all evaluated at the plant output. Since matrix multiplication is not commutative $G(s)K(s) \neq K(s)G(s)$ in general. It is thus necessary to distinguish between the sensitivity functions evaluated at the plant input and at the plant output, i.e. at the actuators and the sensors respectively. The sensitivity functions at the plant input are given by:

$$T_i(s) = K(s)G(s)\,(I + K(s)G(s))^{-1} = (I + K(s)G(s))^{-1}\,K(s)G(s) \tag{4.11}$$
$$S_i(s) = (I + K(s)G(s))^{-1} \tag{4.12}$$
$$M_i(s) = (I + K(s)G(s))^{-1}\,K(s) = M_o(s) \tag{4.13}$$

It is seen that $M_i(s) = M_o(s) = M(s)$, so the control sensitivity is independent of the chosen loop breaking point. The relevance of the input sensitivities will become clear shortly.

Now let $\delta(s) = r(s) - d(s)$ denote the "generic" external disturbance. Then from Equation (4.4)-(4.7) the following observations can be made

- For good disturbance error reduction, i.e for $\delta(s)$ to affect $e(s)$ to the least extent, (4.6) shows that the sensitivity $S_o(s)$ should be small.
- For good sensor noise error reduction, i.e for $n(s)$ to affect $e(s)$ to the least extent, (4.6) shows that the complementary sensitivity $T_o(s)$ should be small.
- For disturbances $\delta(s)$ and noise $n(s)$ to affect the control input $u(s)$ to the least extent, Equation (4.7) shows that the control sensitivity $M(s)$ should be small.

For scalar systems the size of the (scalar) transfer functions $S_o(s)$, $T_o(s)$ and $M(s)$ are naturally measured by the absolute value of the complex valued frequency responses $|S_o(j\omega)|$, $|T_o(j\omega)|$ and $|M(j\omega)|$. However, for multivariable systems the frequency responses $S_o(j\omega)$, $T_o(j\omega)$ and $M(j\omega)$ will be complex valued matrices. Thus some scalar measure of the size of a complex valued matrix is needed. Since eigenvalues are used in Theorem 4.2 it is tempting to use the *spectral radius* ρ

$$\rho(A) = \max_i |\lambda_i(A)| \tag{4.14}$$

as measure. However, it is a well known fact that eigenvalues may give poor indication of the "gain" of a transfer function matrix $G(j\omega)$ if the gain is measured as the 2-norm ratio of the output $y(j\omega)$ to the input $u(j\omega)$, see Example 4.1. As discussed in Chapter 3 if $G(s) \in \mathcal{H}_\infty(\mathbf{C}, \mathbf{C}^{m\times n})$ (a stable multivariable transfer function matrix) and $u(t) \in \mathcal{H}_2(\mathbf{R}, \mathbf{R}^n)$ (a causal time function for which the operator 2-norm $\|\cdot\|_{2,2}$ exists), then $y(t) \in \mathcal{H}_2(\mathbf{R}, \mathbf{R}^m)$ and if both $u(t)$ and $y(t)$ are measured by $\|\cdot\|_{2,2}$, then the gain of the system is bounded by $\|G(s)\|_{\mathcal{H}_\infty} = \sup_\omega \bar{\sigma}(G(j\omega))$. There is thus a sound physical motivation for characterizing the frequency response of $G(s)$ by the singular values of $G(j\omega)$. In fact, we may consider the graphs of the singular values $\sigma_i(G(j\omega))$ for all $\omega \geq 0$ as the multivariable generalization of the SISO amplitude Bode plot if we use the operator 2-norm $\|\cdot\|_{2,2}$ for measuring input and output. Other equally well motivated measures (norms) exists, but the control theory is not yet fully developed for these measures.

Example 4.1 (Eigenvalues and Singular Values). Consider the plant $G(s)$ is given by:

$$G(s) = \begin{bmatrix} 7 & -8 \\ -6 & 7 \end{bmatrix} \begin{bmatrix} \frac{1}{s+1} & 0 \\ 0 & \frac{2}{s+2} \end{bmatrix} \begin{bmatrix} 7 & -8 \\ -6 & 7 \end{bmatrix} \tag{4.15}$$

$$= \frac{1}{(s+1)(s+2)} \begin{bmatrix} 2-47s & -56s \\ 42s & 2+50s \end{bmatrix}. \tag{4.16}$$

In Figure 4.2 the size of the eigenvalues $|\lambda_i(G(j\omega))|$ are compared with the singular values $\sigma_i(G(j\omega))$. Notice that the eigenvalues do not show that the 2-norm gain of the system is very dependent on the direction of the input vector.

The current example is a slightly modified version of Example 1 in [Ber94].

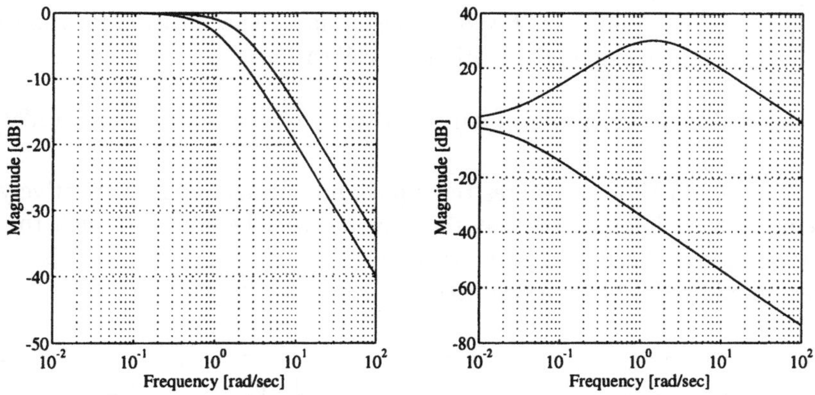

Fig. 4.2. The eigenvalues (left) and singular values (right) of $G(j\omega)$.

■

Let us then return to the sensitivity functions $S_o(s)$, $T_o(s)$ and $M(s)$. Using largest singular value as measure we may now specify standard performance requirements in terms of the maximum singular value. For good rejection of generic disturbances $\delta(s)$ we will eg require that:

$$\bar{\sigma}\left(S_o(j\omega)\right) = \|S_o(j\omega)\|_2 \ll 1 \,. \tag{4.17}$$

Similarly, for good sensor noise rejection we will require:

$$\bar{\sigma}\left(T_o(j\omega)\right) = \|T_o(j\omega)\|_2 \ll 1 \tag{4.18}$$

and for low sensitivity of the input to noise and disturbances

$$\bar{\sigma}\left(M(j\omega)\right) = \|M(j\omega)\|_2 \ll 1 \,. \tag{4.19}$$

However, because $T_o(j\omega) + S_o(j\omega) = I$ the sensitivity $S_o(j\omega)$ and the complementary sensitivity $T_o(j\omega)$ cannot both be small in the same frequency range. Consequently it is demonstrated by (4.17) and (4.18) that optimal tracking or disturbance rejection and optimal sensor noise rejection cannot be obtained *in the same frequency range*! This is a well-known result from classical control. Fortunately, the spectra of disturbances $\delta(s)$ are usually concentrated at low frequencies whereas the spectra of measurement noise $n(s)$ is concentrated at higher frequencies. Thus one may shape the complementary sensitivity $T_o(j\omega)$ and the sensitivity $S_o(j\omega)$ such that $\bar{\sigma}(S_o(j\omega))$ is small at low frequencies and $\bar{\sigma}(T_o(j\omega))$ is small at high frequencies. In modern process control systems, sensor noise will frequently not be of primary concern. However, as we shall see in the next section, robustness to unstructured uncertainty places bounds on the sensitivity functions as well.

Applying weighting functions on the sensitivity functions, we may select the frequency area of interest. A typical performance specification for robust control is given as a weighted sensitivity specification:

$$\sup_{\omega} \bar{\sigma}\left(W_{p2}(j\omega)S_o(j\omega)W_{p1}(j\omega)\right) = \|W_{p2}(s)S_o(s)W_{p1}(s)\|_{\mathcal{H}_\infty} \le 1 \quad (4.20)$$

where $W_{p1}(s)$ and $W_{p2}(s)$ denotes the input and output weight respectively, see Figure 4.3. It is assumed that the weights have been scaled such that a unity bound on the RHS makes sense. The norm $\|\cdot\|_{\mathcal{H}_\infty}$ in (4.20) is the transfer function $\mathcal{H}_\infty$ norm introduced in Section 3.3.2.

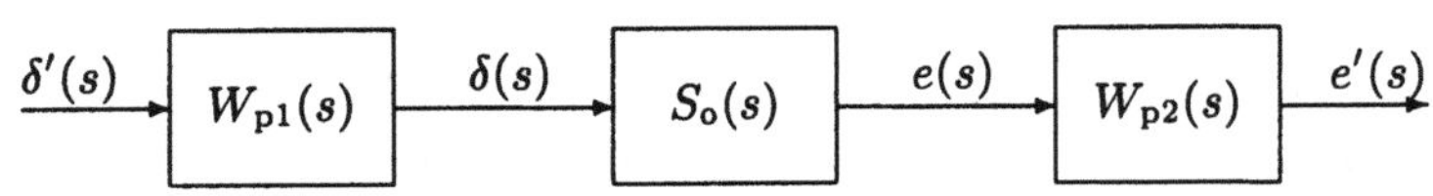

Fig. 4.3. *Output sensitivity $S_o(s)$ with input weight $W_{p1}(s)$ and output weight $W_{p2}(s)$.*

The *normalized input vector* $\delta'(t)$ is assumed to belong to the Hardy space $\mathcal{H}_2(\mathbf{R}, \mathbf{R}^n)$ with norm bounded by 1.

$$\mathcal{D}'_t = \left\{ \delta'(t) \,\middle|\, \|\delta'\|_{2,2} = \sqrt{\int_{-\infty}^{\infty} \delta(t)^T \delta(t) dt} \le 1 \right\} . \quad (4.21)$$

Then the Laplace transform $\delta'(s)$ of $\delta'(t)$ belongs to the set $\mathcal{D}' \in \mathcal{H}_2(\mathbf{C}, \mathbf{C}^n)$ given by

$$\mathcal{D}' = \left\{ \delta'(s) \,\middle|\, \|\delta'\|_{2,2} = \sqrt{\frac{1}{2\pi} \int_{-\infty}^{\infty} \delta(j\omega)^* \delta(j\omega) d\omega} \le 1 \right\} . \quad (4.22)$$

The input weight $W_{p1}(s)$ is used to transform the normalized inputs $\delta'(s)$ to the physical inputs $\delta(s) = W_{p1}(s)\delta'(s)$. For example, disturbances δ are usually expected to have small amplitude at high frequencies. Thus, if disturbance rejection is of primary interest a low pass filter would be a possible choice for $W_{p1}(s)$. If step set-point changes are most important we should choose an input weight such that the physical input $\delta(s) = s^{-1}$, see Example 4.2. Also if the physical inputs are measured in different units, $W_{p1}(s)$ can used to normalize $\delta(s)$ such that $\delta'(s)$ have equal relative magnitude.

Example 4.2 (Performance Input Weights). The following example is from [MZ89]. Assume that we expect steps on $\delta(s)$. Thus we must choose an input weight $W_{p1}(s)$ such that $\delta(s) = s^{-1}$ and $\delta'(s) \in \mathcal{D}'$. For scalar systems a seemingly obvious choice would be $W_{p1}(s) = s^{-1}$ corresponding to an impulse on the normalized input $\delta'(s)$. However, $\delta'(s) = 1$ is *not* a member of the set $\mathcal{D}'$ since the integral becomes unbounded. Thus this will not work. The weight

$$W_{p1}(s) = \frac{s+\beta}{s\sqrt{2\beta}}, \qquad \beta > 0 \tag{4.23}$$

has the desired characteristics since the normalized input

$$\delta'(s) = \frac{\sqrt{2\beta}}{s+\beta} \tag{4.24}$$

satisfies

$$\|\delta'\|_{2,2} = \sqrt{\frac{1}{2\pi}\int_{-\infty}^{\infty} \delta(j\omega)^*\delta(j\omega)d\omega} = \sqrt{\frac{1}{2\pi}\int_{-\infty}^{\infty} \frac{2\beta}{\omega^2+\beta^2}d\omega} = 1 \tag{4.25}$$

so that $\delta'(s) \in \mathcal{D}'$ and $\delta(s)$ will be given by

$$\delta(s) = W_{p1}(s)\delta'(s) = \frac{s+\beta}{s\sqrt{2\beta}}\frac{\sqrt{2\beta}}{s+\beta} = \frac{1}{s}\,. \tag{4.26}$$

■

The physical inputs $\delta(s)$ are consequently assumed to belong to the set

$$\mathcal{D} = \left\{\delta(s) \,\middle|\, \left\|W_{p1}^{-1}\delta\right\|_{2,2} \le 1\right\}\,. \tag{4.27}$$

Treating sets of generic disturbances $\delta(s)$ is attractive because at the design stage it is rarely possible to predict exactly what type of set-point changes $r(s)$ or disturbances $d(s)$ are going to occur during actual operation. Of course, in principle, the control performance could deteriorate significantly if the assumed disturbances for the design is not exactly equal to the input encountered in practice.

The output weight $W_{p2}(s)$ is used to trade off the relative importance of the individual errors in $e(s)$ and to weigh the frequency range of primary interest.

We may now give the following interpretation of the performance specification (4.20).

$$\|W_{p2}(s)S(s)W_{p1}(s)\|_{\mathcal{H}_\infty} \le 1 \tag{4.28}$$

$$\Leftrightarrow \qquad \sup_{\delta'\in\mathcal{H}_2(\mathbf{R},\mathbf{R}^n)\neq 0} \frac{\|e'\|_{2,2}}{\|\delta'\|_{2,2}} \le 1 \tag{4.29}$$

$$\Rightarrow \qquad \sup_{\delta\in\mathcal{D}_t\neq 0} \|e'\|_{2,2} \le 1 \tag{4.30}$$

$$\Leftrightarrow \qquad \sup_{\delta\in\mathcal{D}_t\neq 0} \sqrt{\int_{-\infty}^{\infty} e'(t)^T e'(t)dt} \le 1\,. \tag{4.31}$$

The nominal performance objective is then defined as follows.

Definition 4.1 (Nominal Performance Problem). *The nominal performance problem is, given weighting functions $W_{p1}(s)$ and $W_{p2}(s)$, to design a stabilizing controller $K(s)$ such that the cost function*

$$\mathcal{J}_{np} = \|W_{p2}(s)S_o(s)W_{p1}(s)\|_{\mathcal{H}_\infty} \tag{4.32}$$

is minimized. Thus

$$K(s) = \arg \min_{K(s)\in\mathcal{K}_S} \|W_{p2}(s)S_o(s)W_{p1}(s)\|_{\mathcal{H}_\infty} \tag{4.33}$$

where $\mathcal{K}_S$ denotes the set of all stabilizing controllers and inf(·) *denotes the infimum. If a controller can achieve $\|W_{p2}(s)S_o(s)W_{p1}(s)\|_{\mathcal{H}_\infty} < 1$, we say that the closed loop system has* **nominal performance**.

A very convenient way of formulating the nominal performance problem is by use of the *2×2 Block Problem Formulation.* In Figure 4.4 it is shown how the weights $W_{p1}(s)$ and $W_{p2}(s)$ may be include into the closed loop system in Figure 4.1 if disturbance attenuation is considered. The augmented closed loop system may then be represented as a 2×2 block problem. The *generalized plant* $N(s)$ contains the nominal plant $G(s)$ as well as weighting functions to reflect nominal performance objectives. If set-point changes are of primary importance $d'(s)$ may be replaced by a normalized reference $r'(s)$.

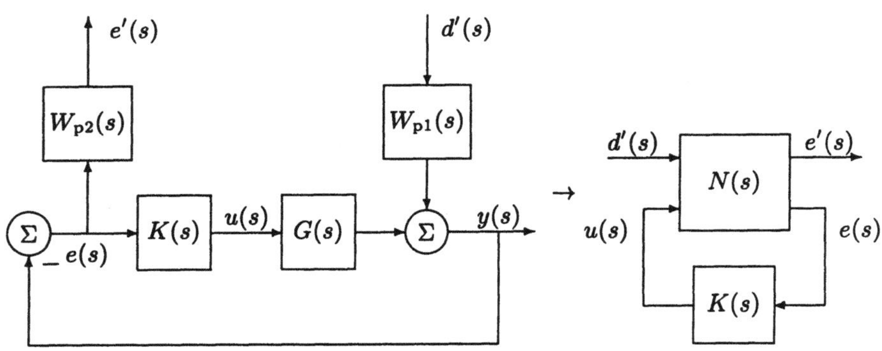

Fig. 4.4. *Nominal performance problem. Augmented closed loop system and corresponding 2 × 2 block problem.*

The transfer function from $d'(s)$ to $e'(s)$ is given by the *linear fractional transformation* (LFT):

$$\begin{aligned} e'(s) &= F_\ell(N(s), K(s))d'(s) && (4.34)\\ &= \left(N_{11}(s) + N_{12}(s)K(s)(I - N_{22}(s)K(s))^{-1}N_{21}(s)\right) d'(s) && (4.35)\\ &= W_{p2}(s)S_o(s)W_{p1}(s)d'(s)\,. && (4.36) \end{aligned}$$

Now the optimal nominal performance problem is one of finding a controller such that:

$$K(s) = \arg \min_{K(s)\in\mathcal{K}_S} \|F_\ell(N(s), K(s))\|_{\mathcal{H}_\infty} . \qquad (4.37)$$

The problem (4.37) is a standard $\mathcal{H}_\infty$ problem which can be solved with well-known techniques, see Section 4.5.

The above formulation of performance objectives are not restricted to "ordinary" sensitivity problems. If eg disturbances enter the loop not only on the output $y(s)$ but also on the input or even as an additional input to $G(s)$ this will merely change the transfer matrix expression in (4.20). Similar rejection of generic disturbances on both error $e(s)$ and control signal $u(s)$ may also easily be incorporated. However, we will then consider the complete transfer matrix from all disturbance inputs $d'(s)$ to all error signals $e'(s)$, thus including all cross-terms. True multi-objective performance measures cannot be considered in the standard $\mathcal{H}_\infty$ framework.

4.3 Robust Stability

In the early 1980'ies it was realized by a number of researchers, see e.g [DS81], that robustness to unmodeled dynamics place upper bounds on the sensitivity functions introduced in Section 4.2.

Let $G(s)$ and $G_T(s)$ denote the nominal model and true system respectively. Then introduce the following perturbation models:

- Additive uncertainty: $G_T(s) = G(s) + \tilde{\Delta}(s)$.
- Input multiplicative uncertainty: $G_T(s) = G(s)(I + \tilde{\Delta}(s))$.
- Output multiplicative uncertainty: $G_T(s) = (I + \tilde{\Delta}(s))G(s)$.
- Inverse input multiplicative uncertainty: $G_T(s) = G(s)(I + \tilde{\Delta}(s))^{-1}$.
- Inverse output multiplicative uncertainty: $G_T(s) = (I + \tilde{\Delta}(s))^{-1}G(s)$.

The perturbation block $\tilde{\Delta}(s)$ is now assumed to be an unstructured full complex block bounded using the matrix 2-norm:

$$\|\tilde{\Delta}(j\omega)\|_2 = \bar{\sigma}(\tilde{\Delta}(j\omega)) \leq \ell(\omega), \qquad \forall\omega \geq 0 . \qquad (4.38)$$

The perturbation $\tilde{\Delta}(j\omega)$ is thus a full complex matrix bounded in magnitude by a frequency dependent scalar. As before, the choice of the bounding norm is a compromise between those that best describe the plant perturbation and those that lead to tractable mathematical problems. Using the 2-norm the perturbation models above describes well the effects of high frequency unmodeled dynamics, infinite-dimensional electro-mechanical resonances and time delays. The perturbation structure (4.38) will, however, lead to conservative descriptions of structured uncertainty like parameter variations. An important reason for using the 2-norm to bound $\tilde{\Delta}(j\omega)$ is that it leads to simple expressions for robust stability.

Usually two diagonal weighting matrices $W_{u1}(s)$ and $W_{u2}(s)$ are introduced such that

$$\tilde{\Delta}(s) = W_{u2}(s)\Delta(s)W_{u1}(s) \tag{4.39}$$

and $\bar{\sigma}(\Delta)(j\omega) \leq 1$, $\forall\omega$. Usually the input weight $W_{u1}(s)$ is used to perform any necessary scaling and $W_{u2}(s)$ is used as a frequency weight to approximate $\ell(\omega)$. There is seldom any reason not to choose $W_{u1}(s)$ and $W_{u2}(s)$ as diagonal matrices.

4.3.1 The Small Gain Theorem

Now the celebrated *Small Gain Theorem*, see e.g. [DV75], will be introduced and applied in connection with the above uncertainty structures. Consider the closed loop system in Figure 4.1 and let $P(s) = G(s)K(s)$ be a square transfer function matrix.

Theorem 4.3 (Small Gain Theorem). *Assume that $P(s)$ is stable. Then the closed loop system is stable if the spectral radius $\rho(P(j\omega)) < 1$, $\forall\omega$.*

Proof (By contradiction). Assume that the spectral radius $\rho(P(j\omega)) < 1$, $\forall\omega$ and that the closed loop system is unstable. Applying Theorem 4.1 instability implies that the map of $\det(I + P(s))$ encircles the origin as s traverses the Nyquist $\mathcal{D}$ contour. Because the Nyquist contour is closed so is the map of $\det(I + P(s))$. Then there exists an $\epsilon \in [0, 1]$ and a frequency ω^* such that

$$\det(I + \epsilon P(j\omega^*)) = 0 \quad \text{i.e the map goes through the origin.} \tag{4.40}$$

$$\Leftrightarrow \quad \prod_i \lambda_i(I + \epsilon P(j\omega^*)) = 0 \tag{4.41}$$

$$\Leftrightarrow \quad 1 + \epsilon\lambda_i(P(j\omega^*)) = 0 \quad \text{for some } i \tag{4.42}$$

$$\Leftrightarrow \quad \lambda_i(P(j\omega^*)) = -\frac{1}{\epsilon} \quad \text{for some } i \tag{4.43}$$

$$\Rightarrow \quad |\lambda_i(P(j\omega^*))| \geq 1 \quad \text{for some } i \tag{4.44}$$

which is a contradiction since we assumed that $\rho(P(j\omega)) < 1$, $\forall\omega$. □

Theorem 4.3 states that for an open-loop stable system a sufficient condition for closed loop stability is to keep the loop "gain" measured by $\rho(P(j\omega))$ less than unity. Fortunately this is only a sufficient condition for stability. Otherwise the usual performance requirement of high controller gain for low frequencies could not be achieved. Theorem 4.3 thus provides only a sufficient, i.e a potentially (very) conservative condition for stability.

The Small Gain Theorem will now be used to assess the closed loop stability under unstructured norm bounded perturbations of the form (4.38). This application is classic in $\mathcal{H}_\infty$ control theory. A famous and often quoted paper on this is [DS81]. Let us eg assume an additive perturbation:

$$G_{\mathrm{T}}(s) = G(s) + W_{\mathrm{u2}}(s)\Delta(s)W_{\mathrm{u1}}(s) \tag{4.45}$$

where $\bar{\sigma}(\Delta(j\omega)) \leq 1$. This can be represented in block-diagram form as in Figure 4.5. Let $P(s)$ denote the transfer matrix "seen" by Δ, see also Figure 4.5. It is easily seen that

$$\begin{aligned} P(s) &= W_{\mathrm{u1}}(s)K(s)(I + G(s)K(s))^{-1}W_{\mathrm{u2}}(s) & (4.46)\\ &= W_{\mathrm{u1}}(s)M(s)W_{\mathrm{u2}}(s)\,. & (4.47)\end{aligned}$$

We then have the following theorem.

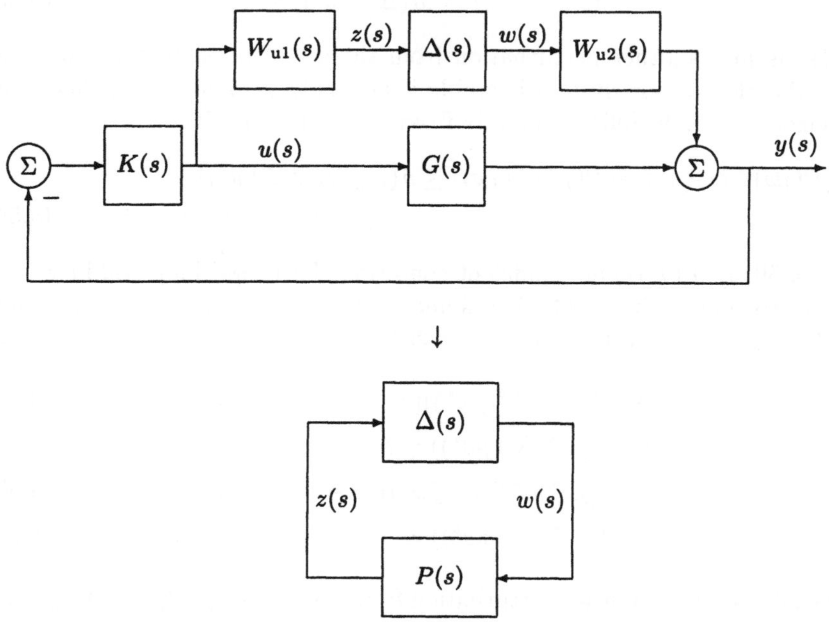

Fig. 4.5. *Closed loop system with additive perturbation. $P(s)$ denotes the transfer function "seen" by the perturbation Δ.*

Theorem 4.4 (Robust Stability). *Assume that the interconnection $P(s)$ is stable and that the perturbation $\Delta(s)$ is of such a form that the perturbed closed loop system is stable if and only if the map of $\det(I - P(s)\Delta(s))$ as s traverses the $\mathcal{D}$ contour does not encircle the origin. Then the closed loop system in Figure 4.5 is stable for all perturbations $\Delta(s)$ with $\bar{\sigma}(\Delta(j\omega)) \leq 1$ if and only if one of the following four equivalent conditions are satisfied:*

$$
\begin{aligned}
& \det(I - P(j\omega)\Delta(j\omega)) \neq 0\,, && \forall\omega,\ \forall\Delta(j\omega) \ni \bar{\sigma}(\Delta(j\omega)) \leq 1. && (4.48)\\
\Leftrightarrow\ & \rho(P(j\omega)\Delta(j\omega)) < 1\,, && \forall\omega,\ \forall\Delta(j\omega) \ni \bar{\sigma}(\Delta(j\omega)) \leq 1. && (4.49)\\
\Leftrightarrow\ & \bar{\sigma}(P(j\omega)) < 1\,, && \forall\omega. && (4.50)\\
\Leftrightarrow\ & \|P(s)\|_{\mathcal{H}_\infty} < 1\,. && && (4.51)
\end{aligned}
$$

Proof. Assume that there exist a perturbation $\Delta^*(s)$ such that $\bar{\sigma}(\Delta^*(j\omega)) \leq 1$ and that the closed loop system is unstable. Then the map of $\det(I - P(s)\Delta(s))$ encircles the origin as s traverses the Nyquist $\mathcal{D}$ contour. Because the Nyquist contour is closed so is the map of $\det(I-P(s)\Delta(s))$. Consequently there exists an $\epsilon \in [0, 1]$ and a frequency ω^* such that

$$\det(I - P(j\omega^*)\epsilon\Delta^*(j\omega^*)) = 0\,. \tag{4.52}$$

Since

$$\bar{\sigma}(\epsilon\Delta^*(j\omega^*)) = \epsilon\bar{\sigma}(\Delta^*(j\omega^*)) \leq 1 \tag{4.53}$$

$\epsilon\Delta^*(s)$ is just another perturbation from the set of possible perturbations. Thus the closed loop system is stable if and only if (4.48) is satisfied. The sufficiency of (4.49) follows directly from Theorem 4.3. Since

$$
\begin{aligned}
\rho(P(j\omega)\Delta(j\omega)) \leq \bar{\sigma}(P(j\omega)\Delta(j\omega)) &\leq \bar{\sigma}(P(j\omega))\bar{\sigma}(\Delta(j\omega))\\
&\leq \bar{\sigma}(P(j\omega)) \leq \|P(s)\|_{\mathcal{H}_\infty}
\end{aligned} \tag{4.54}
$$

both (4.50) and (4.51) are sufficient conditions for closed loop stability.

To prove necessity of (4.49) assume that there exists a $\Delta^*(s)$ for which $\bar{\sigma}(\Delta^*(j\omega)) \leq 1$ and a frequency ω^* such that $\rho(P(j\omega^*)\Delta^*(j\omega^*)) = 1$. Then

$$
\begin{aligned}
& |\lambda_i(P(j\omega^*)\Delta^*(j\omega^*))| = 1\,, && \text{for some } i && (4.55)\\
\Leftrightarrow\ & \lambda_i(P(j\omega^*)\Delta^*(j\omega^*)) = e^{j\theta}\,, && \text{for some } i && (4.56)\\
\Leftrightarrow\ & \lambda_i(P(j\omega^*)e^{-j\theta}\Delta^*(j\omega^*)) = +1\,, && \text{for some } i && (4.57)\\
\Leftrightarrow\ & \lambda_i(P(j\omega^*)\tilde{\Delta}^*(j\omega^*)) = +1\,, && \text{for some } i && (4.58)
\end{aligned}
$$

where $\tilde{\Delta}^*(s)$ is just another perturbation from the set and $\rho(P(j\omega^*)\tilde{\Delta}^*(j\omega^*)) = 1$. Therefore

$$\det(I - P(j\omega^*)\tilde{\Delta}^*(j\omega^*)) = 0 \tag{4.59}$$

and the necessity of (4.49)) has been shown.

To prove necessity of (4.50) we will show that there exists a perturbation $\Delta^*(s)$ for which $\bar{\sigma}(\Delta^*(j\omega)) \leq 1$ such that $\det(I - P(j\omega^*)\Delta^*(j\omega^*)) = 0$ if $\bar{\sigma}(P(j\omega^*)) = 1$. To do so, let $D = \text{diag}\{1, 0, \cdots, 0\}$ and perform a singular value decomposition of $P(j\omega^*)$:

$$P(j\omega^*) = U\Sigma V^H \tag{4.60}$$

where U and V are unitary matrices. Let $\Delta^*(j\omega^*) = VDU^H$. Since U and V are unitary $\bar{\sigma}(\Delta^*) = 1$. We then have that

$$\begin{aligned}\det(I - P(j\omega^*)\Delta^*(j\omega^*)) &= \det(I - U\Sigma V^H V D U^H) = \\ \det(I - U\Sigma D U^H) &= \det\left(U(I - \Sigma D)U^H\right) = \\ \det(U)\det(I - \Sigma D)\det(U^H) &= \det(I - \Sigma D) = 0 \quad (4.61)\end{aligned}$$

since the first row and column in $I - \Sigma D$ are zero. In fact, there exists an infinite number of perturbations for which (4.61) is fulfilled since we may choose $D = \mathrm{diag}\{1, \sigma_2, \cdots, \sigma_k\}$ where $\sigma_i \leq 1$ for $i = 2, \cdots, k$. □

The above proof is due to Lethomaki [Let81]. Theorem 4.4 states that if $\|P(s)\|_{\mathcal{H}_\infty} < 1$, then there is no perturbation $\Delta(s)$ ($\bar{\sigma}(\Delta(j\omega)) \leq 1$) which makes $\det(I - P(s)\Delta(s))$ encircle the origin as s traverses the Nyquist $\mathcal{D}$ contour. Notice that we *assumed* that the absence of encirclements is necessary and sufficient to maintain stability. This is the case, for example, when all perturbations $\Delta(s)$ are stable or when $G_{\mathrm{T}}(s)$ and $G(s)$ has the same number of unstable (right half plane) poles. Any one of these assumptions are standard in robust control.

Notice that the $\mathcal{H}_\infty$-norm constraint (4.51) in Theorem 4.4 is *not* conservative since we have bounded the uncertainty in terms of the spectral norm (maximum singular value). Thus if $\|P(s)\|_{\mathcal{H}_\infty} \geq 1$ there exists a perturbation $\Delta^*(s)$ for which $\bar{\sigma}(\Delta^*(j\omega)) \leq 1$ that will destabilize the closed loop system. If the uncertainty is tightly represented by $\Delta(s)$, then the singular value bound on $P(j\omega)$ is thus a tight robustness bound.

Now let us return to the additively perturbed closed loop system in Figure 4.5. From Theorem 4.4, robust stability of the closed loop system is thus obtained if and only if

$$\bar{\sigma}(P(j\omega)) < 1\,, \qquad \forall\omega. \quad (4.62)$$

$$\Leftrightarrow \quad \bar{\sigma}\left(W_{\mathrm{u1}}(j\omega)K(j\omega)\left(I + G(j\omega)K(j\omega)\right)^{-1} W_{\mathrm{u2}}(j\omega)\right) < 1\,, \quad \forall\omega. \quad (4.63)$$

$$\Leftrightarrow \quad \bar{\sigma}\left(W_{\mathrm{u1}}(j\omega)M(j\omega)W_{\mathrm{u2}}(j\omega)\right) < 1\,, \qquad \forall\omega. \quad (4.64)$$

As for additive uncertainty, we may compute singular value bounds that ensure robust stability under other perturbation models. In Table 4.1 results are given for the uncertainty structures introduced previously.

The robust stability objective is thus, eg given an additive uncertainty specification $W_{\mathrm{u2}}(s)\Delta(s)W_{\mathrm{u1}}(s)$, to design a nominally stabilizing controller $K(s)$ such that the cost function

$$\mathcal{J}_{\mathrm{u}} = \|W_{\mathrm{u1}}(s)M(s)W_{\mathrm{u2}}(s)\|_{\mathcal{H}_\infty} \quad (4.65)$$

is minimized. Thus

$$K(s) = \arg\min_{K(s)\in\mathcal{K}_{\mathrm{S}}} \|W_{\mathrm{u1}}(s)M(s)W_{\mathrm{u2}}(s)\|_{\mathcal{H}_\infty} \quad (4.66)$$

Table 4.1. *Different uncertainty descriptions and their influence on the sensitivity functions.*

Perturbation	Stability Norm Bound
Additive uncertainty	$\|W_{u1}(s)M(s)W_{u2}(s)\|_{\mathcal{H}_\infty} < 1$
Input multiplicative uncertainty	$\|W_{u1}(s)T_i(s)W_{u2}(s)\|_{\mathcal{H}_\infty} < 1$
Output multiplicative uncertainty	$\|W_{u1}(s)T_o(s)W_{u2}(s)\|_{\mathcal{H}_\infty} < 1$
Inverse input mult. uncertainty	$\|W_{u1}(s)S_i(s)W_{u2}(s)\|_{\mathcal{H}_\infty} < 1$
Inverse output mult. uncertainty	$\|W_{u1}(s)S_o(s)W_{u2}(s)\|_{\mathcal{H}_\infty} < 1$

where $\mathcal{K}_S$ denotes the set of all nominally stabilizing controllers. If a controller can achieve $\mathcal{J}_u < 1$, we say that the closed loop system is ***robustly stable.*** Notice then how the structure of the robust stability problem (4.66) equals the structure of the nominal performance problem (4.33). Consequently the robust stability problem may be formulated as a 2×2 block problem as well. Given an additive uncertainty specification, a 2×2 block problem formulation may be derived, see Figure 4.6. Compare also with Figure 4.5.

The transfer function from $w(s)$ to $z(s)$ s given by the LFT

$$z(s) = F_\ell(N(s), K(s))w(s) \tag{4.67}$$

$$= W_{u1}(s)K(s)\left(I + G(s)K(s)\right)^{-1} W_{u2}(s)w(s) \tag{4.68}$$

$$= W_{u1}(s)M(s)W_{u2}(s)w(s)\ . \tag{4.69}$$

The optimal robust stability problem is thus one of finding the controller given by

$$K(s) = \arg \min_{K(s)\in\mathcal{K}_S} \|F_\ell(N(s), K(s))\|_{\mathcal{H}_\infty}\ . \tag{4.70}$$

Like the nominal performance problem, the robust stability problem (4.70) is a standard $\mathcal{H}_\infty$ problem with a known solution.

4.4 Robust Performance

The robust performance objective is derived from (4.32) with nominal sensitivity $S_o(s)$ replaced by perturbed sensitivity $\tilde{S}_o(s)$:

$$\mathcal{J}_{rp} = \left\| W_{p2}(s)\tilde{S}_o(s)W_{p1}(s) \right\|_{\mathcal{H}_\infty}\ . \tag{4.71}$$

Let us once more illustrate with an additive uncertainty model. The robust performance problem can then be formulated as in Figure 4.7.

Let $P(s) = F_\ell(N(s), K(s))$. Then the transfer function from $d'(s)$ to $e'(s)$ is given by the LFT

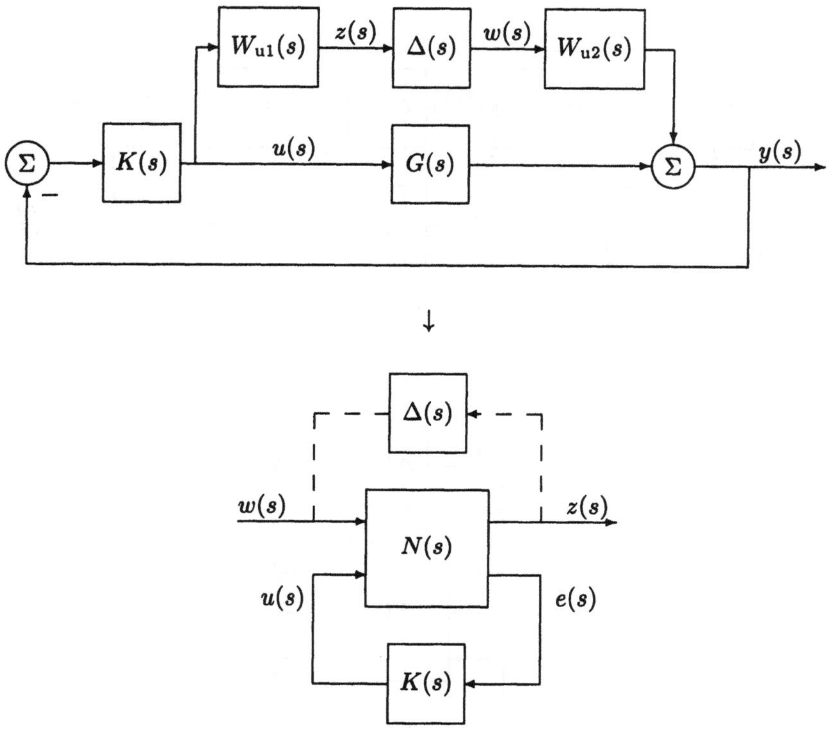

Fig. 4.6. *Uncertain closed loop system and corresponding* 2×2 *block problem.*

$$
\begin{aligned}
e'(s) &= F_{\mathrm{u}}(P(s), \Delta(s))d'(s) && (4.72)\\
&= \left[P_{22}(s) + P_{21}(s)\Delta(s)\left(I - P_{11}(s)\Delta(s)\right)^{-1} P_{12}(s)\right] d'(s) && (4.73)\\
&= W_{\mathrm{p2}}(s)\left(I + G(s)K(s) + G(s)W_{\mathrm{u2}}(s)\Delta(s)W_{\mathrm{u1}}(s)\right)^{-1} W_{\mathrm{p1}}(s) && (4.74)\\
&= W_{\mathrm{p2}}(s)\tilde{S}_{\mathrm{o}}(j\omega)W_{\mathrm{p1}}(s)\ . && (4.75)
\end{aligned}
$$

The optimal robust performance problem can then be formulated

$$
K(s) = \arg \min_{K(s)\in\mathcal{K}_{\mathrm{S}}} \sup_{\Delta(s)\ni\|\Delta(s)\|_{\mathcal{H}_\infty}\le 1} \|F_{\mathrm{u}}(F_\ell(N(s), K(s)), \Delta(s))\|_{\mathcal{H}_\infty} . \quad (4.76)
$$

If

$$
\|F_{\mathrm{u}}(F_\ell(N(s), K(s)), \Delta(s))\|_{\mathcal{H}_\infty} = \|F_{\mathrm{u}}(P(s), \Delta(s))\|_{\mathcal{H}_\infty} < 1 \quad (4.77)
$$

for all $\Delta(s)$ with $\|\Delta(s)\|_{\mathcal{H}_\infty} \le 1$, we say that the closed loop system has *robust performance.* Notice that the robust performance condition (4.77) is similar to the robust stability condition (4.51) in Theorem 4.4. Hence we conclude: *the system* $F_{\mathrm{u}}(P(s), \Delta(s))$ *satisfies the robust performance condition (4.77)* **if and only if** *it is robustly stable for a norm bounded matrix perturbation*

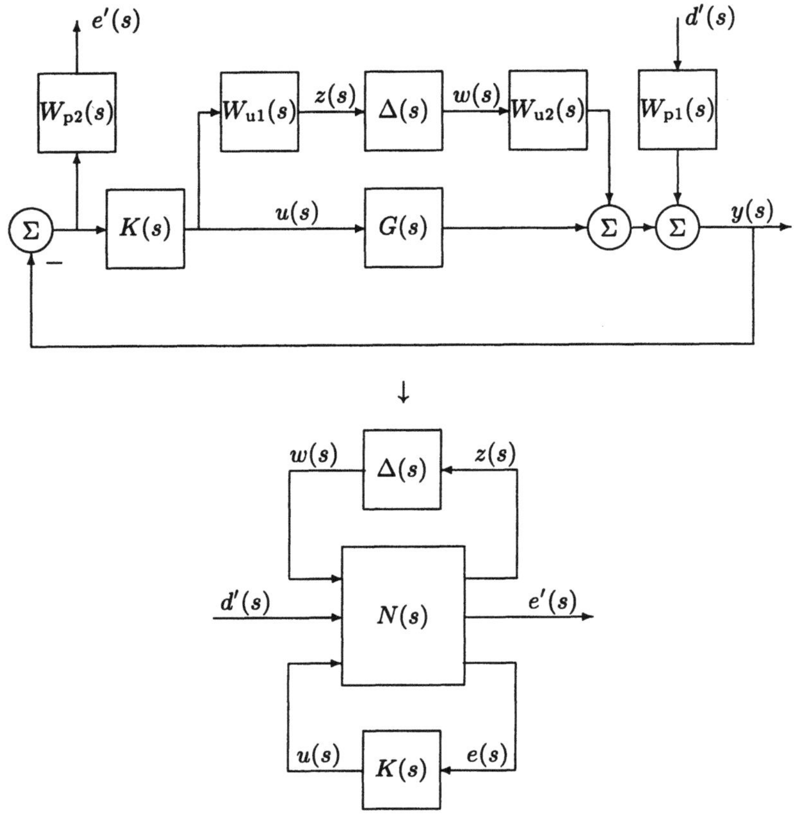

Fig. 4.7. *Robust performance problem with additive uncertainty.*

$\Delta_\mathrm{p}(s)$ *with* $\bar{\sigma}(\Delta_\mathrm{p}(j\omega)) \leq 1$. Thus by augmenting the perturbation structure with a full complex *performance block* $\Delta_\mathrm{p}(s)$ the robust performance condition can be equivalenced with a robust stability condition, see also Figure 4.8.

Let $\tilde{\Delta}(s) = \mathrm{diag}\{\Delta(s), \Delta_\mathrm{p}(s)\}$ denote the augmented perturbation matrix. We then have the following theorem.

Theorem 4.5 (Robust Performance). *Assume that the interconnection* $P(s) = F_\ell(N(s), K(s))$ *is stable and that the perturbation* $\tilde{\Delta}(s)$ *is of such a form that the perturbed closed loop system in Figure 4.8 is stable if and only if the map of* $\det(I - P(s)\tilde{\Delta}(s))$ *as* s *traverses the* $\mathcal{D}$ *contour does not encircle the origin. Then the system* $F_\mathrm{u}(P(s), \Delta(s))$ *will satisfy the robust performance criterion (4.77) if and only if* $P(s)$ *is stable for all perturbations* $\tilde{\Delta}(s)$ *with* $\bar{\sigma}(\tilde{\Delta}(j\omega)) \leq 1$*:*

$$\det(I - P(j\omega)\tilde{\Delta}(j\omega)) \neq 0\,, \quad \forall\omega,\ \forall\tilde{\Delta}(j\omega) \ni \bar{\sigma}(\tilde{\Delta}(j\omega)) \leq 1. \tag{4.78}$$

$$\Leftrightarrow \quad \rho(P(j\omega)\tilde{\Delta}(j\omega)) < 1\,, \quad \forall\omega,\ \forall\tilde{\Delta}(j\omega) \ni \bar{\sigma}(\tilde{\Delta}(j\omega)) \leq 1. \tag{4.79}$$

$$\Leftarrow \quad \|P(s)\|_{\mathcal{H}_\infty} < 1\,. \tag{4.80}$$

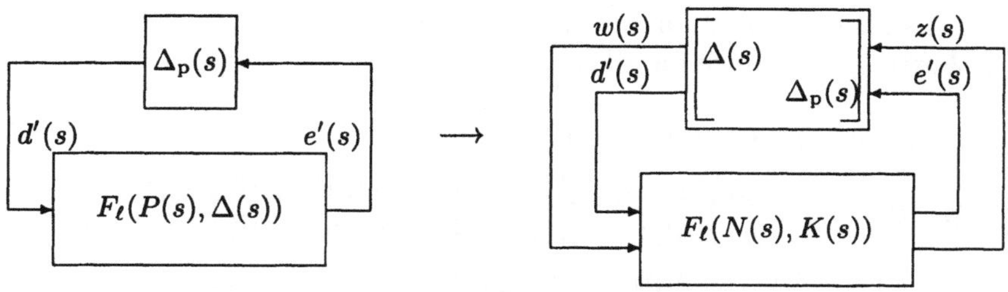

Fig. 4.8. *Block diagram structure for robust performance check. The perturbation structure is augmented with a full complex performance block* $\Delta_p(s)$.

Proof. Follows from Theorem 4.4. Since the structure on $\tilde{\Delta}(s)$ is restricted, (4.80) is a sufficient condition only. □

Clearly robust performance implies both nominal performance and robust stability. Thus a necessary condition for robust performance in connection with additive uncertainty is:

$$\mathcal{J}_{np} = \|W_{p2}(s)S_o(s)W_{p1}(s)\|_{\mathcal{H}_\infty} < 1 \tag{4.81}$$

$$\mathcal{J}_u = \|W_{u1}(s)M(s)W_{u2}(s)\|_{\mathcal{H}_\infty} < 1 \,. \tag{4.82}$$

The sufficient condition for robust performance (4.80) implies that the transfer function $P(s) = F_\ell(N(s), K(s))$ from $[w(s); d'(s)]$ to $[z(s); e'(s)]$ has ∞-norm less than one. We may thus formulated an $\mathcal{H}_\infty$ problem:

$$K(s) = \arg \min_{K(s)\in\mathcal{K}_S} \|F_\ell(N(s), K(s))\|_{\mathcal{H}_\infty} \tag{4.83}$$

with a known solution. If

$$\|F_\ell(N(s), K(s))\|_{\mathcal{H}_\infty} < 1 \tag{4.84}$$

the closed loop system will have robust performance. However, since (4.80) is a sufficient condition only it may be arbitrarily conservative. The next two examples will shed some light on this issue.

Example 4.3 (Robust Performance Problem I). Assume that a performance specification on the output sensitivity function $S_o(s)$ of the form (4.20) is given. Also an additive robust stability specification on the control sensitivity function $M(s)$ of the form (4.64) is assumed. The problem considered is thus the one illustrated in Figure 4.7. The closed loop system $F_\ell(N(s), K(s))$ is then given by

$$F_\ell(N(s), K(s)) = -\begin{bmatrix} W_{u1}(s)M(s)W_{u2}(s) & W_{u1}(s)M(s)W_{p1}(s) \\ W_{p2}(s)S_o(s)W_{u2}(s) & W_{p2}(s)S_o(s)W_{p1}(s) \end{bmatrix} . \tag{4.85}$$

A sufficient conditions for robust performance is thus $\|F_\ell(N(s),K(s))\|_{\mathcal{H}_\infty} < 1$. Furthermore, since for all ω:

$$
\begin{aligned}
&\|F_\ell(N(s),K(s))\|_{\mathcal{H}_\infty} < 1 && (4.86)\\
\Leftrightarrow\ &\bar{\sigma}\left(\begin{bmatrix} W_{u1}(j\omega)M(j\omega)W_{u2}(j\omega) & W_{u1}(j\omega)M(j\omega)W_{p1}(j\omega) \\ W_{p2}(j\omega)S_o(j\omega)W_{u2}(j\omega) & W_{p2}(j\omega)S_o(j\omega)W_{p1}(j\omega)\end{bmatrix}\right) < 1 && (4.87)\\
\Rightarrow\ &\max\left(\bar{\sigma}(W_{u1}(j\omega)M(j\omega)W_{u2}(j\omega)), \bar{\sigma}(W_{u1}(j\omega)M(j\omega)W_{p1}(j\omega)),\right.\\
&\left.\bar{\sigma}(W_{p2}(j\omega)S_o(j\omega)W_{u2}(j\omega)), \bar{\sigma}(W_{p2}(j\omega)S_o(j\omega)W_{p1}(j\omega))\right) < 1 && (4.88)\\
\Rightarrow\ &\bar{\sigma}(W_{u1}(j\omega)M(j\omega)W_{u2}(j\omega)) < 1,\ \bar{\sigma}(W_{p2}(j\omega)S_o(j\omega)W_{p1}(j\omega)) < 1 && (4.89)
\end{aligned}
$$

the robust performance condition will imply both robust stability and nominal performance. However, due to the off-diagonal elements in $F_\ell(N(s),K(s))$, the robust performance criterion may be conservative in the general case. If the weighting functions $W_{p1}(s)$ and $W_{u2}(s)$ are restricted to scalar transfer functions multiplied with a unity matrix of appropriate dimension, i.e if

$$
W_{p1}(s) = w_{p1}(s)\cdot I, \qquad W_{u2}(s) = w_{u2}(s)\cdot I \tag{4.90}
$$

where $w_{p1}(s)$ and $w_{u2}(s)$ are scalar systems, it will now be shown that the sufficient robust performance conditions (4.80) is only slightly conservative. Because of (4.90) the uncertainty weights may be gathered in $\tilde{W}_{u1}(s) = W_{u1}(s)w_{u2}(s)$ and the performance weights in $\tilde{W}_{p2}(s) = W_{p2}(s)w_{p1}(s)$. The robust performance requirement can then be written

$$
\begin{aligned}
&\bar{\sigma}(F_\ell(N(j\omega),K(j\omega))) < 1\,, && \forall\omega \quad (4.91)\\
\Leftrightarrow\ &\bar{\sigma}\left(\begin{bmatrix} \tilde{W}_{u1}(j\omega)M(j\omega) & \tilde{W}_{u1}(j\omega)M(j\omega) \\ \tilde{W}_{p2}(j\omega)S_o(j\omega) & \tilde{W}_{p2}(j\omega)S_o(j\omega)\end{bmatrix}\right) < 1\,, && \forall\omega \quad (4.92)\\
\Leftrightarrow\ &\bar{\sigma}\left(\begin{bmatrix} \tilde{W}_{u1}(j\omega)M(j\omega) \\ \tilde{W}_{p2}(j\omega)S_o(j\omega)\end{bmatrix}\right) < \frac{1}{\sqrt{2}}\,, && \forall\omega \quad (4.93)\\
\Leftarrow\ &\max\{\bar{\sigma}(\tilde{W}_{u1}(j\omega)M(j\omega)), \bar{\sigma}(\tilde{W}_{p2}(j\omega)S_o(j\omega))\} < \frac{1}{2}\,, && \forall\omega. \quad (4.94)
\end{aligned}
$$

The inequality (4.93) follows since

$$
\bar{\sigma}\left(\begin{bmatrix} \tilde{W}_{u1}(j\omega)M(j\omega) & \tilde{W}_{u1}(j\omega)M(j\omega) \\ \tilde{W}_{p2}(j\omega)S_o(j\omega) & \tilde{W}_{p2}(j\omega)S_o(j\omega)\end{bmatrix}\right) = \sqrt{2}\bar{\sigma}\left(\begin{bmatrix} \tilde{W}_{u1}(j\omega)M(j\omega) \\ \tilde{W}_{p2}(j\omega)S_o(j\omega)\end{bmatrix}\right) \tag{4.95}
$$

Thus (4.94) becomes a sufficient condition for robust performance. Consequently if the nominal performance and robust stability criterion are satisfied with some margin – namely a factor 2 – robust performance will be guaranteed. Conversely, for a nominal performance and additive uncertainty specification restricted as in (4.90), robust performance can *not* be arbitrarily poor if nominal performance and robust stability is obtained since $\|F_\ell(N(s), K(s))\|_{\mathcal{H}_\infty}$ then will be bounded by two (implied by (4.94) $\rightarrow$ (4.91)). Fully equivalent results can be found for output multiplicative uncertainty. ■

Example 4.4 (Robust Performance Problem II). Now let us assume a standard performance specification as in Example 4.3 but an input multiplicative robust stability specification on the input complementary sensitivity $T_\mathrm{i}(s)$, see Table 4.1. We may rewrite this as an additive perturbation simply by multiplying $W_{\mathrm{u}2}(s)$ in Example 4.3 with the plant transfer function $G(s)$. The (sufficient) robust performance condition then becomes

$$\bar{\sigma}(F_\ell(N(j\omega), K(j\omega))) < 1 \quad \Leftrightarrow \tag{4.96}$$

$$\bar{\sigma}\left(\begin{bmatrix} W_{\mathrm{u}1}(j\omega)M(j\omega)G(j\omega)W_{\mathrm{u}2}(j\omega) & W_{\mathrm{u}1}(j\omega)M(j\omega)W_{\mathrm{p}1}(j\omega) \\ W_{\mathrm{p}2}(j\omega)S_\mathrm{o}(j\omega)G(j\omega)W_{\mathrm{u}2}(j\omega) & W_{\mathrm{p}2}(j\omega)S_\mathrm{o}(j\omega)W_{\mathrm{p}1}(j\omega) \end{bmatrix}\right) < 1 \tag{4.97}$$

for all $\omega \geq 0$. Note that $M(s)G(s) = T_\mathrm{i}(s)$. Again, due to the off-diagonal elements in $F_\ell(N(s), K(s))$, the condition (4.97) may be conservative. In fact, even if the weightings are restricted as in (4.90) it will be shown that (4.97) may be arbitrarily conservative. With weightings restricted as in (4.90) note that the robust stability criterion becomes

$$\bar{\sigma}\left(\tilde{W}_{\mathrm{u}1}(j\omega)M(j\omega)G(j\omega)\right) < 1\,, \qquad \forall\omega \tag{4.98}$$

$$\Leftarrow \quad \bar{\sigma}\left(\tilde{W}_{\mathrm{u}1}(j\omega)M(j\omega)\bar{\sigma}(G(j\omega))\right) < 1\,, \qquad \forall\omega \tag{4.99}$$

$$\Leftrightarrow \quad \bar{\sigma}\left(\bar{\sigma}(G(j\omega))\tilde{W}_{\mathrm{u}1}(j\omega)M(j\omega)\right) < 1\,, \qquad \forall\omega \tag{4.100}$$

which corresponds to robust stability for an additive perturbation $\tilde{\Delta}(s) = \Delta(s)\bar{\sigma}(G(s))\tilde{W}_{\mathrm{u}1}(s)$. Thus if the system is stable for the additive perturbation $\Delta(s)\bar{\sigma}(G(s))\tilde{W}_{\mathrm{u}1}(s)$ it will also be stable for the input multiplicative perturbation $\Delta(s)\tilde{W}_{\mathrm{u}1}(s)$. Notice that this corresponds to approximating $G(s)$ with its largest singular value and moving it to the uncertainty input weight $\tilde{W}_{\mathrm{u}1}(s)$, see Figure 4.9.

A sufficient condition for robust performance is then that for all $\omega \geq 0$:

$$\bar{\sigma}(F_\ell(N(j\omega), K(j\omega))) < 1 \tag{4.101}$$

$$\Leftarrow \quad \bar{\sigma}\left(\begin{bmatrix} \bar{\sigma}(G(j\omega))\tilde{W}_{\mathrm{u}1}(j\omega)M(j\omega) & \bar{\sigma}(G(j\omega))\tilde{W}_{\mathrm{u}1}(j\omega)M(j\omega) \\ \tilde{W}_{\mathrm{p}2}(j\omega)S_\mathrm{o}(j\omega) & \tilde{W}_{\mathrm{p}2}(j\omega)S_\mathrm{o}(j\omega) \end{bmatrix}\right) < 1 \tag{4.102}$$

$$\Leftrightarrow \quad \bar{\sigma}\left(\begin{bmatrix} \bar{\sigma}(G(j\omega))\tilde{W}_{\mathrm{u}1}(j\omega)M(j\omega) \\ \tilde{W}_{\mathrm{p}2}(j\omega)S_\mathrm{o}(j\omega) \end{bmatrix}\right) < \frac{1}{\sqrt{2}} \tag{4.103}$$

$$\Leftarrow \quad \begin{cases} \bar{\sigma}(G(j\omega))\bar{\sigma}(\tilde{W}_{\mathrm{u}1}(j\omega)T_\mathrm{i}(j\omega)G^{-1}(j\omega)) < \frac{1}{2} \\ \bar{\sigma}(\tilde{W}_{\mathrm{p}2}(j\omega)S_\mathrm{o}(j\omega)) < \frac{1}{2} \end{cases} \tag{4.104}$$

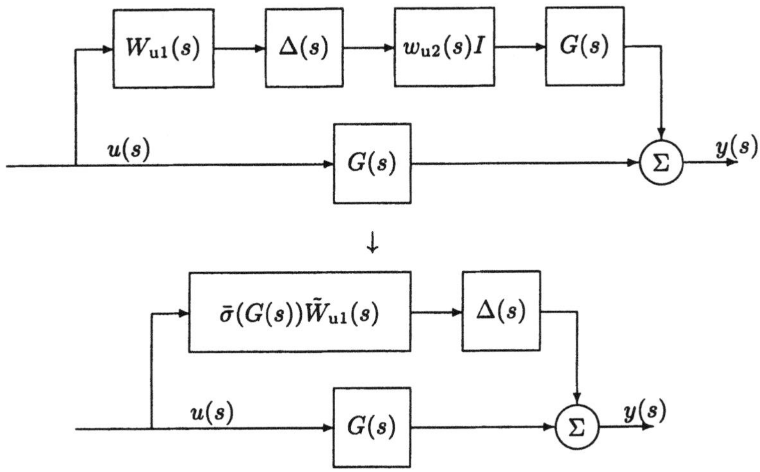

Fig. 4.9. *Approximating $G(s)$ with its largest singular value in the uncertainty specification.*

$$\Leftarrow \quad \begin{cases} \dfrac{\bar{\sigma}(G(j\omega))}{\underline{\sigma}(G(j\omega))}\bar{\sigma}(\tilde{W}_{u1}(j\omega)T_i(j\omega)) < \frac{1}{2} \\ \\ \bar{\sigma}(\tilde{W}_{p2}(j\omega)S_o(j\omega)) < \frac{1}{2} \end{cases} \tag{4.105}$$

$$\Leftrightarrow \quad \begin{cases} \bar{\sigma}(\tilde{W}_{u1}(j\omega)T_i(j\omega)) < \dfrac{1}{2\kappa(G(j\omega))} \\ \\ \bar{\sigma}(\tilde{W}_{p2}(j\omega)S_o(j\omega)) < \frac{1}{2} \end{cases} \tag{4.106}$$

Thus (4.106) becomes a sufficient condition for robust performance. Inequality (4.106) shows that even when nominal performance and robust stability are satisfied with reasonable margin, the (sufficient) robust performance criterion can be violated by an arbitrarily large amount if the plant is ill-conditioned. Conversely, if $\kappa(G(j\omega)) \approx 1$, $\forall\omega$, then robust performance cannot be arbitrarily poor if nominal performance and robust stability are obtained. ■

4.5 Computing the $\mathcal{H}_\infty$ Optimal Controller

Let us finally briefly present a solution to the $\mathcal{H}_\infty$ optimal control problem. The problem of solving minimizations of the form

$$K(s) = \arg \min_{K(s)\in\mathcal{K}_S} \|F_\ell(N(s), K(s))\|_{\mathcal{H}_\infty} \tag{4.107}$$

was probably the single most important research area within the automatic control community in the 1980'ies. At first, only algorithms that produced

$\mathcal{H}_\infty$ optimal controllers of very high order were available, see eg [Fra87]. For polynomial systems numerical algorithms were available which provided $\mathcal{H}_\infty$ optimal controllers of the same order as the augmented plant $N(s)$, see eg [Gri86, Gri88, Kwa85]. However, the numerics were only efficient for scalar systems. Then in early 1988, Doyle, Glover, Khargonekhar and Francis announced an state-space $\mathcal{H}_\infty$ solution which, like the LQG solution, involved only two Ricatti equations and yielded a controller with the same order as the generalized plant. The results were presented at the 1988 American Control Conference and in the 1989 IEEE paper [DGKF89]. This was a major breakthrough for $\mathcal{H}_\infty$ control theory. The parallels now apparent between $\mathcal{H}_\infty$ and LQG theory are pervasive. Both controllers have a state estimator-state feedback structure, two Ricatti equations provides the full state feedback matrix K_c and the output injection matrix K_f in the estimator, respectively. The paper [DGKF89] is now known simply as the DGKF paper.

Given a block 2×2 system $N(s)$ like in Figure 4.4 and a required upper bound γ on the closed loop infinity norm $\|F_\ell(N(s), K(s))\|_{\mathcal{H}_\infty}$ the solution returns a parameterization, which we will denote the DGKF Parameterization, $K(s) = F_\ell(J(s), Q(s))$ of all stabilizing controllers such that $\|F_\ell(N(s), K(s))\|_{\mathcal{H}_\infty} < \gamma$, see Figure 4.10. Any stable transfer function matrix $Q(s)$ satisfying $\|Q(s)\|_{\mathcal{H}_\infty} < \gamma$ will stabilize the closed loop system and cause $\|F_\ell(N(s), K(s))\|_{\mathcal{H}_\infty} < \gamma$. Any $Q(s)$ which is unstable or has $\|Q(s)\|_{\mathcal{H}_\infty} > \gamma$ will destabilize the closed loop system, or cause $\|F_\ell(N(s), K(s))\|_{\mathcal{H}_\infty} > \gamma$ or both. The solution is provided by Theorem 4.6.

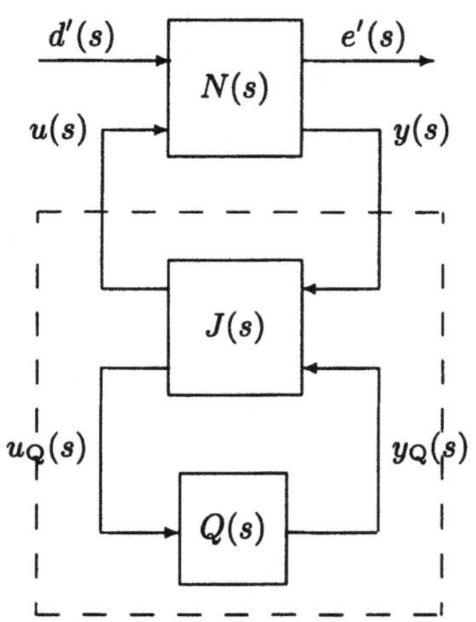

Fig. 4.10. *The DGKF parameterization of all stabilizing controllers.*

Definition 4.2 (Ricatti Solution). *Denote the solution X to the Ricatti equation*

$$A^T X + XA - XRX + Q = 0 \tag{4.108}$$

by $X = \mathbf{Ric}(H)$ *where* H *is the associated Hamiltonian matrix*

$$H = \begin{bmatrix} A & -R \\ -Q & -A^T \end{bmatrix} . \tag{4.109}$$

Theorem 4.6 ($\mathcal{H}_\infty$-Suboptimal Control Problem). *The following solution is taken from [Dai90]. Let* $N(s)$ *be given by its state-space matrices* A, B, C *and* D *and introduce the notation:*

$$N(s) = \left[\begin{array}{c|cc} A & B_1 & B_2 \\ \hline C_1 & D_{11} & D_{12} \\ C_2 & D_{21} & D_{22} \end{array}\right] . \tag{4.110}$$

Make the following assumptions:

1. (A, B_1) *and* (A, B_2) *are stabilizable.*
2. (C_1, A) *and* (C_2, A) *are detectable.*
3. $D_{12}^T D_{12} = I$ *and* $D_{21} D_{21}^T = I$.
4. $D_{11} = D_{22} = 0$.

Let

$$\tilde{D}_{12} = I - D_{12} D_{12}^T, \qquad \tilde{D}_{21} = I - D_{21}^T D_{21} \tag{4.111}$$

and solve the two Ricatti equations:

$$X_\infty = \mathbf{Ric} \begin{bmatrix} A - B_2 D_{12}^T C_1 & \gamma^{-2} B_1 B_1^T - B_2 B_2^T \\ -C_1^T \tilde{D}_{12}^T \tilde{D}_{12} C_1 & -\left(A - B_2 D_{12}^T C_1\right)^T \end{bmatrix} \tag{4.112}$$

$$Y_\infty = \mathbf{Ric} \begin{bmatrix} \left(A - B_1 D_{21}^T C_2\right)^T & \gamma^{-2} C_1^T C_1 - C_2^T C_2 \\ -B_1 \tilde{D}_{21} \tilde{D}_{21}^T B_1^T & -\left(A - B_1 D_{21}^T C_2\right) \end{bmatrix} . \tag{4.113}$$

Form the state feedback matrix K_c*, the output injection matrix* K_f *and the matrices* Z_∞*,* B_Q *and* B_Y*:*

$$K_c = \left(D_{12}^T C_1 + B_2^T X_\infty\right) \tag{4.114}$$

$$K_f = \left(B_1 D_{21}^T + Y_\infty C_2^T\right) \tag{4.115}$$

$$Z_\infty = \left(I - \gamma^{-2} Y_\infty X_\infty\right)^{-1} \tag{4.116}$$

$$B_Q = Z_\infty \left(B_2 + \gamma^{-2} Y_\infty C_1^T D_{12}\right) \tag{4.117}$$

$$B_Y = \gamma^{-2} D_{21} B_1^T X_\infty . \tag{4.118}$$

If $X_\infty \geq 0$ *and* $Y_\infty \geq 0$ *exist and if the spectral radius* $\rho(X_\infty Y_\infty) < \gamma^2$*, then the* $\mathcal{H}_\infty$ *DGKF Parameterization is given by:*

$$J(s) = \left[\begin{array}{c|cc} A_\infty & Z_\infty K_f & B_Q \\ \hline -K_c & 0 & I \\ -\left(C_2 + B_Y\right) & I & 0 \end{array}\right] \tag{4.119}$$

$$= \begin{bmatrix} J_{11}(s) & J_{12}(s) \\ J_{21}(s) & J_{22}(s) \end{bmatrix} \tag{4.120}$$

where A_∞ is given by

$$A_\infty = A - B_2 K_c + \gamma^{-2} B_1 B_1^T X_\infty - Z_\infty K_f (C_2 + B_Y) \; . \qquad (4.121)$$

Stabilizing controllers $K(s)$ may now be constructed by connecting $J(s)$ to any stable transfer function matrix $Q(s)$ with $\|Q(s)\|_{\mathcal{H}_\infty} < \gamma$:

$$\begin{aligned} K(s) &= F_\ell(J(s), Q(s)) & (4.122)\\ &= J_{11}(s) + J_{12}(s)Q(s)\left(I - J_{22}(s)Q(s)\right)^{-1} J_{21}(s) \; . & (4.123) \end{aligned}$$

The ∞-norm of the closed loop system $F_\ell(N(s), F_\ell(J(s), Q(s)))$ satisfy:

$$\|F_\ell\left(N(s), F_\ell(J(s), Q(s))\right)\|_{\mathcal{H}_\infty} < \gamma \; . \qquad (4.124)$$

The controller obtained for $Q(s) = 0$ is known as the central $\mathcal{H}_\infty$ controller.

4.5.1 Remarks on the $\mathcal{H}_\infty$ Solution

Notice that Theorem 4.6 does not provide the optimal $\mathcal{H}_\infty$ control law. Rather it provides a control law satisfying $\|F_\ell(N(s), K(s))\|_{\mathcal{H}_\infty} < \gamma$ after γ has been specified, provided that a control law exists which can do this. Consequently the designer must iterate on γ to approach the optimal ∞-norm γ_0. This is different to the LQG-optimal control problem where the optimal solution is found without iteration. The central controller obtained for $Q(s) = 0$ is not generally the controller achieving the smallest ∞-norm of the closed loop system. However, since it is a valid controller, given the desired bound γ, it is customary to choose this particular controller for implementation. Specifically, in the commercially available software [CS92, BDG+93] this is the controller returned by the $\mathcal{H}_\infty$ control synthesis algorithms.

Furthermore, note that even though the $\mathcal{H}_\infty$ problem is stated in the frequency domain, the solution is presented in state space form. This combination of frequency domain specifications and state space computation has become symptomatic in modern control theory with polynomial systems being an exception.

Consider the necessary assumptions 1-4. Assumption 1 and 2 are clearly reasonable. They simply state that the generalized plant shall be stabilizable and detectable. Assumption 3 can only be met if D_{12} has no more columns than rows (is "tall") and if D_{21} has no more rows than columns (is "fat"). This implies that:

$$\dim e'(s) \geq \dim u(s) \; , \qquad \dim d'(s) \geq \dim y(s) \qquad (4.125)$$

where dim x denotes the dimension of x. Consequently, the number of external outputs (error signals) must be equal to or exceed the number of controlled inputs (actuators) and equivalently, the number of external inputs must be equal to or exceed the number of measured outputs (sensors). For

SISO systems, since dim u = dim $y = 1$ it simply means that both d' and e' should be present. Clearly a fair assumption. For MIMO systems, however, we may form sensible problems where (4.125) is not fulfilled. By adding fictitious inputs or outputs with small weights this may be avoided.

Given a general D matrix where D_{12} and D_{21}^T has full column rank and the desired upper bound γ, a series of scalings and loop shifting operations can be carried out to put D into the form:

$$D_{11} = 0 \tag{4.126}$$

$$D_{12} = \begin{bmatrix} 0 \\ I \end{bmatrix} \tag{4.127}$$

$$D_{21} = \begin{bmatrix} 0 & I \end{bmatrix} \tag{4.128}$$

$$D_{22} = 0 \tag{4.129}$$

so that assumptions 3 and 4 are fulfilled. A suitable loop shifting and scaling algorithm is described in Appendix B. The necessary rank conditions on D_{12} and D_{21} imply that:

$$\text{rank } D_{12} = \dim u\ , \qquad \text{rank } D_{21} = \dim y\ . \tag{4.130}$$

These two conditions are very common in the literature on robust control. In the MATLAB software [CS92, BDG+93] they must be fulfilled both for the $\mathcal{H}_2$ and the $\mathcal{H}_\infty$ problem. The rank condition on D_{12} states that there must be a direct path from the control input $u(s)$ to the error output $e'(s)$. In other words, the open loop transfer function from $u(s)$ to $e'(s)$ must have equal number of poles and zeros. Equivalently, for the rank condition on D_{21} to be fulfilled there must be a direct path from the external input $d'(s)$ to the measured output $y(s)$. However, it is easy to formulate sensible control problems within the 2×2 block structure that do not fulfill the rank conditions (4.130). These problems may usually be reformulated so as to fulfill (4.130), but it does mean that care has to be taken when formulating the 2×2 problem.

The structure of the $\mathcal{H}_\infty$ controller is shown in Figure 4.11. Like the LQG solution it involves a feedback estimator for the estimated state vector $\hat{x}(s)$ and a full state feedback matrix K_c. In contrast to the LQG solution, a scaling matrix Z_∞ appears in series with the output injection matrix K_f and a few extra terms appear for $B_Y = \gamma^{-2} D_{21} B_1^T X_\infty$, $B_X = \gamma^{-2} B_1 B_1^T X_\infty$ and $B_Q = Z_\infty(B_2 + \gamma^{-2} Y_\infty C_1^T D_{12})$.

It can be shown, see eg [TCB95], that for $\gamma \to \infty$ the corresponding *Youla parameterization* for the $\mathcal{H}_2$ (or LQG) problem is recovered from the DGKF Parameterization. Consequently, the $\mathcal{H}_2$ problem is simply a special case of the $\mathcal{H}_\infty$ problem. Comparing the Hamiltonians for the $\mathcal{H}_\infty$ and $\mathcal{H}_2$ Ricatti equations reveal that the only difference is the extra terms $\gamma^{-2} B_1 B_1^T$ and $\gamma^{-2} C_1 C_1^T$ in the upper right corners. The $\mathcal{H}_2$ full state feedback matrix K_c does not depend on B_1, i.e the way in which the external signal $d'(s)$ enter the system does not affect the $\mathcal{H}_2$ state feedback solution. However it

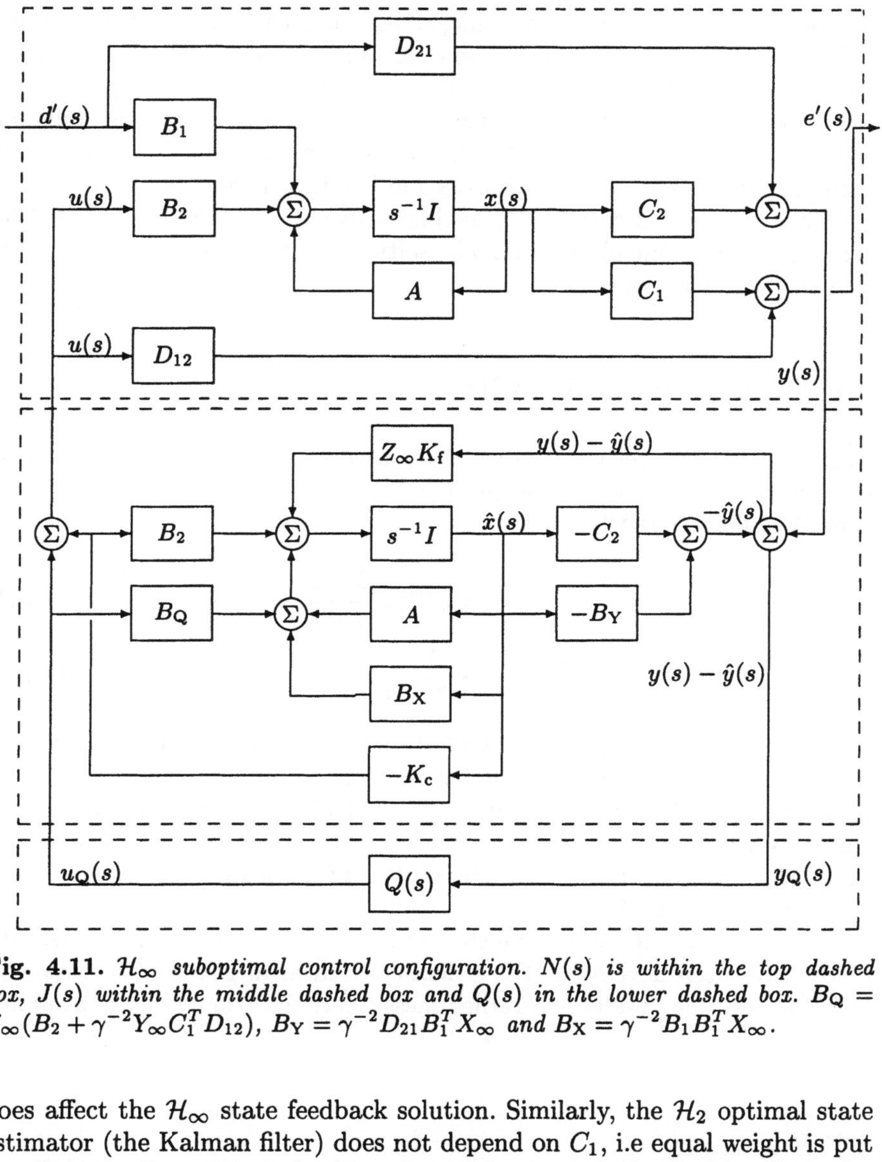

Fig. 4.11. *$\mathcal{H}_\infty$ suboptimal control configuration. $N(s)$ is within the top dashed box, $J(s)$ within the middle dashed box and $Q(s)$ in the lower dashed box. $B_Q = Z_\infty(B_2 + \gamma^{-2}Y_\infty C_1^T D_{12})$, $B_Y = \gamma^{-2}D_{21}B_1^T X_\infty$ and $B_X = \gamma^{-2}B_1 B_1^T X_\infty$.*

does affect the $\mathcal{H}_\infty$ state feedback solution. Similarly, the $\mathcal{H}_2$ optimal state estimator (the Kalman filter) does not depend on C_1, i.e equal weight is put on all states. In contrast, the $\mathcal{H}_\infty$ state estimator is influenced by C_1, i.e by which linear combination of states appears at the external output $e'(s)$.

Most of the robustness problems of LQG optimal control stem from the degradation of the closed loop response when Kalman filter state estimates are substituted for measured states in the state feedback control law. Some states usually contribute more to the loop gain than others, but LQG or $\mathcal{H}_2$ state estimation cannot compensate for this since all states are weighted equally. In the $\mathcal{H}_\infty$ state estimator the designer can tradeoff the contribution of the different states through the C_1 matrix and thus make the closed loop response more insensitive to modeling errors. Similarly, a tradeoff of the external inputs can be made when computing the full state feedback matrix K_c, thus enhancing the closed loop performance. $\mathcal{H}_\infty$ synthesis simply chooses from a wider set of possible stabilizing control laws than $\mathcal{H}_2$, so naturally one would expect to be able to obtain higher performance and robustness levels using $\mathcal{H}_\infty$ given a fixed set of sensors and bandwidth constraints.

There exists commercially available software, e.g. [CS92, BDG+93] for solving the $\mathcal{H}_\infty$ control problem. The software performs an iteration on γ to find the optimal controller.

4.6 Discrete-Time Results

In the current chapter we have only presented continuous-time results. However, it is well-known, see e.g. [KD88], that for sufficiently small sampling time the bilinear transformation:

$$z = e^{sT_s} \approx \frac{1 + s\frac{T_s}{2}}{1 - s\frac{T_s}{2}} \tag{4.131}$$

$$\Leftrightarrow \quad s \approx \frac{2}{T_s}\frac{z-1}{z+1} \tag{4.132}$$

links the continuous-time and discrete-time results. Here T_s denotes the sampling time. Thus if the robust design problem is posed in discrete-time, we may transform $N(z)$ to continuous-time via (4.131), compute the $\mathcal{H}_\infty$ optimal continuous-time control law and back-transform the controller to discrete-time via (4.132). In particular, this is the approach taken in the MATLAB toolboxes [CS92, BDG+93].

Discrete-time algorithms which solves the Z-domain 2×2 problem directly have been developed. For a detailed treatment see [Sto92]. However, the discrete-time results are rather messy and the commercially available software which supports them are yet limited.

4.7 Summary

The classical $\mathcal{H}_\infty$ results in robust control were presented. It was shown how both nominal performance and robust stability can be addressed using a 2×2 block problem formulation. Given any of the unstructured complex perturbations introduced in Section 4.3 the closed loop system may be rewritten in $P\Delta$ form as in Figure 4.5. Then, assuming that every plant in the set described by $\Delta(s)$ with $\bar{\sigma}(\Delta) \leq 1$ can occur in practice, $\|P(s)\|_{\mathcal{H}_\infty} < 1$ is a non conservative condition for robust stability. It was then shown how the robust performance problem can be addressed by augmenting the robust stability problem with a full complex performance block. The robust performance problem then can be formulated as a 2×2 block problem.

However, even though the robust stability criterion (4.51) is non-conservative, in formulating the controller synthesis as a 2×2 block problem some conservatism will always be introduced when robust performance is considered. This is so because the perturbation structure of the robust performance problem is diagonal ($\tilde{\Delta} = \text{diag}\{\Delta, \Delta_p\}$). Unfortunately, with $\mathcal{H}_\infty$ we can only consider full blocks. Thus the off-diagonal elements in this full block will introduce conservatism in the $\mathcal{H}_\infty$ solution. For standard additive and output multiplicative uncertainty the robust performance condition will be reasonably tight, namely up to a factor 2. This applies also for standard input multiplicative uncertainty ***provided the plant is well-conditioned.*** For ill-conditioned plants, robust performance may be arbitrarily poor even though nominal performance and robust stability are obtained. In other words, for ill-conditioned plants the closed loop properties at the plant input may be very different from those at the plant output. For example, the robustness to output multiplicative uncertainty may be satisfactory even though the robustness to input multiplicative uncertainty is very poor.

Generally, it can be concluded that if an unstructured complex perturbation model is tight, i.e if all plants included by the perturbation structure can occur in practice, and if the plant is reasonably well-conditioned, the optimal $\mathcal{H}_\infty$ controller will not be very conservative in the sense that robust performance will not be arbitrarily poor given nominal performance and robust stability.

On the other hand, if an unstructured complex perturbation model is conservative or if the plant is ill-conditioned, an 2×2 block problem $\mathcal{H}_\infty$ optimal controller may be very conservative. In such cases much is to be gained using the structured singular value μ. In the next chapter robust control design with structured singular values will be considered.

CHAPTER 5

ROBUST CONTROL DESIGN USING STRUCTURED SINGULAR VALUES

As we showed in the previous chapter there are two main limitations in connection with $\mathcal{H}_\infty$ control theory. Probably the most important is that we can only handle full complex perturbation structures $\Delta(s) \in \mathbf{C}^{n \times m}$ non-conservatively in an $\mathcal{H}_\infty$ robustness test. The other main limitation is that the robust performance problem can only be considered conservatively *even for a full complex perturbation set* because performance and robustness cannot be separated in the $\mathcal{H}_\infty$ framework. The conservatism introduced depends on the applied perturbation model and the condition number of the plant. In this chapter we shall shown that both these limitations may be overcome using the *structured singular value* μ.

Firstly, we will consider the analysis problem, i.e how do we, given a controller $K(s)$, test for robust stability and robust performance using μ? Secondly the synthesis problem of finding $K(s)$ will be investigated.

5.1 μ Analysis

5.1.1 Robust Stability

We will consider control problems which can be represented in an $N\Delta K$-*framework* as illustrated in Figure 5.1.

Comparing with Figure 4.6 the similarities with the 2×2 block problem are obvious. However, now $\Delta(s)$ will not be restricted to only a full complex block. Rather it is assumed that $\Delta(s)$ has a block diagonal structure as follows. Assume that $\Delta(s)$ is a member of the bounded subset:

$$\mathbf{B\Delta} = \{\Delta(s) \in \mathbf{\Delta} \,|\bar{\sigma}(\Delta(j\omega)) < 1\} \tag{5.1}$$

where $\mathbf{\Delta}$ is defined by:

$$\mathbf{\Delta} = \left\{\operatorname{diag}\left(\delta_1^{\mathrm{r}} I_{\mathrm{r}_1}, \cdots, \delta_{\mathrm{m_r}}^{\mathrm{r}} I_{\mathrm{r}_{\mathrm{m_r}}}, \delta_1^{\mathrm{c}} I_{\mathrm{r}_{\mathrm{m_r}+1}}, \cdots, \delta_{\mathrm{m_c}}^{\mathrm{c}} I_{\mathrm{r}_{\mathrm{m_r}+\mathrm{m_c}}}, \Delta_1, \cdots, \Delta_{\mathrm{m_C}}\right) \right|$$
$$\left. \delta_i^{\mathrm{r}} \in \mathbf{R}, \delta_i^{\mathrm{c}} \in \mathbf{C}, \Delta_i \in \mathbf{C}^{\mathrm{r}_{\mathrm{m_r}+\mathrm{m_c}+i} \times \mathrm{r}_{\mathrm{m_r}+\mathrm{m_c}+i}} \right\} . \tag{5.2}$$

We are thus considering real and complex uncertainty which enters the nominal model in a *linear fractional manner*. Very general classes of robustness problems can be cast into this formulation including, for example, parametric uncertainty, see Example 5.1. Clearly, the block diagonal structure on

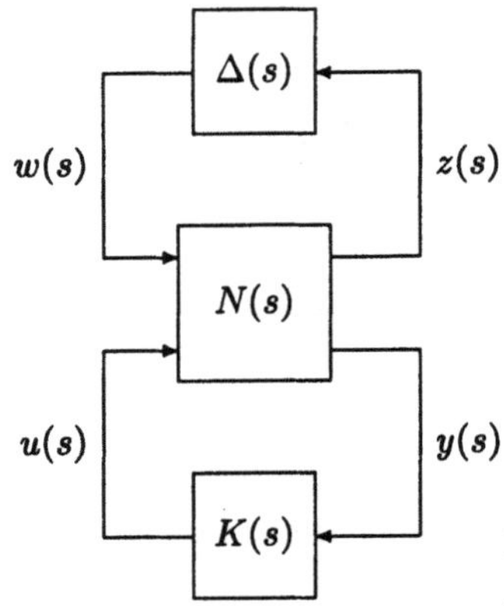

Fig. 5.1. *The $N\Delta K$ formulation of robust stability problem.*

Δ allows much more detailed uncertainty models compared with the $\mathcal{H}_\infty$ approach where $\Delta(s)$ is simply a full complex block. Notice that a full complex block is of course just a special element of the above set $\mathbf{\Delta}$.

Let us also define also the *corresponding complex perturbation set* $\mathbf{\Delta}_\mathrm{c}$:

$$\mathbf{\Delta}_\mathrm{c} = \left\{ \operatorname{diag}\left(\delta_1^\mathrm{c} I_{\mathrm{r}_1}, \cdots, \delta^\mathrm{c}_{\mathrm{m_r+m_c}} I_{\mathrm{r}_{\mathrm{m_r+m_c}}}, \Delta_1, \cdots, \Delta_{\mathrm{m_C}}\right) \mid \right.$$
$$\left. \delta_i^\mathrm{c} \in \mathbf{C}, \Delta_i \in \mathbf{C}^{\mathrm{r}_{\mathrm{m_r+m_c}+i} \times \mathrm{r}_{\mathrm{m_r+m_c}+i}} \right\} \tag{5.3}$$

where the real perturbations δ^r have been replaced by complex perturbations δ^c. This uncertainty set will be use in connection with mixed μ synthesis.

Example 5.1 (Diagonal Perturbation Formulation I). This example is a slightly modified version of that given in [Hol94]. Suppose the plant is given by:

$$G(s) = \frac{\alpha}{\beta s + 1} \tag{5.4}$$

where the DC gain α and the time constant β is only known within $\pm 10\%$, say

$$\alpha = [27.0, 33.0] , \qquad \beta = [0.9, 1.1] . \tag{5.5}$$

We now want to express α and β by their nominal values 30 and 1 and some perturbations Δ_α and Δ_β where $|\Delta_{\alpha,\beta}| \leq 1$. This can for example be accomplished as follows:

$$\alpha = 30 \left(1 + \frac{1}{10}\Delta_\alpha\right) \tag{5.6}$$

$$\beta = 1.0 \left(1 + \frac{1}{10}\Delta_\beta\right) \tag{5.7}$$

with the constraints:

$$\Delta_\alpha \in [-1, +1] , \qquad \Delta_\beta \in [-1, +1] . \tag{5.8}$$

Let us by $\mathbf{B\Delta}$ denote the set $[-1, +1]$. Then the transfer function $G(s)$ may be written:

$$G(s) = \frac{30\,(1 + 0.1\Delta_\alpha)}{(1 + 0.1\Delta_\beta)\,s + 1} \tag{5.9}$$

with

$$\Delta_\alpha, \Delta_\beta \in \mathbf{B\Delta}\,. \tag{5.10}$$

In block diagram form $G(s)$ may be given as in Figure 5.2.

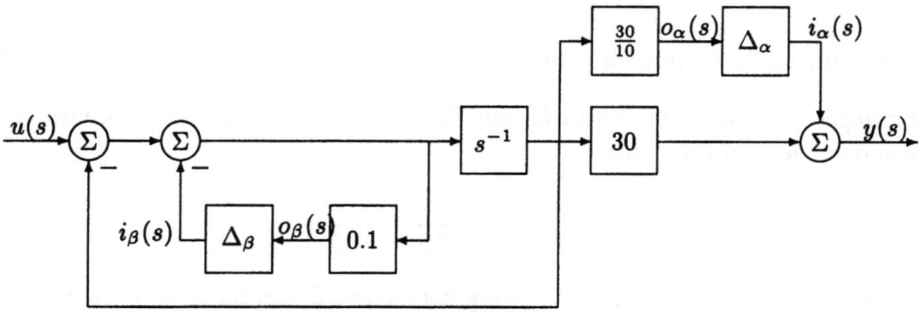

Fig. 5.2. *Example 5.1: Block diagram of $G(s)$.*

Now, in order to find the corresponding $N\Delta K$ form we need to remove the Δ-blocks in Figure 5.2 and find the transfer functions from the three inputs $i_\alpha(s)$, $i_\beta(s)$ and $u(s)$ to the three outputs $o_\alpha(s)$, $o_\beta(s)$ and $y(s)$. In matrix form one may find that

$$\begin{bmatrix} o_\alpha(s) \\ o_\beta(s) \\ y(s) \end{bmatrix} = \begin{bmatrix} 0 & -\frac{30}{10(s+1)} & \frac{30}{10(s+1)} \\ 0 & -\frac{s}{10(s+1)} & \frac{s}{10(s+1)} \\ 1 & -\frac{30}{s+1} & \frac{30}{s+1} \end{bmatrix} \begin{bmatrix} i_\alpha(s) \\ i_\beta(s) \\ u(s) \end{bmatrix}. \tag{5.11}$$

The uncertainty blocks are described by:

$$\begin{bmatrix} i_\alpha(s) \\ i_\beta(s) \end{bmatrix} = \begin{bmatrix} \Delta_\alpha o_\alpha(s) \\ \Delta_\beta o_\beta(s) \end{bmatrix} = \begin{bmatrix} \Delta_\alpha & 0 \\ 0 & \Delta_\beta \end{bmatrix} \begin{bmatrix} o_\alpha(s) \\ o_\beta(s) \end{bmatrix}. \tag{5.12}$$

Now let:

$$w(s) = \begin{bmatrix} i_\alpha(s) \\ i_\beta(s) \end{bmatrix} \tag{5.13}$$

$$z(s) = \begin{bmatrix} o_\alpha(s) \\ o_\beta(s) \end{bmatrix} \tag{5.14}$$

$$N(s) = \begin{bmatrix} 0 & -\frac{30}{10(s+1)} & \frac{30}{10(s+1)} \\ 0 & -\frac{s}{10(s+1)} & \frac{s}{10(s+1)} \\ 1 & -\frac{30}{s+1} & \frac{30}{s+1} \end{bmatrix} \tag{5.15}$$

$$\Delta(s) = \operatorname{diag}\{\Delta_\alpha, \Delta_\beta\} = \begin{bmatrix} \Delta_\alpha & 0 \\ 0 & \Delta_\beta \end{bmatrix} . \tag{5.16}$$

Then the uncertain system is described by the equations:

$$\begin{bmatrix} z(s) \\ y(s) \end{bmatrix} = N(s) \begin{bmatrix} w(s) \\ u(s) \end{bmatrix} \tag{5.17}$$

$$w(s) = \Delta(s) z(s) \tag{5.18}$$

and can readily be put into the $N\Delta K$ structure. ■

Example 5.2 (Diagonal Perturbation Formulation II). Let us next consider a standard second order lag:

$$G(s) = \frac{\alpha \omega_{\mathrm{n}}^2}{s^2 + 2\zeta\omega_{\mathrm{n}} + \omega_{\mathrm{n}}^2} . \tag{5.19}$$

Assume that the gain α, damping coefficient ζ and natural frequency ω_{n} is known only to belong to the following sets

$$\alpha = \alpha_{\mathrm{o}}(1 + \delta_\alpha \Delta_\alpha) \tag{5.20}$$

$$\zeta = \zeta_{\mathrm{o}}(1 + \delta_\zeta \Delta_\zeta) \tag{5.21}$$

$$\omega_{\mathrm{n}} = \omega_{\mathrm{n_o}}(1 + \delta_\omega \Delta_\omega) \tag{5.22}$$

with

$$\Delta_\alpha, \Delta_\zeta, \Delta_\omega \in \mathbf{B\Delta} \qquad \mathbf{B\Delta} = [-1; +1] . \tag{5.23}$$

A state-space representation of $G(s)$ is given in Figure 5.3. Due to the increased complexity of this system compared with the first order system in Example 5.1 we will work with state space representations. Define the states as:

$$x_1 = \dot{y} , \qquad x_2 = y . \tag{5.24}$$

A few manipulations reveal that the augmented system $N(s)$ can be written

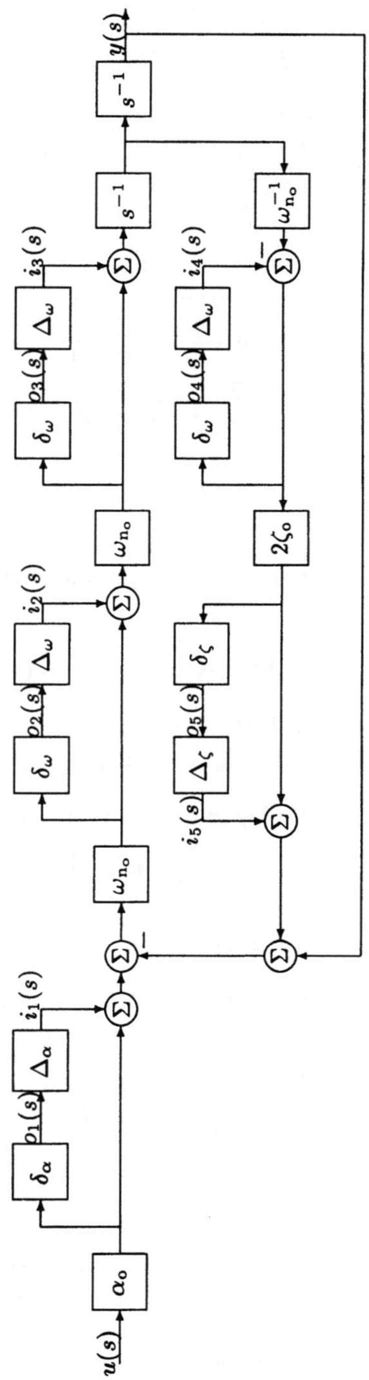

Fig. 5.3. *Transfer function representation of uncertain second order lag.*

$$
N(s) = \left[\begin{array}{c|c} A & B \\ \hline C & D \end{array}\right] = \left[\begin{array}{cc|cccccc}
-2\zeta_o\omega_{n_o} & -\omega_{n_o}^2 & \omega_{n_o}^2 & \omega_{n_o} & 1 & 2\zeta_o\omega_{n_o}^2 & -\omega_{n_o}^2 & \alpha_o\omega_{n_o}^2 \\
1 & 0 & 0 & 0 & 0 & 0 & 0 & 0 \\ \hline
0 & 0 & 0 & 0 & 0 & 0 & 0 & \alpha_o\delta_\alpha \\
-2\delta_\omega\zeta_o & -\delta_\omega\omega_{n_o} & \delta_\omega\omega_{n_o} & 0 & 0 & 2\delta_\omega\zeta_o\omega_{n_o} & -\delta_\omega\omega_{n_o} & \delta_\omega\alpha_o\omega_{n_o} \\
-2\delta_\omega\zeta_o\omega_{n_o} & -\delta_\omega\omega_{n_o}^2 & \delta_\omega\omega_{n_o}^2 & \delta_\omega\omega_{n_o} & 0 & 2\delta_\omega\zeta_o\omega_{n_o}^2 & -\delta_\omega\omega_{n_o}^2 & \delta_\omega\alpha_o\omega_{n_o}^2 \\
\delta_\omega\omega_{n_o}^{-1} & 0 & 0 & 0 & 0 & -\delta_\omega & 0 & 0 \\
2\delta_\zeta\zeta_o\omega_{n_o}^{-1} & 0 & 0 & 0 & 0 & -2\delta_\zeta\zeta_o & 0 & 0 \\
0 & 1 & 0 & 0 & 0 & 0 & 0 & 0
\end{array}\right] \tag{5.25}
$$

where we have used the notation introduced in Theorem 4.6. Proceeding as in Example 5.1, define

$$w(s) = \left[\begin{array}{ccccc} i_1(s) & i_2(s) & i_3(s) & i_4(s) & i_5(s) \end{array}\right]^T \tag{5.26}$$

$$z(s) = \left[\begin{array}{ccccc} o_1(s) & o_2(s) & o_3(s) & o_4(s) & o_5(s) \end{array}\right]^T \tag{5.27}$$

$$\Delta(s) = \operatorname{diag}\{\Delta_\alpha, \Delta_\omega, \Delta_\omega, \Delta_\omega, \Delta_\zeta\}\ . \tag{5.28}$$

Then the uncertain second order lag is described by the equations

$$\left[\begin{array}{c} z(s) \\ y(s) \end{array}\right] = N(s) \left[\begin{array}{c} w(s) \\ u(s) \end{array}\right] \tag{5.29}$$

$$w(s) = \Delta(s) z(s) \tag{5.30}$$

and can readily be put into the $N\Delta K$ structure. Notice that in this example the perturbation block structure includes repeated scalar blocks:

$$\boldsymbol{\Delta} = \left[\begin{array}{ccc} \delta_1 & 0 & 0 \\ 0 & \delta_2 I_{3\times 3} & 0 \\ 0 & 0 & \delta_3 \end{array}\right]\ . \tag{5.31}$$

■

As illustrated by the 2 above examples, highly structured uncertainty descriptions can be represented in the $N\Delta K$ framework in a straightforward manner. However, the extraction of the uncertainty blocks may involve a lot of tedious algebra. Fortunately, in the MATLAB *μ-Analysis and Synthesis Toolbox* a very handy m-function (`sysic.m`) is available which automates the necessary system interconnections.

Dynamic uncertainty as introduced in Section 4.3 may be included through complex blocks of appropriate size.

As before, denote by $F_\ell(N(s), K(s)) = P(s)$ the transfer function obtained by closing the lower loop in Figure 5.1. $P(s)$ is the *generalized closed loop transfer function.* $P(s)$ is given by the LFT

$$P(s) = F_\ell(N(s), K(s)) \tag{5.32}$$

$$= N_{11}(s) + N_{12}(s)K(s)\left(I - N_{22}(s)K(s)\right)^{-1} N_{21}(s) \tag{5.33}$$

Robust stability under structured perturbations $\Delta(s) \in \mathbf{B\Delta}$ is then determined by the following theorem which is an extension of Theorem 4.4.

Theorem 5.1. *Assume that the interconnection $P(s)$ is stable and that the perturbation $\Delta(s)$ is of such a form that the perturbed closed loop system is stable if and only if the map of $\det(I - P(s)\Delta(s))$ as s traverses the Nyquist $\mathcal{D}$ contour does not encircle the origin. Then the closed loop system in Figure 5.1 is stable for all perturbations $\Delta(s) \in \mathbf{B\Delta}$ if and only if*

$$\det(I - P(j\omega)\Delta(j\omega)) \neq 0\,, \qquad \forall\omega, \forall\Delta(j\omega) \in \mathbf{B\Delta} \tag{5.34}$$

$$\Leftrightarrow \quad \rho(P(j\omega)\Delta(j\omega)) < 1\,, \qquad \forall\omega, \forall\Delta(j\omega) \in \mathbf{B\Delta} \tag{5.35}$$

$$\Leftarrow \quad \bar{\sigma}(P(j\omega)) < 1\,, \qquad \forall\omega. \tag{5.36}$$

Proof. The proof follows easily from Theorem 4.4 with $\Delta(s) \in \mathbf{B\Delta}$. □

Note that (5.36) is a sufficient condition only for robust stability. The necessity of the similar condition (4.50) for unstructured perturbations follows from the fact that the unstructured perturbation set include *all* $\Delta(s)$ with $\bar{\sigma}(\Delta(j\omega)) \leq 1$. Here, however, we restrict the set of perturbations to $\Delta(s) \in \mathbf{B\Delta}$ and therefore, in general, condition (5.36) can be arbitrarily conservative. Rather than a singular value constraint we need some measure which takes into account the structure of the perturbations $\Delta(s)$. This is precisely the structured singular value μ.

Given a matrix $P \in \mathbf{C}^{n \times m}$, the positive real-valued function μ is then defined by:

$$\mu_{\mathbf{\Delta}}(P) = \frac{1}{\min\left\{\bar{\sigma}(\Delta) : \Delta \in \mathbf{\Delta}, \det(I - P\Delta) = 0\right\}} \tag{5.37}$$

unless no $\Delta \in \mathbf{\Delta}$ makes $I - P\Delta$ singular, in which case $\mu_{\mathbf{\Delta}}(P) = 0$. Thus $1/\mu_{\mathbf{\Delta}}(P)$ is the "size" of the smallest perturbation Δ, measured by its maximum singular value, which makes $I - P\Delta$ singular ($\det(I - P\Delta) = 0$). If $P(s)$ is a transfer function we can interpret $1/\mu_{\mathbf{\Delta}}(P(j\omega))$ as the size of the smallest perturbation $\Delta(j\omega)$ which shifts the characteristic loci of the transfer matrix $P(s)$ to the Nyquist point -1 at the frequency ω.

From the definition of μ and Theorem 5.1 we now have the following theorem for assessing robust stability, see also [DP87, PD93]:

Theorem 5.2 (Robust Stability with μ). *Assume that the interconnection $P(s)$ is stable and that the perturbation $\Delta(s)$ is of such a form that the perturbed closed loop system is stable if and only if the map of* $\det(I - P(s)\Delta(s))$ *as s traverses the Nyquist $\mathcal{D}$ contour does not encircle the origin. Then the closed loop system in Figure 5.1 is stable for all perturbations $\Delta(s) \in \mathbf{B\Delta}$ if and only if :*

$$\|\mu_{\mathbf{\Delta}}(P(s))\|_{\infty} \leq 1 \tag{5.38}$$

where:

$$\|\mu_{\mathbf{\Delta}}(P(s))\|_{\infty} = \sup_{\omega} \mu_{\mathbf{\Delta}}(P(j\omega)) \ . \tag{5.39}$$

5.1.2 Robust Performance

For robust performance we will include the normalized exogenous disturbances $d'(s)$ and the normalized error signals (controlled outputs) $e'(s)$ into the $N\Delta K$ formulation. Then we may obtain the general framework for robustness analysis and synthesis of linear systems illustrated as in Figure 5.4. Any linear interconnection of control inputs u, measured outputs y, disturbances d', controlled outputs (error signals) e', perturbations w and a controller K can be expressed within this framework.

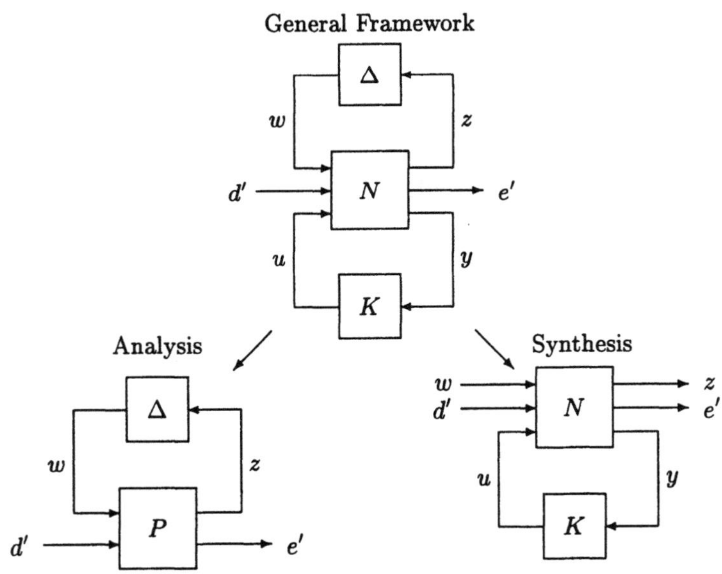

Fig. 5.4. *A general framework for analysis and synthesis of linear control systems.*

Within the general framework analysis and synthesis constitutes two special cases as illustrated in Figure 5.4. Like the 2×2 block problem, scalings and weights are conventionally absorbed into the transfer function $N(s)$ in order to normalize $d'(s)$, $e'(s)$ and $\Delta(s)$ to norm 1. Notice that if we partition $P(s)$ into four blocks consistent with the dimensions of the two input (w, d') and two output (z, e') vectors we can identify $P_{11}(s)$ as the transfer matrix $P(s)$ introduced in the previous section, see e.g. Theorem 5.2. Context will determine which $P(s)$ we refer to.

For robust performance the transfer function F_u from d' to e' may be partitioned as a linear fractional transformation:

$$e'(s) = F_u(P(s), \Delta(s))d'(s) \tag{5.40}$$

$$= \left[P_{22}(s) + P_{21}(s)\Delta(s)\left(I - P_{11}(s)\Delta(s)\right)^{-1} P_{12}(s)\right] d'(s) \,. \tag{5.41}$$

Here $P_{22}(s)$ is the weighted nominal performance function and $F_u(P(s), \Delta(s))$ is thus the weighted perturbed performance function. Now from (5.41) the robust performance objective becomes

$$\|F_u(P(s), \Delta(s))\|_\infty = \sup_\omega \bar{\sigma}\left(F_u(P(j\omega), \Delta(j\omega))\right) < 1\,, \quad \forall \Delta \in \mathbf{B\Delta}. \tag{5.42}$$

As in Section 4.4 we notice that the condition for robust performance is a singular value constraint similar to the robust stability condition (4.50) and conclude: *the robust performance objective (5.42) is satisfied if and only if the interconnection $F_u(P(s), \Delta(s))$ is robustly stable for a norm bounded matrix perturbation $\Delta_\mathrm{p}(s)$ with $\bar{\sigma}(\Delta_\mathrm{p}(j\omega)) \le 1$.* Hence by augmenting the perturbation structure with a full complex performance block $\Delta_\mathrm{p}(s)$ the robust performance condition can be equivalenced with an unstructured robust stability condition, see Figure 4.8. However, in the μ-framework we can perform this augmentation naturally since the allowed perturbation structure is block diagonal.

We thus have the following theorem for assessing robust performance, see [DP87, PD93]:

Theorem 5.3 (Robust Performance with μ). *Let an $\mathcal{H}_\infty$ performance specification be given on the transfer function from d' to e' – typically a weighted sensitivity specification – of the form:*

$$\|F_u(P(s), \Delta(s))\|_\infty = \sup_\omega \bar{\sigma}\left(F_u(P(j\omega), \Delta(j\omega))\right) < 1\,. \tag{5.43}$$

Then $F_u(P(s), \Delta(s))$ is stable and $\|F_u(P(s), \Delta(s))\|_\infty < 1$, $\forall \Delta(s) \in \mathbf{B\Delta}$ if and only if

$$\left\|\mu_{\tilde{\mathbf{\Delta}}}(P(s))\right\|_\infty \le 1 \tag{5.44}$$

where the perturbation set is augmented with a full complex performance block:

$$\tilde{\mathbf{\Delta}} = \left\{ diag\left(\Delta, \Delta_\mathrm{p}\right) \middle| \Delta \in \mathbf{\Delta}, \Delta_\mathrm{p} \in \mathbf{C}^{k \times k} \right\} \,. \tag{5.45}$$

Theorem 5.3 is the real payoff for measuring performance in terms of the $\mathcal{H}_\infty$-norm and bounding model uncertainty in the same manner. Using μ it is then possible to test for both robust stability and robust performance in a non-conservative manner. Indeed, if the uncertainty is modeled exactly by $\Delta(s)$, i.e if all plants in the norm-bounded set can really occur in practice, then the μ condition for robust performance is necessary and sufficient. Notice how the μ theorems provide much tighter conditions for robust performance compared with the $\mathcal{H}_\infty$ results in Section 4.4. Performance and stability conditions are now separated, much tighter uncertainty descriptions may be given due to the diagonal structure on $\mathbf{\Delta}$ and non-conservative results are provided for all perturbation models and even for ill-conditioned systems.

Equation (5.44) provides a very simple test for checking robust performance. Plotting $\mu_{\tilde{\mathbf{\Delta}}}(P(j\omega))$ versus frequency ω will reveal whether the conditions of Theorem 5.3 are met.

Since $\Delta_1 = \mathrm{diag}\{\Delta, 0\}$ and $\Delta_2 = \mathrm{diag}\{0, \Delta_\mathrm{p}\}$ are special cases of $\Delta \in \tilde{\mathbf{\Delta}}$ it is clear that

$$\mu_{\tilde{\mathbf{\Delta}}}(P(j\omega)) \geq \max\{\mu_{\mathbf{\Delta}}(P_{11}(j\omega)), \mu_{\mathbf{\Delta}_\mathrm{p}}(P_{22}(j\omega)) = \bar{\sigma}(P_{22}(j\omega))\} \quad (5.46)$$

which implies that for robust performance it is necessary that the closed loop system is robustly stable and satisfies the nominal performance objective.

5.1.3 Computation of μ

As we have shown above μ is a powerful tool for assessing robust stability and robust performance under structured and/or unstructured perturbations. Unfortunately, the computation of μ is a very difficult and yet unsolved mathematical problem. Equation (5.37) is not suitable for computing μ since the implied optimization problem may have multiple local maxima [DP87, FTD91]. However, tight upper and lower bounds for μ may be effectively computed for both complex and mixed perturbations sets. Algorithms for computing these bounds have been documented in several papers, see e.g. [DP87, YND91]. For simplicity, assume in the following that $P \in \mathbf{C}^{n\times n}$ is square.

μ with Complex Perturbations. Let us first consider computation of μ when the perturbation structure consists of complex elements only, i.e $m_r = 0$ in (5.2). Then we can relate $\mu_{\mathbf{\Delta}}(P)$ to familiar linear algebra quantities when $\mathbf{\Delta}$ is one of two extreme sets [DPZ91]:

- If $\mathbf{\Delta} = \{\delta^\mathrm{c} I_n \,|\, \delta^\mathrm{c} \in \mathbf{C}\}$ ($m_\mathrm{r} = 0$, $m_\mathrm{c} = 1$, $m_\mathrm{C} = 0$ in (5.2)), then $\mu_{\mathbf{\Delta}}(P) = \rho(P)$, the spectral radius.
- If $\mathbf{\Delta} = \{\Delta \,|\, \Delta \in \mathbf{C}^{n\times n}\}$ ($m_\mathrm{r} = 0$, $m_\mathrm{c} = 0$, $m_\mathrm{C} = 1$ in (5.2)), then $\mu_{\mathbf{\Delta}}(P) = \bar{\sigma}(P)$, the maximum singular value.

For a general complex perturbation $\mathbf{\Delta}$ we have that

$$\{\delta^\mathrm{c} I_n \,|\, \delta^\mathrm{c} \in \mathbf{C}\} \subset \mathbf{\Delta} \subset \{\Delta \,|\, \Delta \in \mathbf{C}^{n\times n}\} \ . \quad (5.47)$$

Thus, directly from the definition of μ (5.37) we conclude that

$$\rho(P) \leq \mu_{\mathbf{\Delta}}(P) \leq \bar{\sigma}(P) . \tag{5.48}$$

However, these bounds are insufficient since the gap between $\rho(P)$ and $\bar{\sigma}(P)$ can be arbitrarily large. Thus the bounds (5.48) must be tightened. This is done through transformations on P that *do not affect* $\mu_{\mathbf{\Delta}}(P)$ but *do affect* $\rho(P)$ *and* $\bar{\sigma}(P)$. Define the following two subsets of $\mathbf{C}^{n \times n}$:

$$\mathbf{Q} = \left\{ Q \in \mathbf{\Delta} \middle| m_{\mathrm{r}} = 0, \delta_i^{\mathrm{c}^*} \delta_i^{\mathrm{c}} = 1, \Delta_i^* \Delta_i = I_{\mathrm{r}_{m_\mathrm{c}+i}} \right\} \tag{5.49}$$

$$\mathbf{D} = \Big\{ \mathrm{diag}\left(D_1, \cdots, D_{m_\mathrm{c}}, d_1 I_{\mathrm{r}_{m_\mathrm{c}+1}}, \cdots, d_{m_\mathrm{C}} I_{\mathrm{r}_{m_\mathrm{c}+m_\mathrm{C}}} \right) \\ \Big| D_i \in \mathbf{C}^{\mathrm{r}_i \times \mathrm{r}_i}, D_i^* = D_i > 0, d_i \in \mathbf{R}, d_i > 0 \Big\} . \tag{5.50}$$

It can then be shown, see eg the original paper on μ by Doyle [Doy82] that for any $\Delta \in \mathbf{\Delta}$ (with $m_\mathrm{r} = 0$), $Q \in \mathbf{Q}$ and $D \in \mathbf{D}$,

$$Q^* \in \mathbf{Q}, \quad Q\Delta \in \mathbf{\Delta}, \quad \Delta Q \in \mathbf{\Delta}, \quad \bar{\sigma}(Q\Delta) = \bar{\sigma}(\Delta Q) = \bar{\sigma}(\Delta), \tag{5.51}$$

$$D\Delta = \Delta D . \tag{5.52}$$

From (5.51) and (5.52) the next theorem follow.

Theorem 5.4 (Upper and Lower Bounds on Complex μ). *For all $Q \in \mathbf{Q}$ and $D \in \mathbf{D}$*

$$\mu_{\mathbf{\Delta}}(PQ) = \mu_{\mathbf{\Delta}}(QP) = \mu_{\mathbf{\Delta}}(P) = \mu_{\mathbf{\Delta}}(DPD^{-1}) . \tag{5.53}$$

Thus the bounds in Equation (5.48) may be tightened to

$$\max_{Q \in \mathbf{Q}} \rho(QP) \leq \mu_{\mathbf{\Delta}}(P) \leq \inf_{D \in \mathbf{D}} \bar{\sigma}(DPD^{-1}) . \tag{5.54}$$

The lower bound $\max_{Q \in \mathbf{Q}} \rho(QP)$ is in fact an equality [Doy82], but $\rho(QP)$ is unfortunately non-convex with multiple local maxima which are not global. Hence, local search cannot be guaranteed to obtain μ, but can only yield a lower bound. Computation of the upper bound is a convex problem so the global minimum $\inf_{D \in \mathbf{D}} \bar{\sigma}(DPD^{-1})$ can, in principle at least, be found. Unfortunately, the upper bound is not always equal to μ. It can be shown that for block structures $\mathbf{\Delta}$ with $m_\mathrm{r} = 0$ and $2m_\mathrm{c} + m_\mathrm{C} \leq 3$, the upper bound is always equal to $\mu_{\mathbf{\Delta}}(P)$. However, for block structures with $2m_\mathrm{c} + m_\mathrm{C} > 3$, there exists matrices for which μ is less than the infimum. Nevertheless, even in the case where the upper bound do not equal μ exactly numerical experience suggests that the upper bound in general is reasonably tight.

With the release of the MATLAB *μ-Analysis and Synthesis Toolbox* [BDG+93] in 1993 commercially available routines exist for computing the bounds in Theorem 5.4. In practical control design (with purely complex perturbations), the mathematical problems with computation of μ may thus be considered rather theoretical.

μ with Mixed Real and Complex Perturbations. Computation of the mixed real and complex μ problem has been the focus of intensive research over the last decade, see eg [FTD91, YND91, YND92, You93]. It is beyond the scope of this book to treat the bounds on mixed μ[1] in any great detail. A complete treatment may be found in the thesis by Young [You93]. Here we will merely state some of the important results on the subject. Let us start by defining the following sets:

$$\mathbf{Q} = \left\{ Q \in \mathbf{\Delta} \middle| \delta_i^{\mathrm{r}} \in [-1;1], \delta_i^{\mathrm{c}^*} \delta_i^{\mathrm{c}} = 1, \Delta_i^* \Delta_i = I_{\mathrm{r}_{\mathrm{m_r+m_c}+i}} \right\} \tag{5.55}$$

$$\mathbf{D} = \left\{ \mathrm{diag}\left(D_1, \cdots, D_{\mathrm{m_r+m_c}}, d_1 I_{\mathrm{r}_{\mathrm{m_r+m_c}+1}}, \cdots, d_{\mathrm{m_C}} I_{\mathrm{r_m}}\right) \middle| D_i \in \mathbf{C}^{\mathrm{r}_i \times \mathrm{r}_i}, D_i^* = D_i > 0, d_i \in \mathbf{R}, d_i > 0 \right\} \tag{5.56}$$

$$\mathbf{G} = \left\{ \mathrm{diag}\left(G_1, \cdots, G_{\mathrm{m_r}}, O_{\mathrm{r}_{\mathrm{m_r}+1}}, \cdots, O_{\mathrm{r_m}}\right) \middle| G_i \in \mathbf{C}^{\mathrm{r}_i \times \mathrm{r}_i}, G_i = G_i^* \right\} \tag{5.57}$$

$$\hat{\mathbf{D}} = \left\{ \mathrm{diag}\left(D_1, \cdots, D_{\mathrm{m_r+m_c}}, d_1 I_{\mathrm{r}_{\mathrm{m_r+m_c}+1}}, \cdots, d_{\mathrm{m_C}} I_{\mathrm{r_m}}\right) \middle| D_i \in \mathbf{C}^{\mathrm{r}_i \times \mathrm{r}_i}, \det(D_i) \neq 0, d_i \in \mathbf{C}, d_i \neq 0 \right\} \tag{5.58}$$

$$\hat{\mathbf{G}} = \left\{ \mathrm{diag}\left(g_1, \cdots, g_{\mathrm{n_r}}, O_{\mathrm{n_c}}\right) \middle| g_i \in \mathbf{R} \right\} \tag{5.59}$$

where $\mathrm{r_m}=\mathrm{r}_{\mathrm{m_r+m_c+m_C}}$, $\mathrm{n_r}=\sum_{i=1}^{\mathrm{m_r}} \mathrm{r}_i$ and $\mathrm{n_c}=\mathrm{n}-\mathrm{n_r}$. Notice that for compatibility with $P(s)$ we must have that $\sum_{i=1}^{\mathrm{m}} \mathrm{r}_i = n$.

We then have the upper and lower bounds as in Theorem 5.5 [FTD91].

Theorem 5.5 (Upper and Lower Bounds on Mixed μ [FTD91]). *Let $\bar{\lambda}_{\mathrm{R}}$ denote the largest real eigenvalue and let $\rho_{\mathrm{R}}(P)$ denote the real spectral radius of P:*

$$\rho_{\mathrm{R}}(P) = \max\left\{|\lambda_{\mathrm{R}}(P)| \ : \ \lambda_{\mathrm{R}}(P) \textit{ is a real eigenvalue of } P\right\} . \tag{5.60}$$

If P has no real eigenvalues then $\rho_{\mathrm{R}}(P) = 0$. Furthermore suppose that α_* is the result of the minimization problem

$$\alpha_* = \inf_{D \in \mathbf{D}, G \in \mathbf{G}} \min_{\alpha \in \mathbf{R}} \left\{ \alpha \middle| \bar{\lambda}_{\mathrm{R}}\left(P^* D P + j\left(GP - P^* G\right) - \alpha D\right) \leq 0 \right\} . \tag{5.61}$$

Then

$$\rho_{\mathrm{R}}(P) \leq \mu_{\mathbf{\Delta}}(P) \leq \sqrt{\max(0, \alpha_*)} . \tag{5.62}$$

---

[1] We will often refer to the mixed real and complex μ problem simply as the mixed μ problem.

Notice that computation of the upper bound involves a Linear Matrix Inequality (LMI). A variety of numerical techniques exist to tackle such minimizations. Furthermore, numerical recipes for solving LMIs is an area of intensive research.

However, even for comparatively medium size problems ($n \leq 100$), the optimization over the D and G scaling matrices could involve thousands of parameters. Thus, in order to tackle such problems with reasonably computation times, straightforward application of brute force optimization techniques will not suffice. Therefore, the above bounds have been reformulated exploiting the specific structure of the mixed μ upper bound problem, see e.g. [YND92]. Several forms of the upper bound can be found as stated in Theorem 5.6.

Theorem 5.6 (Reformulation of Mixed μ Upper Bound). *Suppose we have a matrix $P \in \mathbf{C}^{n\times n}$ and a real scalar $\beta > 0$. Furthermore for any $D \in \mathbf{C}^{n\times n}$ denote $P_{\mathrm{D}} = DPD^{-1}$. Then the following statements are equivalent:*

1. *There exists matrices $D_1 \in \mathbf{D}$, $G_1 \in \mathbf{G}$ such that:*
$$\bar{\lambda}_{\mathrm{R}}\left(P^* D_1 P + j(G_1 P - P^* G_1) - \beta^2 D_1\right) \leq 0\,. \tag{5.63}$$
2. *There exists matrices $D_2 \in \mathbf{D}$, $G_2 \in \mathbf{G}$ (or $D_2 \in \hat{\mathbf{D}}$, $G_2 \in \hat{\mathbf{G}}$) such that:*
$$\bar{\lambda}_{\mathrm{R}}\left(P^*_{\mathrm{D}_2} P_{\mathrm{D}_2} + j(G_2 P_{\mathrm{D}_2} - P^*_{\mathrm{D}_2} G_2)\right) \leq \beta^2\,. \tag{5.64}$$
3. *There exists matrices $D_3 \in \mathbf{D}$, $G_3 \in \mathbf{G}$ (or $D_3 \in \hat{\mathbf{D}}$, $G_3 \in \hat{\mathbf{G}}$) such that:*
$$\bar{\sigma}\left[\left(\frac{P_{\mathrm{D}_3}}{\beta} - jG_3\right)\left(I + G_3^2\right)^{-\frac{1}{2}}\right] \leq 1\,. \tag{5.65}$$
4. *There exists matrices $D_4 \in \mathbf{D}$, $G_4 \in \mathbf{G}$ (or $D_4 \in \hat{\mathbf{D}}$, $G_4 \in \hat{\mathbf{G}}$) such that:*
$$\bar{\sigma}\left[\left(I + G_4^2\right)^{-\frac{1}{4}}\left(\frac{P_{\mathrm{D}_4}}{\beta} - jG_4\right)\left(I + G_4^2\right)^{-\frac{1}{4}}\right] \leq 1\,. \tag{5.66}$$

Proofs may be found in [You93]. Using the equivalence relations from the above Theorem we may easily form alternative upper bounds on μ. In particular, the bound implemented in the MATLAB μ toolbox is derived from (5.66). Define β^* as:

$$\beta^* = \inf_{\beta\in\mathbf{R}_+, G\in\hat{\mathbf{G}}, D\in\hat{\mathbf{D}}} \{\beta \,|\, \bar{\sigma}\left(P_{\mathrm{DG}}\right) \leq 1\} \tag{5.67}$$

with P_{DG} given by

$$P_{\mathrm{DG}} = \left(I + G^2\right)^{-\frac{1}{4}}\left(\frac{DPD^{-1}}{\beta} - jG\right)\left(I + G^2\right)^{-\frac{1}{4}}\,. \tag{5.68}$$

Then

$$\max_{Q\in\mathbf{Q}} \rho(QP) \leq \mu_{\mathbf{\Delta}}(P) \leq \beta^*\,. \tag{5.69}$$

5.2 μ Synthesis

For robust synthesis the transfer function F_ℓ from $[w\ d']^T$ to $[z\ e']^T$ may be partitioned as the linear fractional transformation:

$$\begin{bmatrix} z(s) \\ e'(s) \end{bmatrix} = F_\ell(N(s), K(s)) \begin{bmatrix} w(s) \\ d'(s) \end{bmatrix} =$$
$$\left[N_{11}(s) + N_{12}(s)K(s)\left(I - N_{22}(s)K(s)\right)^{-1} N_{21}(s)\right] \begin{bmatrix} w(s) \\ d'(s) \end{bmatrix} . \quad (5.70)$$

Noticing that $F_\ell(N(s), K(s)) = P(s)$ and using Theorem 5.3 a stabilizing controller $K(s)$ achieves robust performance if and only if for each frequency $\omega \in [0, \infty]$, the structured singular value satisfies:

$$\mu_{\tilde{\Delta}}\left(F_\ell(N(j\omega), K(j\omega))\right) < 1 . \quad (5.71)$$

Consequently, the robust performance problem becomes one of synthesizing a nominally stabilizing controller $K(s)$ that minimizes $\mu_{\tilde{\Delta}}\left(F_\ell(N, K)\right)$ across frequency:

$$K(s) = \arg \min_{K(s)\in\mathcal{K}_S} \left\| \mu_{\tilde{\Delta}}\left(F_\ell(N(s), K(s))\right) \right\|_\infty \quad (5.72)$$

where $\mathcal{K}_S$ denotes the set of all nominally stabilizing controllers.

5.2.1 Complex μ Synthesis – *D-K* Iteration

Unfortunately, (5.72) is not tractable since μ cannot be directly computed. Rather the upper bound on μ is used to formulate the control problem. For purely complex perturbations, $m_r = 0$ in (5.2), the upper bound problem becomes

$$K(s) = \arg \min_{K\in\mathcal{K}_S} \sup_{\omega} \inf_{D\in\mathbf{D}} \left\{ \bar{\sigma}\left(D(\omega)F_\ell\left(N(j\omega), K(j\omega)\right)D^{-1}(\omega)\right)\right\} . \quad (5.73)$$

Unfortunately, it is also not known how to solve (5.73). However, an approximation to complex μ synthesis can be made by the following iterative scheme. For a fixed controller $K(s)$, the problem of finding $D(\omega)$ at a set of chosen frequency points ω is just the complex μ upper bound problem which is a convex problem with known solution. Having found these scalings we may fit a real rational stable minimum phase transfer function matrix $D(s)$ to $D(\omega)$ by fitting each element of $D(\omega)$ with a real rational stable minimum phase SISO transfer function. We may impose the extra constraint that the approximations $D(s)$ should be minimum phase (so that $D^{-1}(s)$ is stable too) since any phase in $D(s)$ is absorbed into the complex perturbations. For a given magnitude of $D(\omega)$, the phase corresponding to a minimum phase transfer function system may be computed using complex cepstrum techniques. Accurate transfer function estimates may then be generated using standard frequency domain least squares techniques.

For given scalings $D(s)$ the problem of finding a controller $K(s)$ which minimizes the norm $\|F_\ell(D(s)N(s)D^{-1}(s), K(s))\|_{\mathcal{H}_\infty}$ will be reduced to a standard $\mathcal{H}_\infty$ problem. Repeating this procedure, denoted *D-K* iteration, several times will yield the complex μ optimal controller provided the algorithm converges. Even though the computation of the D scalings and the optimal $\mathcal{H}_\infty$ controller are both convex problems, the *D-K* iteration procedure is *not jointly convex* in $D(s)$ and $K(s)$ and counter examples of convergence has been given [Doy85]. However, *D-K* iteration seems to work quite well in practice and has been successfully applied to a large number of applications. The *D-K* iteration procedure may be outlined as below.

Procedure 5.1 (*D-K* Iteration).

1. *Given an augmented system $N(s)$, let $i = 1$ and $D_i^\star(\omega) = I, \forall\omega$.*
2. *Fit a stable minimum phase transfer function matrix $D_i(s)$ to the pointwise scalings $D_i^\star(\omega)$. Augment $D_i(s)$ with a unity matrix of appropriate size such that $D_i(s)$ is compatible with $N(s)$. Construct the interconnection $N_{\mathrm{D}_i}(s) = D_i(s)N(s)D_i^{-1}(s)$.*
3. *Find the $\mathcal{H}_\infty$ optimal controller $K_i(s)$:*

$$K_i(s) = \arg \min_{K(s)\in\mathcal{K}_S} \|F_\ell(N_{\mathrm{D}_i}(s), K(s))\|_{\mathcal{H}_\infty} \; . \tag{5.74}$$

4. *Compute the new scalings $D_{i+1}^\star(\omega)$ solving the complex μ upper bound problem*

$$D_{i+1}^\star(\omega) = \arg \min_{D(\omega)\in\mathbf{D}} \left\{\bar{\sigma}\left(D(\omega)F_\ell(N(j\omega), K_i(j\omega))D^{-1}(\omega)\right)\right\} \tag{5.75}$$

   pointwise across frequency ω.
5. *Compare $D_{i+1}^\star(\omega)$ and $D_i^\star(\omega)$. Stop if they are close. Otherwise let $i = i + 1$ and repeat from 2.*

Notice that we use the $\mathcal{H}_\infty$ optimal control solution to synthesize the controller in step 3. The K-step of the procedure may be illustrated as in Figure 5.5.

With the release of the MATLAB *μ-Analysis and Synthesis Toolbox* commercially available software now exists to support complex μ synthesis using *D-K* iteration. The procedure 5.1 may be implemented quite easily with the aid of the toolbox. In the current release of the μ toolbox (Version 2.0), repeated scalar complex blocks are not supported in connection with *D-K* iteration. For repeated scalar blocks the corresponding D scaling is a full matrix. Thus the number of SISO transfer function approximations needed in connection with *D-K* iteration grows quadratically with the number of repeated scalar perturbations. An approximation to *D-K* iteration with repeated scalar blocks can of course be made simply by considering them as

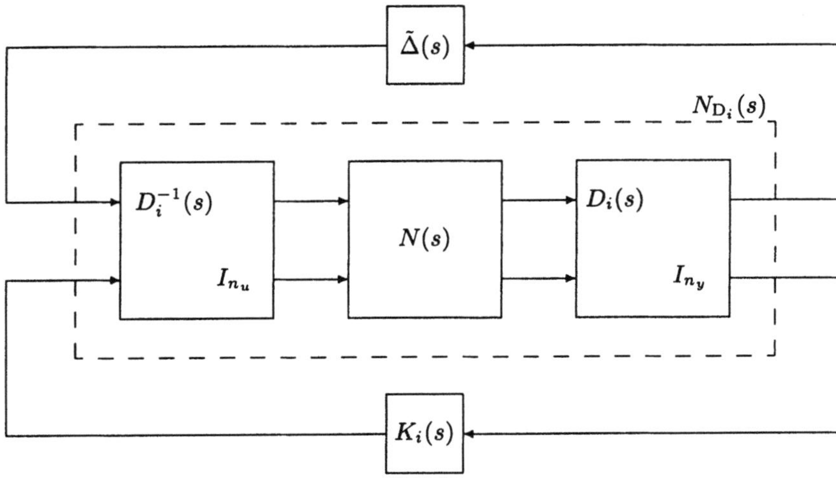

Fig. 5.5. *K-step in D-K iteration.* n_u *and* n_y *are the number of manipulated and controlled variables respectively.*

uncorrelated. However, the design will then not be optimal. By analyzing the final design with the true perturbation structure, the amount of conservatism introduced can be assessed.

Furthermore, notice that an approximation to mixed real and complex μ synthesis can be achieved with *D-K* iteration by approximating the mixed perturbation set $\mathbf{\Delta}$ with the corresponding complex set $\mathbf{\Delta}_c$, see (5.3). Thus all real perturbations are approximated with complex ones. Again, the design will not be optimal but through μ analysis the design may be tested against the true perturbation structure.

5.2.2 Mixed μ Synthesis – *D,G-K* Iteration

The general mixed real and complex μ problem is unfortunately much more difficult than the purely complex μ problem. In fact, until recently it was an unsolved problem. In his thesis [You93] and the papers [YÅ94, You94] Young has presented a solution, denoted *D,G-K iteration*, to the mixed μ synthesis problem. This approach is strongly motivated by the *D-K* procedure for purely complex perturbations. Young chooses the formulation 3 from Theorem 5.6 to pose the mixed μ upper bound problem

$$K(s) = \arg \min_{K \in \mathcal{K}_S} \sup_{\omega} \inf_{D(\omega)\in\mathbf{D}, G(\omega)\in\mathbf{G}} \inf_{\beta(\omega)\in\mathbf{R}_+} \{\beta(\omega) \,|\bar{\sigma}(\Gamma(\omega)) \leq 1\} \quad (5.76)$$

where

$$\Gamma(\omega) = \left(\frac{D(\omega)F_\ell(N(j\omega), K(j\omega))D^{-1}(\omega)}{\beta(\omega)} - jG(\omega)\right)\left(I + G^2(\omega)\right)^{-\frac{1}{2}} . \quad (5.77)$$

We will denote this a *direct* upper bound problem emphasizing that the problem is posed directly in line with the way the upper bound is computed. Again no known solution to (5.77) exists. Rather it must be solved iteratively similarly to *D-K* iteration. For fixed $K(s)$ the problem of finding $D(\omega)$, $G(\omega)$ and $\beta(\omega)$ is just the mixed μ upper bound problem. Having found these scalings we may fit real rational transfer function matrices to them such that the interconnection is stable. For given β, $D(s)$ and $G(s)$ transfer matrices the problem of finding the controller $K(s)$ will be reduced to a standard $\mathcal{H}_\infty$ problem.

However, the upper bound problem in (5.76)-(5.77) is not in the form of a singular value minimization such as (5.73), but rather a minimization subject to a singular value constraint. Thus we cannot proceed as before for purely complex perturbations. Instead we need the following 2 theorems which tell us how the mixed μ upper bound scales. Easy proofs may be found in [You93].

Theorem 5.7 (Scaling of Mixed μ Upper Bound I). *Suppose we have matrices $P \in \mathbf{C}^{n\times n}$, $D \in \mathbf{D}$, $G \in \mathbf{G}$ and a real scalar $\beta > 0$ such that*

$$\bar{\sigma}\left(\left(\frac{DPD^{-1}}{\beta} - jG\right)\left(I+G^2\right)^{-\frac{1}{2}}\right) \leq 1 . \tag{5.78}$$

Thus β is an upper bound on $\mu_{\mathbf{\Delta}}(P)$. Then for any real $\hat{\beta} > \beta$ there exists matrices $\hat{D} \in \mathbf{D}$, $\hat{G} \in \mathbf{G}$ such that

$$\bar{\sigma}\left(\left(\frac{\hat{D}P\hat{D}^{-1}}{\hat{\beta}} - j\hat{G}\right)\left(I+\hat{G}^2\right)^{-\frac{1}{2}}\right) < 1 . \tag{5.79}$$

Theorem 5.8 (Scaling of Mixed μ Upper Bound II). *Suppose we have matrices $P \in \mathbf{C}^{n\times n}$, $D \in \mathbf{D}$, $G \in \mathbf{G}$ and real scalars $\beta > 0$ and $0 < r \leq 1$ such that*

$$\bar{\sigma}\left(\left(\frac{DPD^{-1}}{\beta} - jG\right)\left(I+G^2\right)^{-\frac{1}{2}}\right) \leq r . \tag{5.80}$$

Then there exists matrices $\hat{D} \in \mathbf{D}$, $\hat{G} \in \mathbf{G}$ such that

$$\bar{\sigma}\left(\left(\frac{\hat{D}P\hat{D}^{-1}}{r\beta} - j\hat{G}\right)\left(I+\hat{G}^2\right)^{-\frac{1}{2}}\right) \leq 1 \tag{5.81}$$

so that $r\beta$ is an upper bound on $\mu_{\mathbf{\Delta}}(P)$.

Both these results are obviously true in the complex case for $G = O_n$ since

$$\bar{\sigma}\left(\frac{DPD^{-1}}{\beta}\right) \leq 1 \Leftrightarrow \frac{\beta}{\hat{\beta}}\bar{\sigma}\left(\frac{DPD^{-1}}{\beta}\right) \leq \frac{\beta}{\hat{\beta}} \Leftrightarrow \bar{\sigma}\left(\frac{DPD^{-1}}{\hat{\beta}}\right) < 1 \tag{5.82}$$

$$\bar{\sigma}\left(\frac{DPD^{-1}}{\beta}\right) \leq r \Leftrightarrow \bar{\sigma}\left(\frac{DPD^{-1}}{r\beta}\right) \leq 1 . \tag{5.83}$$

However, the results are not so obvious in the mixed case. In particular, note that Theorem 5.8 applies *only* for $r \leq 1$.

Young then proposes the following iterative scheme for mixed μ synthesis.

Procedure 5.2 (*D,G-K* Iteration [You93]).

1. *Given an augmented system $N(s)$, let $i = 1$, $D_i^\star(w) = I_n, \forall\omega$, $G_i^\star(\omega) = O_n, \forall\omega$ and $\beta_i^\star=1$.*
2. *Fit transfer function matrices $D_i(s)$ and $G_i(s)$ to the pointwise scalings $D_i^\star(\omega)$ and $jG_i^\star(\omega)$ so that $D_i(j\omega)$ approximates $D_i^\star(\omega)$ and $G_i(j\omega)$ approximates $jG_i^\star(\omega)$. Replace $D_i(s)$ and $G_i(s)$ with appropriate factors so that $D_i(s)$, $D_i^{-1}(s)$, $G_{\mathrm{h}_i}(s)$ and $G_i(s)G_{\mathrm{h}_i}(s)$ are all stable, where $G_{\mathrm{h}_i}(s)$ is a spectral factor satisfying $(I + G_i(s)G_i(-s)^T)^{-1} = G_{\mathrm{h}_i}(s)G_{\mathrm{h}_i}(-s)^T$. Augment $D_i(s)$ and $G_{\mathrm{h}_i}(s)$ with identity matrices, and $G_i(s)$ with a zero matrix of appropriate dimensions so that $D_i(s)$, $G_i(s)$ and $G_{\mathrm{h}_i}(s)$ are all compatible with $N(s)$. Construct the interconnection*

$$N_{\mathrm{DG}_i}(s) = (D_i(s)N(s)D_i^{-1}(s) - \beta_i^\star G_i(s))G_{\mathrm{h}_i}(s)\ . \tag{5.84}$$

3. *Find the $\mathcal{H}_\infty$ optimal controller $K_i(s)$:*

$$K_i(s) = \arg \min_{K(s)\in\mathcal{K}_{\mathrm{S}}} \|F_\ell(N_{\mathrm{DG}_i}(s), K(s))\|_{\mathcal{H}_\infty}\ . \tag{5.85}$$

4. *Compute the maximum upper bound $\beta_{i+1}^\star = \|\mu_{\tilde{\mathbf{\Delta}}}(F_\ell(N(s), K(s)))\|_\infty$:*

$$\beta_{i+1}^\star = \sup_{\omega\in\mathbf{R}} \inf_{D(\omega)\in\mathbf{D}, G(\omega)\in\mathbf{G}} \inf_{\beta(\omega)\in\mathbf{R}_+} \{\beta(\omega)\,|\bar{\sigma}(\Gamma(\omega)) \leq 1\} \tag{5.86}$$

   where $\Gamma(\omega)$ is given by

$$\Gamma(\omega) = \left(\frac{D(\omega)F_\ell(N(j\omega), K_i(j\omega))D^{-1}(\omega)}{\beta(\omega)} - jG(\omega)\right) \cdot \left(I + G^2(\omega)\right)^{-\frac{1}{2}}\ . \tag{5.87}$$

5. *Calculate the new scalings $D_{i+1}^\star(\omega)$ and $G_{i+1}^\star(\omega)$ solving the minimization problem*

$$\inf_{D_{i+1}^\star(\omega)\in\mathbf{D}, G_{i+1}^\star(\omega)\in\mathbf{G}} \bar{\sigma}\left(\left(\frac{D_{i+1}^\star(\omega)F_\ell(N(j\omega), K_i(j\omega))D_{i+1}^{\star^{-1}}(\omega)}{\beta_{i+1}^\star} - jG_{i+1}^\star(\omega)\right)\left(I + G_{i+1}^{\star^2}(\omega)\right)^{-\frac{1}{2}}\right) \tag{5.88}$$

   pointwise across frequency.
6. *Compare the new scalings $D_{i+1}^\star(\omega)$ and $G_{i+1}^\star(\omega)$ with the previous ones. Stop if they are close. Otherwise let $i = i + 1$ and repeat from 2.*

The similarities between the *D-K* iteration, see Procedure 5.1, and the above *D,G-K* iteration are obvious. However, there is one notable difference, namely the determination of the scaling matrices in step 5 of the *D,G-K* iteration. Notice that $D^\star_{i+1}(\omega)$ and $G^\star_{i+1}(\omega)$ are *not* the scalings from the μ upper bound computation in step 4. Rather it is a similar minimization with *constant* $\beta = \beta^\star_{i+1}$. As we shall soon see, this "twist" is necessary to avoid "pop-up" type phenomena where for some frequencies a small increase in the maximum singular value creates a very large increase in μ.

D,G-K iteration, as the corresponding *D-K* iteration for complex perturbations, cannot be guaranteed to converge to the global minimum of (5.76) since the procedure is not jointly convex in the scalings and the controller. However, we can show that is it monotonically non-increasing in $\mu_{\tilde{\mathbf{\Delta}}}(F_\ell(N(s),K(s)))$ given perfect realizations of the scaling matrices. Then under weak conditions the procedure will converge to a local minimum of μ. Having converged to such a point, the controller $K_i(s)$ from step 3 in Procedure 5.2 is the final mixed μ controller which satisfies

$$\sup_{\omega\in\mathbf{R}} \mu_{\tilde{\mathbf{\Delta}}}(F_\ell(N(j\omega),K(j\omega))) \leq \beta^\star_{i+1} \,. \tag{5.89}$$

Again we have used the $\mathcal{H}_\infty$ optimal control solution to synthesize the controller in step 3. The *K*-step of *D,G-K* iteration can be illustrated as in Figure 5.6

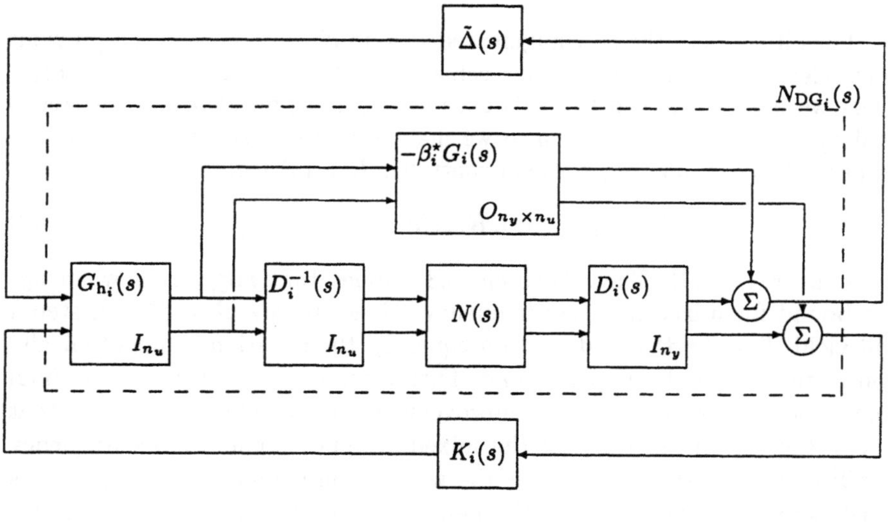

Fig. 5.6. *K-step of D,G-K iteration.* n_u *and* n_y *are the number of manipulated and controlled variables respectively.*

In order to show that Procedure 5.2 is non-increasing in μ we will need the results from Theorem 5.7 and 5.8. Note that $\beta^\star_{i+1}$ was defined in step 4 as the maximum of the μ upper bound across frequency. Thus if $\beta_{i+1}(\omega)$ denotes the μ upper bound across frequency ω, then $\beta^\star_{i+1} \geq \beta_{i+1}(\omega)$ for all frequencies, with $\beta^\star_{i+1} = \beta_{i+1}(\omega)$ only for those frequency points where the maximum is achieved. Then Theorem 5.7 guarantees that in step 5 we can achieve:

$$\bar{\sigma}\left(\left(\frac{D^\star_{i+1}(\omega)F_\ell(N(j\omega),K_i(j\omega))D^{\star^{-1}}_{i+1}(\omega)}{\beta^\star_{i+1}} - jG^\star_{i+1}(\omega)\right)\left(I + G^{\star^2}_{i+1}(\omega)\right)^{-\frac{1}{2}}\right) < 1 \quad (5.90)$$

for all frequency points where $\beta^\star_{i+1} > \beta_{i+1}(\omega)$ and

$$\bar{\sigma}\left(\left(\frac{D^\star_{i+1}(\omega)F_\ell(N(j\omega),K_i(j\omega))D^{\star^{-1}}_{i+1}(\omega)}{\beta^\star_{i+1}} - jG^\star_{i+1}(\omega)\right)\left(I + G^{\star^2}_{i+1}(\omega)\right)^{-\frac{1}{2}}\right) = 1 \quad (5.91)$$

for points where $\beta^\star_{i+1} = \beta_{i+1}$. Then note that for perfect realizations of the (new) scalings $D_{i+1}(s)$ and $G_{i+1}(s)$ in step 2 of the next iteration we will have that

$$\left\|F_\ell(N_{DG_{i+1}}(s), K_i(s))\right\|_{\mathcal{H}_\infty} = \beta^\star_{i+1} \,. \quad (5.92)$$

Thus if we close the new augmented system $N_{\mathrm{DG}_{i+1}}(s)$ with the previous controller, the maximum singular value of the interconnection equals $\beta^\star_{i+1}$. Thus we are guaranteed the existence of a stabilizing controller in step 3 achieving at least this level since we may simply choose the previous controller. Thus the new $\mathcal{H}_\infty$ optimal controller $K_{i+1}(s)$ will satisfy

$$\left\|F_\ell(N_{DG_{i+1}}(s), K_{i+1}(s))\right\|_{\mathcal{H}_\infty} = r\beta^\star_{i+1} \quad (5.93)$$

for some $r \in [0,1]$. Then Theorem 5.8 implies that $r\beta^\star_{i+1}$ is now an upper bound for mixed μ across frequency. Thus the new $\beta = \beta^\star_{i+2}$ in step 4 will yield $\beta^\star_{i+2} \leq r\beta^\star_{i+1} \leq \beta^\star_{i+1}$. Consequently, the iteration is monotonically non-increasing in the upper bound. Furthermore, if the new $\mathcal{H}_\infty$ optimal controller $K_{i+1}(s)$ achieves any improvement over the old one $K_i(s)$ so that $r < 1$ then the peak value of the mixed μ upper bound across frequency ω will be strictly decreased. Thus the way the interconnection $N_{\mathrm{DG}_{i+1}}(s)$ is constructed guarantees that if the new controller $K_{i+1}(s)$ reduces the $\mathcal{H}_\infty$ norm of $F_\ell(N_{\mathrm{DG}_{i+1}}(s), K_{i+1}(s))$, then the peak value of the mixed μ upper bound across frequency is reduced. It is important to have this guarantee because it ensures that we do not suffer from any "pop-up" type phenomena as discussed above. The possibility of this type of behavior is a direct consequence of the fact that Theorem 5.8 does not hold for $r > 1$. The *D,G-K* iteration procedure explicitly inhibits such "pop-up" type of behavior.

The necessary factorizations required to carry out step 2 in the D,G-K iteration above are described in [You93]. However, for the general upper bound problem the scalings cannot in general be restricted to be minimum phase since the phase is not absorbed into the (real) perturbations. Thus the scalings must be fitted both in magnitude and phase. One exception is the diagonal elements of the D-scalings. For the G scalings we must require that the phase of the diagonal elements of $G_i(j\omega)$ is 90° *for all frequencies* ω. The fitting of these purely imaginary diagonal elements can only be obtained using high order all pass structures. Since the order of $G_i^\star(s)$ adds to the order of the controller $K(s)$, the final controller will usually be of very high order.

Successful results on D,G-K iteration has been reported for simple mixed μ problems, see e.g. [You94]. However, the need for fitting purely complex scalings with high order all-pass transfer functions as well as the general need for fitting both in phase and magnitude severely hamper the practical use of D,G-K iteration for mixed μ synthesis. In many applications the G scalings seem to change quite rapidly. Thus it may be very difficult to obtain any reasonable fit for G.

5.2.3 Mixed μ Synthesis – μ-K Iteration

As discussed above the general D,G-K approach for controller synthesis with mixed perturbation sets $\boldsymbol{\Delta}$ provides a procedure which is monotonically non-increasing in the mixed μ upper bound. Unfortunately, the D,G-K iteration is much more complicated than the corresponding D-K iteration for purely complex perturbations. In particular, the fitting of scaling matrices is much more complex in the mixed case since we are forced to fit both in phase and magnitude. Especially, fitting the purely imaginary diagonal elements of $jG^\star(\omega)$ with high order all-pass structures will be difficult and furthermore result in augmented systems $N_{\mathrm{DG}_i}(s)$ of very high order. Even though the commercially available software tools for $\mathcal{H}_\infty$ optimal controller synthesis works quite well also for high order systems, when the order exceeds 100 the algorithms slow down rather dramatically even on high-performance UNIX installations.

In the following we will propose a new approach to mixed μ synthesis, denoted *μ-K iteration*, which sacrifices some of the guaranteed convergence properties of D,G-K iteration, but which on the other hand only requires that scalings are fitted in magnitude.

Whereas the D,G-K approach is a *direct* upper bound minimization, μ-K iteration is an *indirect* upper bound minimization in the sense that the augmented system matrix corresponding to $N_{\mathrm{DG}}(s)$ above does not directly reflect the structure of the μ upper bound as in (5.77).

The main idea of the proposed μ-K iteration scheme is to perform a scaled D-K iteration where the ratio between mixed and complex μ is taken into account through an additional scaling matrix $\Gamma(s)$. Given the augmented system $N(s)$ and any stabilizing controller $K(s)$, we may compute upper bounds for μ across frequency given both the "true" mixed perturbation set $\tilde{\mathbf{\Delta}}$ and a fully complex approximation $\tilde{\mathbf{\Delta}}_c$, see Equation (5.3). In order to "trick" the $\mathcal{H}_\infty$ optimal controller in the next iteration to concentrate more on mixed μ, we will construct a system $N_{D\Gamma}(s)$ such that $\bar{\sigma}(F_\ell(N_{D\Gamma}(j\omega), K(j\omega)))$ approximates the mixed μ upper bound just computed. This is fully equivalent to D,G-K iteration. In μ-K iteration, however, the structure of the approximation is different. $N_{D\Gamma}(s)$ is constructed by applying two scalings to the original system $N(s)$. A D scaling such that $\bar{\sigma}(F_\ell(N_{D\Gamma}(j\omega), K(j\omega)))$ approximates the complex μ upper bound and a Γ scaling to shift from complex to mixed μ. In order to avoid the "pop-up" type phenomena discussed in connection with D,G-K iteration, the Γ scalings are filtered through a first order filter with filter constant α. The iteration is performed as follows:

Procedure 5.3 (μ-K Iteration).

1. *Given the augmented system $N(s)$, let $\gamma_0(s) = 1$, $D_0(s) = I$ and $N_{D\Gamma_0}(s)$ be given by*

$$N_{D\Gamma_0}(s) = \begin{bmatrix} \gamma_0(s) I_{n_{ze}} & 0 \\ 0 & I_{n_y} \end{bmatrix} D_0(s) N(s) D_0^{-1}(s) \tag{5.94}$$

$$= \Gamma_0(s) D_0(s) N(s) D_0^{-1}(s) = N(s) \,. \tag{5.95}$$

   n_y denotes the number of controlled outputs $y(s)$ and n_{ze} denotes the number of external outputs $z(s)+e'(s)$. Let $K_0(s) = K_1(s)$ be a stabilizing controller and $i = 1$.

2. *Compute the mixed and corresponding complex μ upper bounds*

$$\bar{\mu}_{\tilde{\mathbf{\Delta}}}\left(F_\ell(N(j\omega), K_i(j\omega))\right) \,, \qquad \bar{\mu}_{\tilde{\mathbf{\Delta}}_c}\left(F_\ell(N(j\omega), K_i(j\omega))\right)$$

   at each frequency ω. Note that the complex μ upper bound equals

$$\bar{\sigma}(D_i^\star(\omega) F_\ell(N(j\omega), K_i(j\omega)) D_i^{\star^{-1}}(\omega))$$

   where the scalings $D_i^\star(\omega)$ are found solving the minimizations

$$D_i^\star(\omega) = \arg\min_{D \in \mathbf{D}} \bar{\sigma}\left(D F_\ell(N(j\omega), K_i(j\omega)) D^{-1}\right) \,, \qquad \forall \omega \geq 0 \,. \tag{5.96}$$

3. *Compute $\beta_i(\omega)$ given by:*

$$\beta_i(\omega) = \frac{\bar{\mu}_{\tilde{\mathbf{\Delta}}}\left(F_\ell(N(j\omega), K_i(j\omega))\right)}{\bar{\mu}_{\tilde{\mathbf{\Delta}}_c}\left(F_\ell(N(j\omega), K_i(j\omega))\right)} \frac{1}{|\gamma_{i-1}(j\omega)|} - 1 \,. \tag{5.97}$$

4. *Fit, in magnitude, a stable minimum phase transfer function matrix $D_i(s)$ to $D_i^\star(\omega)$ so that $D_i(j\omega)$ approximates $D_i^\star(\omega)$ across frequency ω. Augment $D_i(s)$ with a unity matrix of appropriate size such that $D_i(s)$ is compatible with $N(s)$.*

5. *Determine an upper bound for the constant* $\alpha_i \in [0;1]$ *according to*

$$\bar{\alpha}_i(\omega) = \begin{cases} \min\{1, \xi_i(\omega)\}\ , & \text{if } \beta_i(\omega) > 0 \\ 1\ , & \text{if } \beta_i(\omega) \leq 0 \end{cases} \tag{5.98}$$

where $\xi_i(\omega)$ *is given by*

$$\xi_i(\omega) = \left(\frac{\|F_\ell(N_{D\Gamma_{i-1}}(s), K_{i-1}(s))\|_{\mathcal{H}_\infty}}{\bar{\sigma}\left(F_\ell(D_i(j\omega)N(j\omega)D_i^{-1}(j\omega), K_i(j\omega))\right)|\gamma_{i-1}(j\omega)|} - 1 \right) \cdot \frac{1}{\beta_i(\omega)}\ . \tag{5.99}$$

6. *Choose a constant* $\alpha_i = \kappa \inf_\omega \bar{\alpha}_i(\omega)$ *where* $\kappa \in [0;1]$ *and compute for all* ω*:*

$$\gamma_i^\star(\omega) = (1 - \alpha_i)|\gamma_{i-1}(j\omega)| + \alpha_i \frac{\bar{\mu}_{\tilde{\mathbf{\Delta}}}\left(F_\ell(N(j\omega), K_i(j\omega))\right)}{\bar{\mu}_{\tilde{\mathbf{\Delta}}_c}\left(F_\ell(N(j\omega), K_i(j\omega))\right)}\ . \tag{5.100}$$

Fit, in magnitude, a stable minimum phase scalar transfer function $\gamma_i(s)$ *to* $\gamma_i^\star(\omega)$ *such that* $\gamma_i(j\omega)$ *approximates* $\gamma_i^\star(\omega)$ *across frequency* ω.

7. *Construct*

$$N_{D\Gamma_i}(s) = \begin{bmatrix} \gamma_i(s)I_{n_{ze}} & 0 \\ 0 & I_{n_y} \end{bmatrix} D_i(s)N(s)D_i^{-1}(s) \tag{5.101}$$

$$= \Gamma_i(s)D_i(s)N(s)D_i^{-1}(s) \tag{5.102}$$

and compute the optimal $\mathcal{H}_\infty$ *controller:*

$$K_{i+1}(s) = \arg \min_{K(s)\in\mathcal{K}_S} \|F_\ell(N_{D\Gamma_i}(s), K(s))\|_{\mathcal{H}_\infty} \tag{5.103}$$

8. *Compute the mixed and corresponding complex* μ *upper bounds*

$$\bar{\mu}_{\tilde{\mathbf{\Delta}}}\left(F_\ell(N(j\omega), K_{i+1}(j\omega))\right)\ , \qquad \bar{\mu}_{\tilde{\mathbf{\Delta}}_c}\left(F_\ell(N(j\omega), K_{i+1}(j\omega))\right)$$

at each frequency ω. *Computation of* $\bar{\mu}_{\tilde{\mathbf{\Delta}}_c}\left(F_\ell(N(j\omega), K_{i+1}(j\omega))\right)$ *provides the scalings* $D_{i+1}^\star(\omega)$ *for the next iteration.*

9. *Compute* $\beta_{i+1}(\omega)$ *given by:*

$$\beta_{i+1}(\omega) = \frac{\bar{\mu}_{\tilde{\mathbf{\Delta}}}\left(F_\ell(N(j\omega), K_{i+1}(j\omega))\right)}{\bar{\mu}_{\tilde{\mathbf{\Delta}}_c}\left(F_\ell(N(j\omega), K_{i+1}(j\omega))\right)} \frac{1}{|\gamma_i(j\omega)|} - 1\ . \tag{5.104}$$

10. *If* $\sup_\omega |\beta_{i+1}(\omega)| > \sup_\omega |\beta_i(\omega)|$ *return to 6 and reduce* κ. *Otherwise let* $i = i + 1$.
11. *Compare* $D_{i+1}(s)$ *and* $\gamma_i(s)$ *with the previous scalings* $D_i(s)$ *and* $\gamma_{i-1}(s)$. *Stop if they are close and* $\sup_\omega |\beta_{i+1}(\omega)| \approx 0$. *Otherwise repeat from step 4.*

Clearly, μ-K iteration is not so conceptually straight forward as D-K iteration for purely complex perturbation nor D,G-K iteration for mixed perturbation sets. The main obstacle is to avoid the "pop-up" phenomena.

In order to clarify matters let us assume perfect realizations of the D and Γ scalings and that we may choose $\alpha_1 = 1$. Then

$$\begin{aligned}
&\bar{\sigma}\left(F_\ell\left(N_{D\Gamma_1}(j\omega), K_1(j\omega)\right)\right) \\
&= F_\ell\left(\Gamma_1(j\omega)D_1(j\omega)N(j\omega)D_1^{-1}(j\omega), K_1(j\omega)\right) && (5.105)\\
&= |\gamma_1(j\omega)|\bar{\sigma}\left(F_\ell\left(D_1(j\omega)N(j\omega)D_1^{-1}(j\omega), K_1(j\omega)\right)\right) && (5.106)\\
&= |\gamma_1(j\omega)|\bar{\mu}_{\tilde{\Delta}_c}\left(F_\ell\left(N(j\omega), K_1(j\omega)\right)\right) && (5.107)\\
&= \bar{\mu}_{\tilde{\Delta}}\left(F_\ell\left(N(j\omega), K_1(j\omega)\right)\right), \quad \text{for } \alpha_1 = 1 && (5.108)
\end{aligned}$$

and the controller $K_2(s)$ will minimize the ∞-norm of an augmented system which closed with the previous controller $K_1(s)$ has maximum singular value approximating mixed μ. New mixed and complex μ bounds may then be computed and the procedure may be repeated. Unfortunately, it is not possible to choose $\alpha_i = 1$ in general since we may then suffer from "pop-up" type phenomena where for some frequencies a small increase in the maximum singular value of $N_{D\Gamma_i}(s)$ creates a very large increase in μ. However, by reducing α, i.e by filtering γ through a stable first order filter the "pop-up" type phenomena may be avoided with proper choice of α. Thus filtering $\gamma(s)$ in μ-K iteration is the equivalent of step 5 in D,G-K iteration. Filtering $\gamma(s)$ does *not* provide the same guarantee for a monotonically non-increasing μ upper bound as we can achieve in D,G-K iteration. However, numerical experience suggests that it works well in practice.

As before we use the $\mathcal{H}_\infty$ optimal control solution to synthesize the controller in step 7. The K-step of μ-K iteration can be illustrated as in Figure 5.7.

As shown in Equation (5.108), $\bar{\sigma}\left(F_\ell\left(N_{D\Gamma_i}(j\omega), K_i(j\omega)\right)\right)$ equals the upper bound $\bar{\mu}_{\tilde{\Delta}}\left(F_\ell\left(N(j\omega), K_i(j\omega)\right)\right)$ for perfect realizations of the scalings $\gamma_i(s)$ and $D_i(s)$ and for $\alpha_i = 1$. For $\alpha_i \neq 1$ this is not true. However, if

$$|\gamma_{i-1}(j\omega)| \rightarrow \frac{\bar{\mu}_{\tilde{\Delta}}\left(F_\ell(N(j\omega), K_i(j\omega))\right)}{\bar{\mu}_{\tilde{\Delta}_c}\left(F_\ell(N(j\omega), K_i(j\omega))\right)}, \qquad \text{for } i \rightarrow \infty \qquad (5.109)$$

then from (5.100) and perfect realization of $\gamma_i(j\omega)$:

$$|\gamma_{i-1}(j\omega)| \rightarrow |\gamma_i(j\omega)|, \qquad \text{for } i \rightarrow \infty \qquad (5.110)$$

and

$$\bar{\sigma}\left(F_\ell(\Gamma_i(j\omega)D_i(j\omega)N(j\omega)D_i^{-1}(j\omega), K_i(j\omega))\right) \rightarrow \bar{\mu}_{\tilde{\Delta}}\left(F_\ell(N(j\omega), K_i(j\omega))\right) \qquad (5.111)$$

for $i \rightarrow \infty$. Notice that Equation (5.109) implies that $\beta_i(\omega) \rightarrow 0$. Thus in order to reach a local minimum for $\|\bar{\mu}_{\tilde{\Delta}}(F_\ell(N(s), K(s)))\|_\infty$ the necessary conditions are:

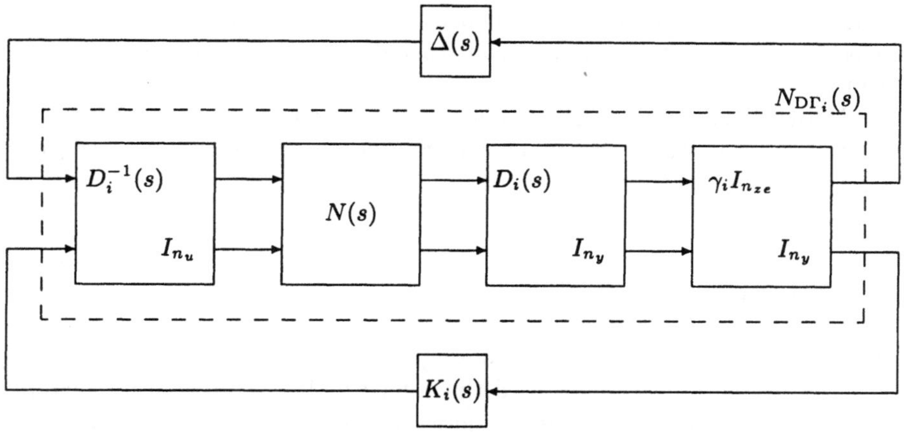

Fig. 5.7. *K-step of μ-K iteration. n_u, n_y and n_{ze} are the number of manipulated inputs $u(s)$, measured outputs $y(s)$ and exogenous outputs $z(s)$ and $e'(s)$ respectively.*

1. The iteration must be monotonically non-increasing in the $\mathcal{H}_\infty$ norm $\|F_\ell(N_{D\Gamma_i}(s), K_i(s))\|_{\mathcal{H}_\infty}$, i.e

$$\|F_\ell(N_{D\Gamma_i}(s), K_i(s))\|_{\mathcal{H}_\infty} \leq \|F_\ell(N_{D\Gamma_{i-1}}(s), K_{i-1}(s))\|_{\mathcal{H}_\infty} \quad \forall i\,. \quad (5.112)$$

2. Furthermore the iteration must be monotonically non-increasing in the supremum $\|\beta_i(\omega)\|_\infty$, i.e

$$\|\beta_i(\omega)\|_\infty \leq \|\beta_{i-1}(\omega)\|_\infty\,, \quad \forall i\,. \quad (5.113)$$

In Appendix C it is shown that the criteria (5.112) can be met by choosing $\bar{\alpha}_i(\omega)$ as in (5.98). If κ then can be chosen to fulfill the criteria (5.113), then under weak conditions the iteration will converge to a local minimum for $\|\bar{\mu}_{\tilde{\mathbf{\Delta}}}(F_\ell(N(s), K(s)))\|_\infty$. We then have the following lemma:

Lemma 5.1. *The μ-K iteration procedure described above is monotonically non-increasing in $\|F_\ell(N_{D\Gamma_i}(s), K_i(s))\|_{\mathcal{H}_\infty}$ given perfect realizations of the $D(s)$ and $\gamma(s)$ scalings. Provided $\|\beta(\omega)\|_\infty$ converges to zero then under weak conditions the procedure will converge to a local minimum for $\|\bar{\mu}_{\tilde{\mathbf{\Delta}}}\left(F_\ell(N(j\omega), K_i(j\omega))\right)\|_\infty$.*

Numerical evidence suggests that κ can indeed be chosen to fulfill (5.113). For $\alpha_i = 0$, $\gamma_i(s) = \gamma_{i-1}(s)$ and we will only concentrate on reducing the $\mathcal{H}_\infty$ bound in (5.112). Conversely, if $\beta_i(\omega) > 0$, for some ω, then for $\kappa = 1$ it can be shown, see Appendix C, that

$$\|F_\ell(N_{D\Gamma_i}(s), K_i(s))\|_{\mathcal{H}_\infty} = \|F_\ell(N_{D\Gamma_{i-1}}(s), K_{i-1}(s))\|_{\mathcal{H}_\infty}\,. \quad (5.114)$$

Thus in this situation we will only concentrate on the criteria (5.113) during the i'th step of the iteration. By choosing α (or κ) in between these two extremes we may attempt to fulfill both criteria (5.112) and (5.113). However, it is likely that examples may be constructed where no κ fulfilling (5.113) can be found.

Notice that for a purely complex perturbation set, if we choose $\kappa = 1$ then $\beta_i(\omega) = 0, \forall\omega, \forall i$ and $\gamma_i(s) = 1, \forall i$. Then μ-K iteration will reduce to standard D-K iteration.

5.3 Summary

An introduction to the structured singular value μ has been given. Using a $N\Delta K$ framework we addressed the robust stability problem for possibly repeated real and complex uncertainties which enter the nominal model in a linear fractional manner. This uncertainty description is superior to the assumption in connection with $\mathcal{H}_\infty$ control where only full complex blocks can be considered. Using μ we then derived simple expressions for both robust stability and robust performance with an ∞ norm bounded performance cost function. Thus we can consider robust performance non-conservatively for a very large class of perturbation structures including parametric uncertainty and mixed dynamic and parametric uncertainty.

Unfortunately, μ cannot be directly computed since the implied optimization problem is non-convex and may have multiple local minima. This fact has naturally hampered the application of μ theory to real control systems design. Rather than trying to compute μ itself it is then customary to look for upper and lower bounds on μ. From the definition of μ it is easy to see that μ is bounded from below by the spectral radius and from above by the singular value. The idea is to utilize transformations which are μ-invariant to refine these bounds. In the original paper on μ [Doy82], Doyle derived such bounds for purely complex perturbation sets. The complex μ upper bound is generally very close to the true μ value.

However, one of the powerful features of μ is that it can handle real perturbations (parametric uncertainty) as well. Unfortunately, the purely real (for robust stability analysis with parametric uncertainty only) or the mixed real and complex (for robust stability analysis for mixed uncertainty or for robust performance analysis in general) upper bound is more complicated. During the past 5 years, formulation of mixed μ upper and lower bounds have received considerable attention within the automatic control community. Today, some of these bounds are commercially available through the MATLAB μ toolbox. However, in general the bounds on mixed μ seem to be much more conservative than the purely complex μ bounds. In particular, for a large number of real perturbations, the gap between the upper and lower bound may be quite large.

Nevertheless, for a limited number (less than 10-15) of non-repeated real uncertainties, the bounds on μ seem in general to be quite tight. Thus from a control engineer's point of view the mathematical subtleties in μ is of minor importance in connection with μ analysis.

This is unfortunately not true when it comes to μ synthesis. Since we cannot compute μ (unless in some very simple cases where the upper bound is equal to μ) it is clear that controller design with μ will be difficult. The approach usually taken is to consider an upper bound problem instead. However, even for purely complex perturbations this is still an unsolved mathematical problem despite the fact that it has received much attention. For complex perturbation sets an iterative scheme, known as *D-K iteration*, has been known for some time. This approach possesses no guaranteed convergence properties. Nevertheless, it seems to work quite successfully in practice and has been applied to a large number of applications. Furthermore, with the release of the MATLAB *μ-Analysis and Synthesis Toolbox* commercially available software exists to support *D-K* iteration.

Conversely, for mixed real and complex perturbations until recently no known methodology was available for solving the synthesis problem. Then in 1993 Young [You93] presented a iterative solution procedure, denoted *D,G-K iteration*, to the mixed μ problem. The *D,G-K* iteration is a more or less straightforward generalization of *D-K* iteration. Unfortunately, since the problem is posed directly in line with the way μ is computed we now need to fit scalings both in magnitude and phase. In complex μ synthesis we need only to fit in magnitude since any phase is absorbed into the complex perturbations. In mixed real and complex μ synthesis the phase is not absorbed since the perturbations are real; thus the need for fitting both in phase and magnitude. This considerably complicates matters, in particular since one of the scaling matrices are purely imaginary for all frequencies.

A novel approach for mixed μ synthesis is then proposed. The new approach considers a corresponding complex problem where we approximate all real perturbations with complex ones. Then we only need to fit scalings in magnitude. We then introduce an additional scaling matrix to shift from complex to mixed μ. With this new approach, denoted *μ-K iteration*, we cannot obtain the same guarantee for monotonically non-increasing behavior as for *D,G-K* iteration. However, as we shall show in the next Chapter, *μ-K* iteration seems to work quite well in practice.

CHAPTER 6

MIXED μ CONTROL OF A COMPACT DISC SERVO DRIVE

The main purpose of this chapter is to illustrate how the proposed μ-K iteration performs on a practical design example. A compact disc (CD) servo drive has be chosen as application partly because it is a simple and illustrative example and partly because it enables us to compare the results with reported results from D,G-K iteration by Young [You93, YÅ94].

The servo arm for a computer disc drive or a compact disc player is essentially a small flexible structure, whose dynamics will depend on various physical quantities. Provided the product is to be mass produced, the control design for the servo should work for any product from the assembly line and thus the design should be insensitive to variations in the parameters of the servo model.

Here a very simple idealized model of the servo is considered. The control problem formulation is identical to that in [YÅ94]. It is assumed that the low frequency plant dynamics can be approximated by a double integrator with an uncertain gain:

$$G_T(s) = \frac{k_p}{s^2}, \qquad 0.1 \le k_p \le 10 \,. \tag{6.1}$$

Notice that the ratio of $k_{p,max}$ to $k_{p,min}$ is as large as 100. The model is assumed valid up to the frequency $\omega_o = 100$ rad/s. The parameters k_p and ω_o do not necessarily reflects physical meaningful numbers. This is, however, of minor importance here since the main objective is to illustrate different design techniques. The robust performance control problem is thus to design a fixed dynamic controller which stabilizes the plant and complies with performance demands for all possible plants. To ensure that the controller rolls off at frequencies above ω_o a complementary sensitivity specification is included into the design. Performance requirements are given as a standard sensitivity specification. The control problem can then be specified as in Figure 6.1.

Notice that the perturbed low frequency plant (6.1) can be written

$$G_T(s) = \frac{1}{s^2}\left(5.05 + 4.95\delta^r\right) = \frac{5.05}{s^2}\left(1 + \frac{4.95}{5.05}\delta^r\right) = G(s)\left(1 + W_r\delta^r\right) \tag{6.2}$$

where $\delta^r \in \mathbf{R}$ is a real scalar perturbation $\delta^r \in [-1, 1]$. The sensitivity weights on $S(s)$ and $T(s)$ are standard weights taken from [YÅ94]:

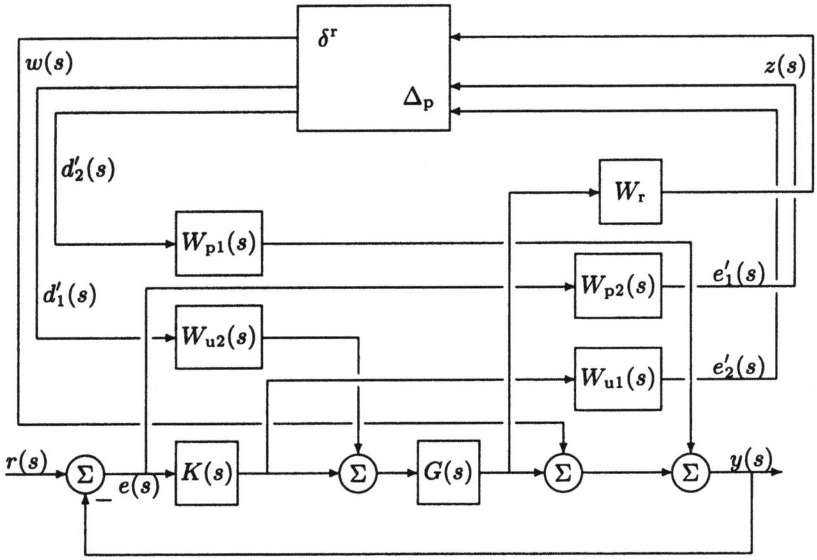

Fig. 6.1. *Control problem set-up for the compact disc servo drive.*

$$W_{u1}(s) = W_{u2}(s) = \frac{5(s+0.001)}{s+5} \tag{6.3}$$

$$W_{p1}(s) = W_{p2}(s) = \frac{0.03}{s+0.05} \tag{6.4}$$

Notice that we have gathered the weighted sensitivity functions $W_{p2}SW_{p1}$ and $W_{u1}TW_{u2}$ into a single performance specification. We in fact believe that this is a somewhat awkward way of specifying the control problem since the complementary sensitivity weights $W_{u1}(s)$ and $W_{u2}(s)$ introduced to make the controller roll off at high frequencies reflect an uncertainty specification rather than a performance specification. Thus it could have been introduced via a separate scalar uncertainty block δ^c. In this way we would decouple the specifications on the sensitivity and complementary sensitivity. However, in order to compare our results with those reported in [You93, YÅ94] we will keep the formulation in Figure 6.1.

6.1 Complex μ Design

Our first approach will be to design a complex μ optimal controller:

$$K_{\mu_c}(s) = \arg \min_{K(j\omega)\in\mathcal{K}_S} \sup_{\omega} \mu_{\tilde{\mathbf{\Delta}}_c}\left(F_\ell(N(j\omega), K(j\omega))\right) . \tag{6.5}$$

Hence, at first we approximate the real perturbation δ^r with a complex one δ^c. The complex μ problem was solved using D-K iteration as outlined in Procedure 5.1. The results from the iteration is shown in Figure 6.2 and tabulated in Table 6.1.

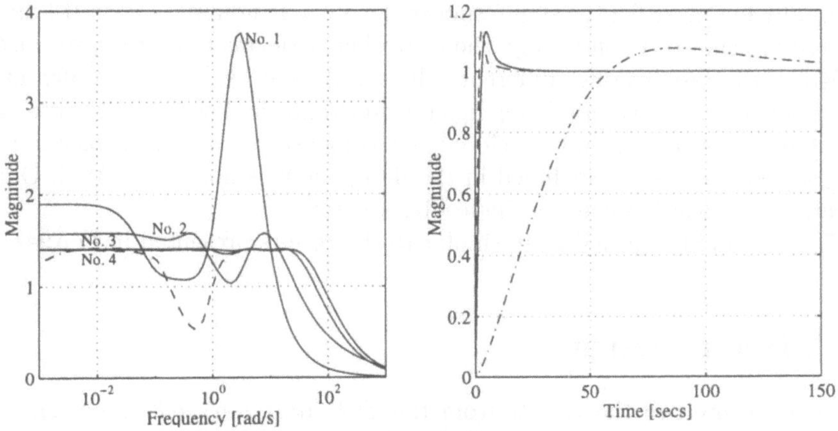

Fig. 6.2. *Results from D-K iteration on the CD player servo. Shown are on the left* $\mu_{\tilde{\Delta}_c}(F_\ell(N(j\omega), K(j\omega)))$ *for each iteration 1-4 and for the 7th order reduced controller (dash-dotted). Also on the left are* $\mu_{\tilde{\Delta}}(F_\ell(N(j\omega), K_{\mu_c}(j\omega)))$ *(dashed). On the right are the step responses for* $k_p = 1$ *(solid),* $k_p = 0.1$ *(dash-dotted) and* $k_p = 10$ *(dashed).*

In Figure 6.2 $\mu_{\tilde{\Delta}_c}(F_\ell(N(j\omega), K(j\omega)))$ for each iteration is shown across frequency. As seen the iteration converges rapidly with a final complex μ peak at 1.40. Using a third order approximation for $D(s)$, the final complex μ controller had 12 states. Using the standard model reduction facilities in the MATLAB μ toolbox it was possible to reduce the order to 7 with very little increase in $\|\mu_{\tilde{\Delta}_c}(F_\ell(N(j\omega), K(j\omega)))\|_\infty$.

Table 6.1. *Results from D-K iteration on the CD player servo.*

	Iteration No.				
	1	2	3	4	7th order K
$\|\mu_{\tilde{\Delta}_c} F_\ell(N(s), K_i(s))\|_\infty$	3.75	1.60	1.41	1.40	1.42
$\|\mu_{\tilde{\Delta}} F_\ell(N(s), K_i(s))\|_\infty$	–	–	–	1.40	1.42

	Overshoot	Settling time
$k_p = 0.1$	1.14	6.4 secs
$k_p = 1$	1.13	13.6 secs
$k_p = 10$	1.08	∞

In Figure 6.2 also the mixed μ result for the final controller $K_{\mu_c}(s)$ is shown. Notice how there is a noticeable dive in mixed μ around 0.5 rad/s. It thus seem possible that a significant improvement in control performance can be obtained by applying mixed μ synthesis.

Finally in Figure 6.2 the step responses for $k_p = 1$ (nominal), $k_p = 0.1$ and $k_p = 10$ are shown. Even though the controller performs quite well, see also Table 6.1, the responses for different values of the gain k_p are very different, with the high gain response being much faster than the low gain response as expected. If it is desired that the closed loop response is invariant to variations in k_p a prefilter may be included in the design as it is done in [YÅ94]. Our objective here will however be the feedback design.

The results in Figure 6.2 are similar to the results presented in [YÅ94].

6.2 Mixed μ Design

In order to improve the results from the *D-K* iteration, μ-*K* iteration as outlined in Procedure 5.3 was applied to the servo problem. The complex μ optimal controller $K_{\mu_c}(s)$ is then an obvious choice for the initial controller $K_0(s)$. Let us illustrate the first run through of the μ-*K* iteration. Thus let $N(s)$ be the augmented open loop plant derived from Figure 6.1. Then $N_{D\Gamma_0}(s) = D_0(s)N(s)D_0^{-1}(s)$ where $D_0(s)$ is the scaling matrix from step 2 in the final *D-K* iteration. Furthermore $K_0(s) = K_1(s) = K_{\mu_c}(s)$. The complex and mixed μ upper bounds in step 2 of Procedure 5.3 are the bounds shown in Figure 6.2. The scalings $D_1^\star(\omega)$ are returned by the MATLAB *μ-Analysis and Synthesis Toolbox* `mu.m` command. We may then compute $\beta_1(\omega)$ as

$$\beta_1(\omega) = \frac{\bar{\mu}_{\tilde{\Delta}}\left(F_\ell(N(j\omega), K_1(j\omega))\right)}{\bar{\mu}_{\tilde{\Delta}_c}\left(F_\ell(N(j\omega), K_1(j\omega))\right)} \frac{1}{|\gamma_0(j\omega)|} - 1 \tag{6.6}$$

$$= \frac{\bar{\mu}_{\tilde{\Delta}}\left(F_\ell(N(j\omega), K_1(j\omega))\right)}{\bar{\mu}_{\tilde{\Delta}_c}\left(F_\ell(N(j\omega), K_1(j\omega))\right)} - 1 \tag{6.7}$$

since $\gamma_0(\omega) = 1, \forall\omega$. $\beta_1(\omega)$ is shown in Figure 6.3. Note that $\beta_1(\omega)$ is simply the ratio between mixed and complex μ minus 1. Next we may use the μ toolbox command `musynfit.m` to fit the scalings $D_1^\star(\omega)$ with a real rational stable minimum phase transfer function matrix as required in step 4 of μ-*K* iteration. `musynfit.m` furthermore automatically augment with an appropriate unity matrix. Since mixed μ is always smaller than the corresponding complex μ:

$$\bar{\mu}_{\tilde{\Delta}}\left(F_\ell(N(j\omega), K_1(j\omega))\right) \leq \bar{\mu}_{\tilde{\Delta}_c}\left(F_\ell(N(j\omega), K_1(j\omega))\right), \quad \forall\omega \tag{6.8}$$

we will always have that $\beta_1(\omega) \leq 0, \forall\omega$ in the first run through of μ-*K* iteration. Thus in step 5 of Procedure 5.3 we will have that $\bar{\alpha}_i(\omega) = 1, \forall\omega$. Consequently, there is no need for calculating $\xi_1(\omega)$, see Equation (5.98).

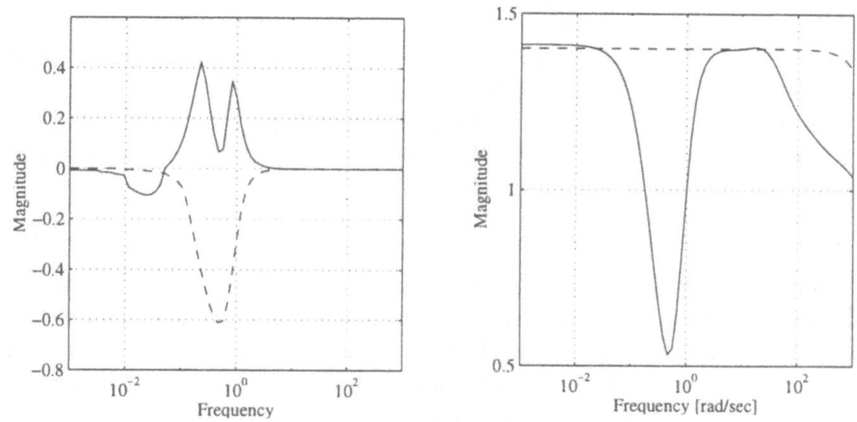

Fig. 6.3. *Results from first iteration on the CD player servo. On the left is shown $\beta_2(\omega)$ and $\beta_1(\omega)$ (dashed) and on the right is shown $\bar{\sigma}(F_\ell(N_{\mathrm{D}\Gamma_0}(j\omega), K_0(j\omega)))$ (dashed) and $\bar{\sigma}(F_\ell(N_{\mathrm{D}\Gamma_1}(j\omega), K_1(j\omega)))$.*

Then in step 6 let us choose $\kappa = 1$. Then $\gamma_1^\star(\omega)$ will be equal to the mixed to complex μ ratio:

$$\gamma_1^\star(\omega) = \frac{\bar{\mu}_{\tilde{\mathbf{\Delta}}}\left(F_\ell(N(j\omega), K_1(j\omega))\right)}{\bar{\mu}_{\tilde{\mathbf{\Delta}}_c}\left(F_\ell(N(j\omega), K_1(j\omega))\right)} \frac{1}{|\gamma_0(j\omega)|} = \beta_1(\omega) + 1 \,. \tag{6.9}$$

A scalar transfer function $\gamma_1(s)$ is then fitted, in magnitude, to $\gamma_1^\star(\omega)$. Here we may use the fitting routine `fitmag.m` from the μ toolbox. Next the augmented system $N_{\mathrm{D}\Gamma_1}(s)$ is formed:

$$N_{\mathrm{D}\Gamma_1}(s) = \begin{bmatrix} \gamma_1(s) I_3 & 0 \\ 0 & 1 \end{bmatrix} D_1(s) N(s) D_1^{-1}(s) \tag{6.10}$$

and the $\mathcal{H}_\infty$ optimal controller $K_2(s)$:

$$K_2(s) = \arg \min_{K(s)\in\mathcal{K}_S} \|F_\ell(N_{\mathrm{D}\Gamma_1}(s), K(s))\|_{\mathcal{H}_\infty} \tag{6.11}$$

is computed, eg using the `hinfsyn.m` function from the μ toolbox.

Then in step 8 of the μ-K iteration we compute upper bounds for mixed and complex μ, the latter providing the D-scalings $D_2^\star(\omega)$ for the next iteration. In step 9 we finally find $\beta_2(\omega)$ given by:

$$\beta_2(\omega) = \frac{\bar{\mu}_{\tilde{\mathbf{\Delta}}}\left(F_\ell(N(j\omega), K_2(j\omega))\right)}{\bar{\mu}_{\tilde{\mathbf{\Delta}}_c}\left(F_\ell(N(j\omega), K_2(j\omega))\right)} \frac{1}{|\gamma_1(j\omega)|} - 1 \,. \tag{6.12}$$

$\beta_2(\omega)$ is also shown in Figure 6.3. Since

$$\sup_\omega |\beta_2(\omega)| < \sup_\omega |\beta_1(\omega)| \tag{6.13}$$

the condition in step 10 for terminating the first run through is fulfilled. Since $\sup_\omega |\beta_2(\omega)|$ is not close to zero we will repeat the iteration from step 4.

The second parameter besides β which we use to judge the convergence properties of the μ-K iteration is the ∞-norm $\|F_\ell(N_{D\Gamma_i}(s), K_i(s))\|_{\mathcal{H}_\infty}$. Note that we need this norm to compute $\xi_i(\omega)$ in step 5. In Figure 6.3 we have plotted $\bar{\sigma}(F_\ell(N_{D\Gamma_0}(j\omega), K_0(j\omega)))$ and $\bar{\sigma}(F_\ell(N_{D\Gamma_1}(j\omega), K_1(j\omega)))$ across frequency ω. Note that Appendix C guarantees that

$$\|F_\ell(N_{D\Gamma_i}(s), K_i(s))\|_{\mathcal{H}_\infty} \leq \|F_\ell(N_{D\Gamma_{i-1}}(s), K_{i-1}(s))\|_{\mathcal{H}_\infty} \tag{6.14}$$

for perfect realizations of the D and γ scalings. The results in Figure 6.3 confirms this except for a little overshoot at low frequencies due to imperfect scalings. This is in fact trivial for the first iteration since $K_0(s) = K_1(s)$ and $|\gamma_1(j\omega)| \leq 1 = |\gamma_0(j\omega)|$ for all frequencies ω. However, for the next iterations this result is not so trivial, see Appendix C.

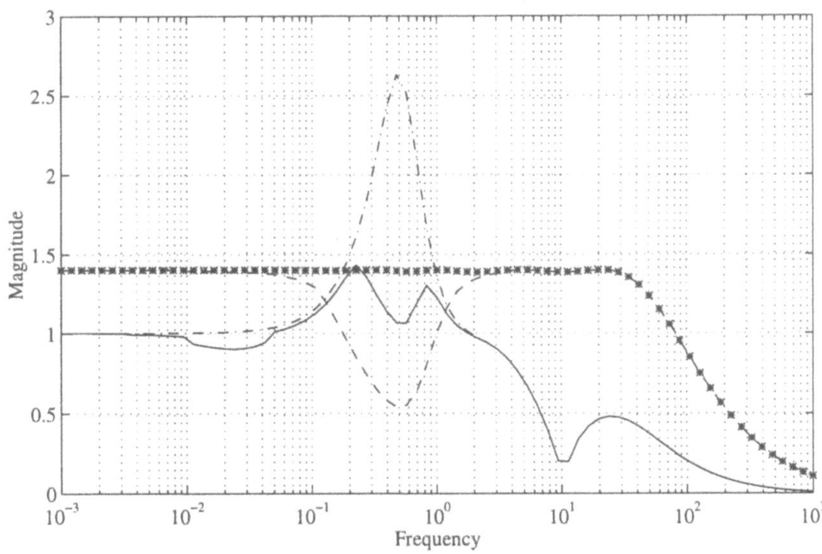

Fig. 6.4. *μ results from first μ-K iteration on the CD servo. New (solid) and previous (dashed) mixed μ upper bound and new (dash-dotted) and previous (*) complex μ upper bound.*

Finally let us have a look at μ. In Figure 6.4 $\mu_{\tilde{\Delta}}(F_\ell(N(j\omega), K_1(j\omega)))$ (initial mixed μ), $\mu_{\tilde{\Delta}}(F_\ell(N(j\omega), K_2(j\omega)))$ (new mixed μ), $\mu_{\tilde{\Delta}_c}(F_\ell(N(j\omega), K_1(j\omega)))$ (initial complex μ) and $\mu_{\tilde{\Delta}_c}(F_\ell(N(j\omega), K_1(j\omega)))$ (new complex μ) are shown versus frequency.

Note that with the γ scaling we have put less emphasis on the mid-frequency range in the $\mathcal{H}_\infty$ optimization and thus have provided the opportunity for reducing μ outside this frequency range. Then, of course, the new mixed μ value will increase in the mid frequency range. Here we have not obtained any decrease in the maximum μ upper bound across frequency since $\|\mu_{\tilde{\Delta}}(F_\ell(N(s), K_1(s)))\|_\infty \approx \|\mu_{\tilde{\Delta}}(F_\ell(N(s), K_2(s)))\|_\infty$. Note that we have not postulated that this would be the case in every iteration. However, we have shifted the emphasis in the $\mathcal{H}_\infty$ optimization from complex to mixed μ and during the next iterations $\sup_\omega \beta_i(\omega) \to 0$ and the mixed μ upper bound will decrease.

In Table 6.2 the results of the μ-K iteration are given. Notice how $\sup_\omega \beta(\omega)$ converges to zero indicating that the γ scalings approximates the mixed to complex μ ratio. The ∞-norm $\|F_\ell(P_i(s), K_i(s))\|_{\mathcal{H}_\infty}$ is monotonically decreasing as shown in Appendix C. κ had to be reduced from iteration 3+ in order to reduce the β peaks (mixed μ pop-up phenomena).

Table 6.2. *Results from μ-K iteration on the CD player servo.*

	μ-K Iteration No.						
	1	2	3	4	5	6	7
$\left\|\mu_{\tilde{\Delta}_c}(F_\ell(N(s), K_i(s)))\right\|_\infty$	1.40	–	–	–	–	–	2.46
$\|\mu_{\tilde{\Delta}}(F_\ell(N(s), K_i(s)))\|_\infty$	1.40	1.43	1.15	1.06	1.08	1.06	1.02
$\sup_\omega \beta_i(\omega)$	0.61	0.42	0.30	0.13	0.07	0.06	0.04
$\|F_\ell(P_i(s), K_i(s))\|_{\mathcal{H}_\infty}$	1.40	1.41	1.32	1.12	1.05	1.024	1.019
$\inf_\omega \bar{\alpha}_i(\omega)$	–	1	0.95	1	1	0.62	0.32
κ	–	1	1	0.5	0.5	0.5	1

	Overshoot	Settling time
$k_p = 0.1$	1.32	13.9 secs
$k_p = 1$	1.36	20.6 secs
$k_p = 10$	1.10	102 secs

In Figure 6.5 mixed and complex μ is shown for the initial (complex μ optimal) and final mixed μ optimal controller. Mixed μ peaks at approximately 1.02 and a 25% improvement has thus been obtained through μ-K iteration. Notice how this is achieved at the expense of complex μ which peaks at almost 2.5 for the mixed μ optimal controller $K_\mu(s)$.

In Figure 6.5 also the step responses are shown for different values of the uncertain gain k_p. Comparing with the results for the the complex μ optimal controller $K_{\mu_c}(s)$, see Figure 6.2, we note that the mixed μ controller is more invariant to changes in k_p. The worst-case rise- and settling time (for $k_p = 0.1$) has been much improved. The expense has been an increase in overshoot. Using a pre-filter, these overshoots may be removed.

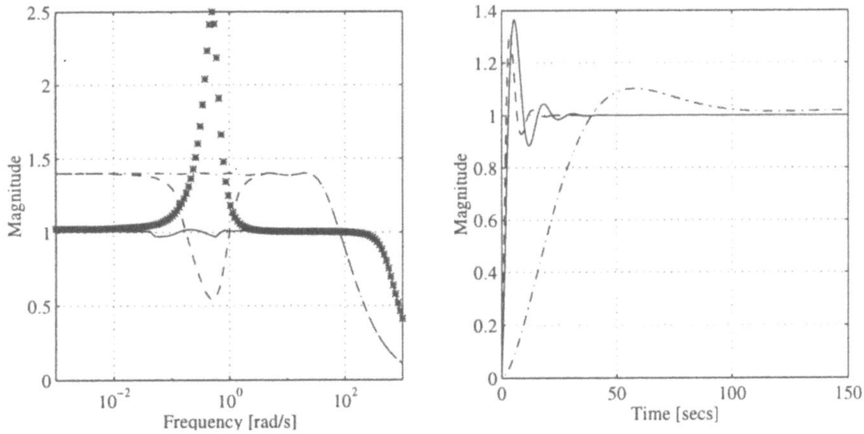

Fig. 6.5. *Results from μ-K iteration on the CD player servo. Shown are on the left $\mu_{\tilde{\Delta}}(F_\ell(N(j\omega), K(j\omega)))$ (dashed and solid) and $\mu_{\tilde{\Delta}_c}(F_\ell(N(j\omega), K(j\omega)))$ (dash-dotted and *) for the initial (complex μ optimal) and final mixed μ optimal controller respectively. On the right are the step responses for $k_p = 1$ (solid), $k_p = 0.1$ (dash-dotted) and $k_p = 10$ (dashed).*

In [YÅ94] D,G-K iteration is used to find a mixed μ controller for the CD player servo. Here mixed μ was reported to peak at 1.25 for a ninth order realization of the final mixed μ controller. In order to compare this result with the μ-K iteration result, the final controller $K_\mu(s)$ which had 26 states must be reduced using model reduction. Using the routines provided in the MATLAB μ toolbox $K_\mu(s)$ was reduced to ninth order with virtually no increase in μ. With the reduced order controller $K_{\mu_r}(s)$, $\mu_{\tilde{\Delta}}(F_\ell(N(j\omega), K_{\mu_r}(j\omega)))$ also peaked at 1.02. For lower order realizations there were, however, a significant increase in mixed μ.

In this case it thus seems that μ-K iteration performs better than D,G-K iteration. We believe that the reason for this is problems in fitting the G scalings. In Figure 6.6 $G(\omega)$ is shown for the final mixed μ controller $K_\mu(s)$. Notice that $G(\omega)$ is identically zero for a large part of the frequency range (namely at those frequency points where mixed and complex μ are identical) but rapidly rise to approximately $2 \cdot 10^4$ at those frequencies where mixed and complex μ differs. Fitting an all-pass transfer function to this kind of data will be very difficult. The corresponding γ scaling in μ-K iteration behaves much better as can be seen in Figure 6.6.

6.3 Summary

The control of compact disc servo drive was considered. A very simple model of a CD drive is a double integrator with uncertain gain. Here the ratio of the maximum to minimum gain was 100. Using weighting functions from [YÅ94]

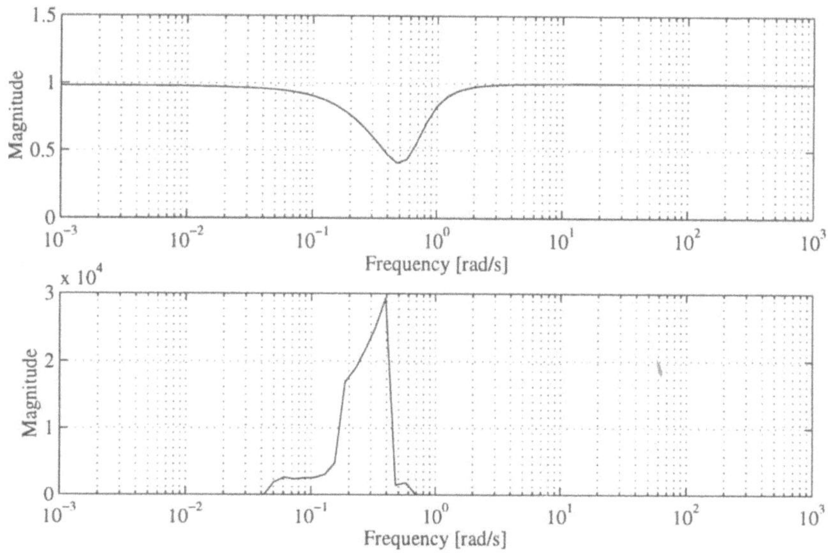

Fig. 6.6. *γ scalings to be fitted in μ-K iteration (top) and G scalings to be fitted in D, G-K iteration (bottom).*

the control design was formulated as a $N\Delta K$ problem. Initially a complex μ design was performed by treating the real perturbation as a complex one. The result (after model reduction) was a 7th order controller which achieved real μ peaking at 1.42. The complex μ value $\mu_{\tilde{\Delta}_c}(F_\ell(N(j\omega), K(j\omega)))$ peaked also at 1.42[1]. However, whereas complex μ was flat across frequency ω there was a noticeable dive in the mixed μ response. Thus significant improvement should be possible with a mixed μ design. This was confirmed using μ-K iteration. The final controller had 9 states (after model reduction) and the corresponding mixed μ response peaked at 1.02. Thus the complex μ result was improved by 25%. This was achieved at the expense of complex μ which peaked at 2.5 for the mixed μ controller. Compared with the results reported by Young & Åström on D,G-K iteration for the same system (mixed μ peaked at 1.25) we obtained better results using μ-K iteration.

[1] Note that $\mu_{\tilde{\Delta}}(F_\ell(N(j\omega), K(j\omega)))$ and $\mu_{\tilde{\Delta}_c}(F_\ell(N(j\omega), K(j\omega)))$ are equal for $w = 0$ and ∞ because then $F_\ell(N(j\omega), K(j\omega))$ is real.

CHAPTER 7

μ CONTROL OF AN ILL-CONDITIONED AIRCRAFT

In this chapter the methods introduced in Chapter 5 will be applied in control of an Advanced Short Take-Off and Vertical Landing (ASTOVL) aircraft. The model applied for the vehicle is valid at low speeds in the transition zone from jet-borne to fully wing-borne flight. In this flight condition the aircraft is unstable in the longitudinal axis and there is very poor decoupling between pilot commands and aircraft response. Furthermore the aircraft is extremely ill-conditioned as the gain of the system decays to zero at steady-state for one particular input direction. The limits this assign on the closed loop system will be thoroughly investigated. The control objectives for the aircraft will be discussed and a performance specification will be derived with due attention to the special gain characteristics of the system. For robustness to dynamic uncertainty, a diagonal dynamic multiplicative perturbation models will be applied at both input and output. A classical control configuration will then be investigated. For a modern design approach the general $N\Delta K$ framework becomes a very natural way of formulating the control problem. This furthermore possesses the advantage that the uncertainty specification may easily be augmented with a parametric uncertainty description as well. This has been investigated in [TCABG95]. Here we will however confine ourselves to considering dynamic uncertainty. A complex μ optimal controller will be designed for the aircraft using *D-K* iteration.

7.1 The Aircraft Model

The aircraft model describes a complete pitch axis system for a generic canard-delta configuration with flap and foreplane angles fixed. The aircraft is controlled using vectored thrust produced from engine nozzles at the front and the rear of the aircraft. The thrust is split with one part being directed through the rear nozzle at a variable angle, θ_{R} to the horizontal, producing a thrust, T_{R}, and the remainder through the front nozzle inclined at a fixed angle, θ_{F} to the horizontal, producing a thrust, T_{F}. The forward thrust can be augmented with plenum chamber burning when required. The aircraft is controlled by varying the real nozzle angle, θ_{R}, and the magnitude of the thrusts, T_{R} and T_{F}. The pilot commands for pitch attitude θ, height rate

$\dot{h}$ and longitudinal acceleration a_x are processed by the flight control computer hardware $G_c(s)$ which also converts pilot commands into demands on the controlled variables θ_R, T_F and T_R through a 3×3 axis transformation matrix. The thrusts produced by the engine, $G_E(s)$, are then resolved via a force transformation matrix, F_{mat}, into an axial force, X_F, a normal force, Z_F, and a pitching moment, M, which are the inputs to the rigid aircraft model, denoted $G_A(s)$.

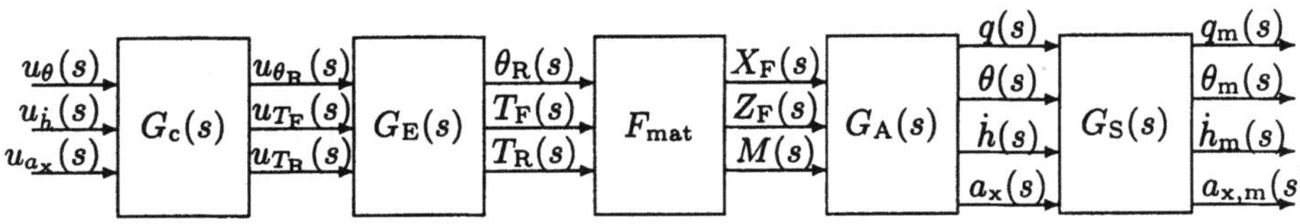

Fig. 7.1. *The rigid-body model of the aircraft $G(s)$.*

The model for the rigid aircraft, $G_A(s)$, represents the linearized, rigid-body dynamics of a canard-delta configured aircraft. The model is linearized for straight and level flight at 100 ft, Mach no. 0.151 and 6 deg angle of attack. The dynamics at this particular operating point are unstable.

The pitch rate q [deg/s], pitch attitude θ [deg], height rate $\dot{h}$ [ft/s] and longitudinal acceleration a_x [g] are measured and available for feedback. The diagonal transfer function matrix modeling the sensors is denoted $G_S(s)$.

The complete model of the aircraft is thus given as in Figure 7.1. A detailed description of the rigid body aircraft model is given in Appendix D. The open loop transfer function $G(s) = G_S(s)G_A(s)F_{mat}G_E(s)G_c(s)$ has 26 states, 4 outputs and 3 inputs. The aircraft model is thus a high order non-square system.

7.1.1 Plant Scaling

Notice that the input/outputs of the aircraft model are measured in different units. If norm-based control design methods like $\mathcal{H}_\infty$ or μ are applied, the control problem will be skewed if not due attention is taken to scale the plant so that the relative magnitude (or importance) of the different input/outputs will be approximately equal. This is done by multiplication on the left and right with 2 diagonal matrices $N_u(s)$ and $N_y(s)$:

$$G_s(s) = N_y(s)G(s)N_u^{-1}(s) . \tag{7.1}$$

$N_u(s)$ and $N_y(s)$ should be chosen so that the relative magnitude of the inputs/outputs equal unity and so that the input and output are scaled to norm 1. A popular method for choosing the scaling matrices is the use of expected maximum magnitude. In Table 7.1 the expected maximum magnitude for each input/output are given.

Table 7.1. *Expected maximum magnitude for each input/output component.*

Signal	u_θ	$u_{\dot{h}}$	u_{a_x}	q	θ	$\dot{h}$	a_x
Units	deg	ft/s	g	deg/s	deg	ft/s	g
Expected max. value	± 30	± 100	± 2	± 30	± 30	± 100	± 2

From Table 7.1 the scalings are then chosen as

$$N_u = \text{diag}\left\{\frac{1}{30}, \frac{1}{100}, \frac{1}{2}\right\}, \qquad N_y = \text{diag}\left\{\frac{1}{30}, \frac{1}{30}, \frac{1}{100}, \frac{1}{2}\right\}. \tag{7.2}$$

Now the system "seen" by the controller will have input and output with maximum magnitude scaled to norm one. If the controller for the scaled system is denoted $K_s(s)$, the controller for the true system $K(s)$ will be given by:

$$K(s) = N_u^{-1}(s)K_s(s)N_y(s)\ . \tag{7.3}$$

Throughout the remainder of this chapter we will almost exclusively deal with the scaled plant $G_s(s)$ and controller $K_s(s)$. In order not to lengthen the notation excessively, unless any confusion may occur we will from now on use $G(s)$ and $K(s)$ to denote the scaled plant and controller respectively.

7.1.2 Dynamics of the Scaled Aircraft Model

The dynamics of the scaled aircraft model will now be investigated using singular values. In Figure 7.2 the singular value Bode plot of the (scaled) plant $G(s)$ is shown. The plant is seen to be very ill-conditioned. Notice how one of the singular values decays to zero at steady-state ($\omega = 0$) thus making the condition number of the plant infinite at steady-state. In other words, the plant model $G(j\omega)$ looses rank at steady-state. This of course impose some unusual limitations on the achievable performance of the closed loop system. Let the singular value decomposition of $G(j\omega)$ be

$$G(j\omega) = Y(\omega)\Sigma(\omega)U^*(\omega) \tag{7.4}$$

$$= \begin{bmatrix} y_1 & y_2 & y_3 & y_4 \end{bmatrix} \begin{bmatrix} \sigma_1 & 0 & 0 \\ 0 & \sigma_2 & 0 \\ 0 & 0 & \sigma_3 \\ 0 & 0 & 0 \end{bmatrix} \begin{bmatrix} u_1^* \\ u_2^* \\ u_3^* \end{bmatrix}. \tag{7.5}$$

Here y_i and u_i are the output and input principal directions respectively, see Section 3.2.1. Since U is unitary u_1, u_2 and u_3 will be mutually orthogonal and we may partition the input $u(j\omega)$ after the input principal directions:

$$u(j\omega) = \alpha_1(\omega)u_1(\omega) + \alpha_2(\omega)u_2(\omega) + \alpha_3(\omega)u_3(\omega)\ . \tag{7.6}$$

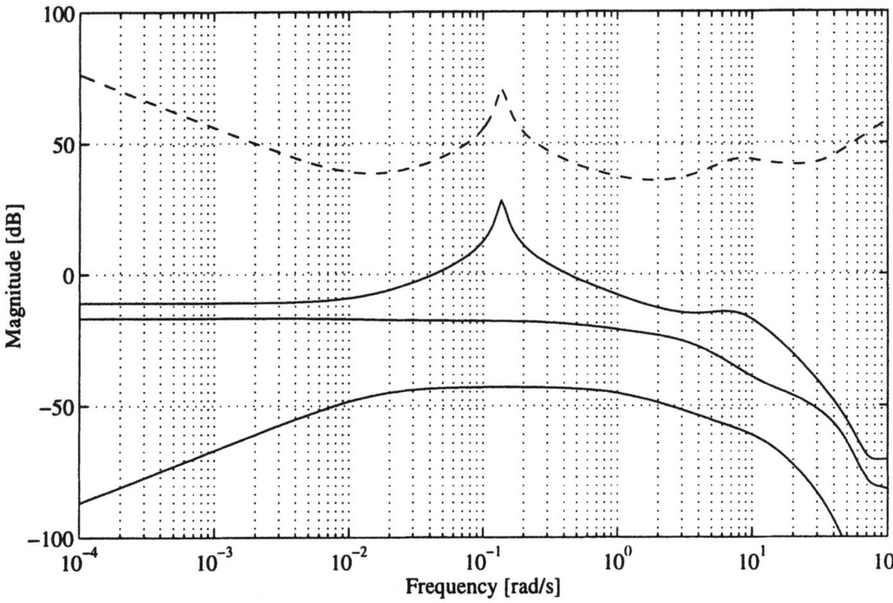

Fig. 7.2. *Singular value Bode plot of the plant $G(s)$ (solid) and plant condition number (dashed).*

In other words we express $u(j\omega)$ in the coordinate system defined by U. The output is then given by:

$$y(j\omega) = Y(\omega)\Sigma(\omega)U^*(\omega)\left(\alpha_1(\omega)u_1(\omega) + \alpha_2(\omega)u_2(\omega) + \alpha_3(\omega)u_3(\omega)\right) \tag{7.7}$$

$$= \sum_{i=1}^{3} y_i(\omega)\sigma_i(\omega)\alpha_i(\omega) \tag{7.8}$$

$$= \sigma_1(\omega)\alpha_1(\omega)y_1(\omega) + \sigma_2(\omega)\alpha_2(\omega)y_2(\omega) + \sigma_3(\omega)\alpha_3(\omega)y_3(\omega)\ . \tag{7.9}$$

Thus the output $y(j\omega)$ will be confined to the region spanned by $y_1(\omega)$, $y_2(\omega)$ and $y_3(\omega)$. We cannot require outputs in direction of $y_4(\omega)$ since we can only control as many output directions as we have inputs.

Furthermore, at steady-state σ_3 tends to zero and the output $y(j0)$ becomes confined to the region spanned by $y_1(0)$ and $y_2(0)$. It turns out that for frequencies below 10^{-4} rad/s, $Y(\omega)$ remains fairly constant. A good constant approximation is

$$Y(\omega) \approx \tilde{Y} = \begin{bmatrix} \tilde{y}_1 & \tilde{y}_2 & \tilde{y}_3 & \tilde{y}_4 \end{bmatrix}, \qquad \text{for } \omega < 10^{-4} \tag{7.10}$$

$$= \begin{bmatrix} 0 & 0 & 0 & -1j \\ 0.923 & -0.291 & 0.252j & 0 \\ 0.300 & -0.954 & 0 & 0 \\ -0.240 & -0.076 & -0.968j & 0 \end{bmatrix}, \qquad \text{for } \omega < 10^{-4} \tag{7.11}$$

where the imaginary parts are due to plant transmission zeros at $s = 0$. Since $\tilde{y}_{11} = 0$ and $\tilde{y}_{21} = 0$ the pitch rate q cannot be controlled in the low frequency area. However, we have no intentions of trying to do so. In fact, in the existing classical PI design the pitch rate is used to stabilize the plant, but it is not controlled itself.

Now, $\tilde{y}_3$ is mainly in the direction of the longitudinal acceleration $a_x(s)$ and therefore *independent* low frequency control of the longitudinal acceleration becomes difficult. That is, we cannot achieve acceleration in the low frequency range without changing either the height rate $\dot{h}(s)$ or the pitch attitude $\theta(s)$. This is not surprising since our model is valid in the transition zone between jet-borne and wing-borne flight where any change in longitudinal velocity strongly affects the aircraft lift.

The above singular value analysis has revealed that care must be taken when formulating the control objectives for the system. In the next section it will be shown how the singular value decomposition of $G(s)$ may be utilized to choose adequate weighting functions for robust control design.

7.2 Control Objectives

The singular value analysis made above illustrated that independent control of the pitch rate $q(s)$ will be difficult in the low frequency area. It is furthermore obvious that control of pitch attitude and pitch rate cannot be accomplished independently. It thus seems reasonable to omit direct control of the pitch rate and just use $q(s)$ as feedback (without reference).

A control configuration for the aircraft can thus be formulated as illustrated in Figure 7.3. The pilot commands for pitch attitude, $r_\theta(s)$, height rate, $r_{\dot{h}}(s)$, and longitudinal acceleration, $r_{a_x}(s)$ are compared with the measured values and the corresponding error signals, $e_\theta(s)$, $e_{\dot{h}}(s)$ and $e_{a_x}(s)$ are fed into the controller $K(s)$ together with the measured pitch rate $q_m(s)$.

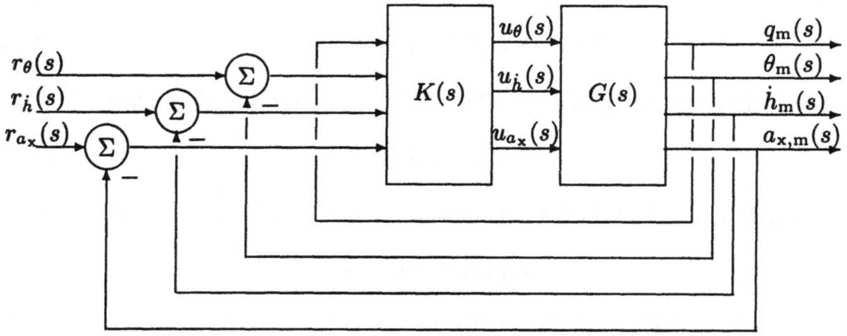

Fig. 7.3. *Block diagram of controlled aircraft.*

There are four main control objectives that will be considered when designing a controller for the ASTOVL vehicle.

1. Stability. The controller must stabilize the aircraft. The closed loop poles must thus all be located in the left half s-plane.
2. Robust stability. The controller must be robustly stable to unmodeled dynamics and parametric uncertainty.
3. Nominal performance. There are two main performance demands to the controlled aircraft. It must have adequate response to pilot commands for pitch attitude, height rate and longitudinal acceleration. Furthermore, rejection of gust disturbances on the airframe must be satisfactory. Here we will consider pilot command response only.
4. Robust performance. Nominal performance specifications must be met by uncertain closed loop system as well.

7.2.1 Robustness

It is critical that the controller stabilizes not only the nominal plant, but also a family of plants covering the inevitable parametric and dynamic uncertainty in the model. Here we will lump all the effects of uncertainty into a dynamic uncertainty description.

The robustness towards unmodeled dynamics can be investigated using the singular value methods in Section 4.3. If, for example, an unstructured input multiplicative uncertainty model is used, then the true system is assumed to be given by:

$$G_{\mathrm{T}}(s) = G(s)\left(I + \tilde{\Delta}(s)\right) \tag{7.12}$$

where $\tilde{\Delta}(s)$ is bounded in magnitude by:

$$\bar{\sigma}(\tilde{\Delta}(j\omega)) < \ell(\omega)\,, \qquad \forall\omega \geq 0\,. \tag{7.13}$$

From Theorem 4.4 the perturbed system will remain stable if and only if:

$$\bar{\sigma}(T_{\mathrm{i}}(j\omega)) < \ell^{-1}(\omega)\,, \qquad \forall\omega \geq 0\,. \tag{7.14}$$

Usually a diagonal weighting function $W_{\mathrm{u}}(s)$ is introduced so that:

$$\tilde{\Delta}(s) = W_{\mathrm{u}}(s)\Delta(s) \tag{7.15}$$

where $\bar{\sigma}(\Delta(j\omega)) < 1$ and $\bar{\sigma}(W_{\mathrm{u}}(j\omega)) = \ell(\omega)$, $\forall\omega \geq 0$. $W_{\mathrm{u}}(s)$ is often given as:

$$W_{\mathrm{u}}(s) = w(s) \cdot I \tag{7.16}$$

where $w(s)$ is a SISO transfer function satisfying $|w(j\omega)| = \ell(\omega)$ and I is the identity matrix of appropriate dimension.

Unfortunately completely unstructured uncertainty descriptions as above may be very conservative for ill-conditioned or poorly scaled systems. To see this let us assume an unstructured input multiplicative uncertainty description like (7.12) with $\tilde{\Delta}(s)$ given by (7.15). It is then easy to show that the perturbed plant $G_{\mathrm{T}}(s)$ is given by:

$$G_{\mathrm{T}}(s) = G(s) + G(s)\Delta(s)W_{\mathrm{u}}(s) = G(s) + G_{\Delta}(s) \tag{7.17}$$

where $G_{\Delta}(s)$ is given by

$$G_{\Delta}(s) = G(s)\Delta(s)W_{\mathrm{u}}(s) = w(s) \cdot \begin{bmatrix} G_{11}\Delta_{11} + G_{12}\Delta_{21} + G_{13}\Delta_{31} & G_{11}\Delta_{12} + G_{12}\Delta_{22} + G_{13}\Delta_{32} & G_{11}\Delta_{13} + G_{12}\Delta_{23} + G_{13}\Delta_{33} \\ G_{21}\Delta_{11} + G_{22}\Delta_{21} + G_{23}\Delta_{31} & G_{21}\Delta_{12} + G_{22}\Delta_{22} + G_{23}\Delta_{32} & G_{21}\Delta_{13} + G_{22}\Delta_{23} + G_{23}\Delta_{33} \\ G_{31}\Delta_{11} + G_{32}\Delta_{21} + G_{33}\Delta_{31} & G_{31}\Delta_{12} + G_{32}\Delta_{22} + G_{33}\Delta_{32} & G_{31}\Delta_{13} + G_{32}\Delta_{23} + G_{33}\Delta_{33} \\ G_{41}\Delta_{11} + G_{42}\Delta_{21} + G_{43}\Delta_{31} & G_{41}\Delta_{12} + G_{42}\Delta_{22} + G_{43}\Delta_{32} & G_{41}\Delta_{13} + G_{42}\Delta_{23} + G_{43}\Delta_{33} \end{bmatrix} . \tag{7.18}$$

Here Δ_{ij} denotes the (i,j)'th entry of $\Delta(s)$. The dependency on s has been omitted for clarity. Now, if there is large differences in the gains of $G_{ij}(s)$ *row-wise*, i.e if, for example,

$$|G_{11}(j\omega)| \ll \{|G_{12}(j\omega)|, |G_{13}(j\omega)|\} \ , \quad \text{for some frequency range,} \tag{7.19}$$

then we will have for those frequencies

$$|G_{\Delta_{11}}(j\omega)| = |w(j\omega)|\,|G_{11}(j\omega)\Delta_{11} + G_{12}(j\omega)\Delta_{21} + G_{13}(j\omega)\Delta_{31}| \tag{7.20}$$

$$\gg |G_{11}(j\omega)| \tag{7.21}$$

unless $|w(j\omega)|$ is very small. Consequently, the uncertainty assumption on $G_{11}(s)$ can be very conservative. Generally, if we have large row-wise gain variations in $G(s)$ an unstructured input multiplicative uncertainty description will be potentially very conservative. Equivalently, it can be shown that if significantly *column-wise* gain variations exist in $G(s)$ an unstructured output multiplicative uncertainty description will be potentially very conservative.

Such gain variations occur typically for poorly scaled or ill-conditioned systems. For the aircraft model it can be checked that there are considerable row-wise gain variations across the entire frequency range and column-wise gain variations in the low frequency area. Thus unstructured perturbation models will surely result in very conservative control designs.

How can we now avoid this conservatism? One remedy is to diagonalize Δ(s):

$$\Delta(s) = \mathrm{diag}\{\Delta_{11}, \Delta_{22}, \cdots, \Delta_{nn}\} \ , \quad |\Delta_{ii}(j\omega)| < 1, \forall\omega \geq 0 \,. \tag{7.22}$$

Then for the case of input multiplicative uncertainty the plant uncertainty $G_{\Delta}(s)$ becomes

$$G_\Delta(s) = G(s)\Delta(s)W_u(s) = w(s)I\begin{bmatrix} G_{11}\Delta_{11} & G_{12}\Delta_{22} & G_{13}\Delta_{33} \\ G_{21}\Delta_{11} & G_{22}\Delta_{22} & G_{23}\Delta_{33} \\ G_{31}\Delta_{11} & G_{32}\Delta_{22} & G_{33}\Delta_{33} \\ G_{41}\Delta_{11} & G_{42}\Delta_{22} & G_{43}\Delta_{33} \end{bmatrix} \tag{7.23}$$

which is less potentially conservative. The diagonalization of $\Delta(s)$ corresponds to assuming that the uncertainty acts individually on each input (or output for output multiplicative uncertainty). We will assume this diagonal uncertainty structure for the aircraft control design. Notice that the control design will now be a μ-problem since we have enforced a structure on $\Delta(s)$. We can thus *not* use the singular value tests from Table 4.1[1]. Rather we must use the results from Chapter 5 to assess whether the closed loop system is robustly stable. From Theorem 5.2 the closed loop system will remain stable under perturbations $\Delta(s)$ if and only if

$$\sup_\omega \mu_\Delta(P(j\omega)) \leq 1 \tag{7.24}$$

where $P(s)$ is the transfer function matrix "seen" by the perturbation.

Another well-known design problem in connection with ill-conditioned systems is that the closed loop properties can be very different when evaluated at different loop breaking points. In [TCB95] it is shown that for the current aircraft the robustness to diagonal multiplicative uncertainty like (7.22) may be adequate when the uncertainty acts at the plant output, but very poor if the uncertainty acts at the plant input. Thus the robustness properties may be very different at the plant input and output. In order to avoid this situation we will assume a diagonal multiplicative perturbation at *both* plant input and output. It is thus assumed that the true plant belongs to the set given by

$$G_T(s) = (I + \tilde{\Delta}_o(s))G(s)(I + \tilde{\Delta}_i(s)) \tag{7.25}$$

where $\tilde{\Delta}_{o/i}(s)$ is a diagonal perturbation:

$$\tilde{\Delta}_{i/o}(s) = W_{i/o}(s)\Delta(s) = w_{i/o}(s)I\Delta(s) \tag{7.26}$$

where $w_{i/o}(s)$ is a scalar transfer function specifying the expected maximum magnitude of $\bar{\sigma}(\tilde{\Delta}_{i/o}(j\omega))$ and $\Delta(s)$ is given by (7.22). Since the plant has been scaled it is assumed that $w_i(s) = w_o(s) = w_u(s)$. Since the nominal model is believed to be structurally correct in the low frequency range, the dynamic uncertainty at low frequencies will be quite small – say 5%. However, at higher frequencies the nominal model becomes more and more uncertain and in the neighborhood of the natural frequency of the airframe the dynamic uncertainty will be large, probably more that 100%. $w_u(s)$ was chosen as:

$$w_u(s) = \frac{10^{15/20}(s+1)}{s+112}\,. \tag{7.27}$$

[1] They will at least be potentially conservative.

In Figure 7.4 the singular value Bode plot of $w_u(s)$ is shown. Notice that the steady-state gain is $0.05 \approx -26$ dB corresponding to 5% uncertainty. The uncertainty then rise to 100% around 20 rad/sec and continues to increase above that frequency in order to illustrate the large uncertainty of the rigid-body airframe model around the natural frequency of the airframe.

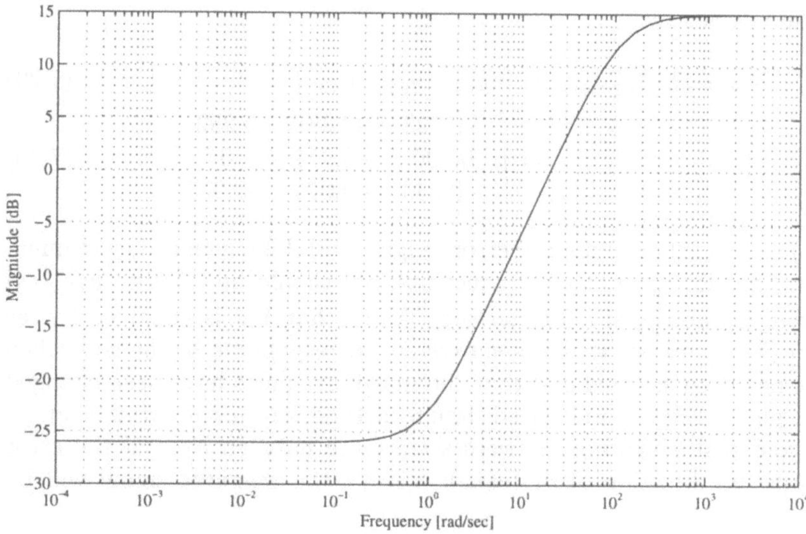

Fig. 7.4. *Singular value Bode plot of the uncertainty specification* $W_u(s) = w_u(s)I$.

7.2.2 Performance

The major performance demand on the controlled aircraft is probably that the pilot command response should be satisfactory. Thus the transfer function from references $r(s)$ to the control error $e(s)$ should be small across the controlled frequency range (up to the closed loop bandwidth ω_b). Since we have not included the pitch rate $q(s)$ into the control error $e(s)$ this transfer matrix will *not* be equal to the standard output sensitivity $S_o(s)$. Nevertheless we will use the term output sensitivity since the implied transfer matrix measures the sensitivity to disturbances (references) at the plant output. However, as was shown in Section 7.1.2 independent control of the longitudinal acceleration $a_x(s)$ in the low frequency range is not feasible. If we attempt to obtain this by specifying standard performance weightings with uniform high gain for low frequencies, the controller will use a lot of effort in trying to accomplish an impossible task. This will surely deteriorate the overall control performance.

To avoid this we must specify a control error weight $W_e(s)$ which does *not* penalize error signals in the direction of $\tilde{y}_3$ in the low frequency range.

Fortunately, such a weighting matrix can be easily constructed from $\tilde{Y}$. Let $W_e(s)$ be given by:

$$W_e(s) = \begin{bmatrix} w_1(s) & 0 & 0 \\ 0 & w_2(s) & 0 \\ 0 & 0 & w_3(s) \end{bmatrix} Y_W^T \tag{7.28}$$

where Y_W is given by:

$$Y_W = \Re e\ (\tilde{Y}_{2:4,1:3}) = \begin{bmatrix} y_{w1} & y_{w2} & y_{w3} \end{bmatrix} \tag{7.29}$$

$$= \begin{bmatrix} -0.923 & -0.291 & 0.252 \\ 0.300 & -0.954 & 0 \\ -0.240 & -0.076 & -0.968 \end{bmatrix} . \tag{7.30}$$

$w_1(s)$ and $w_2(s)$ are standard error weights with high gain at low frequencies. $w_3(s)$ has low gain at low frequencies, but should be equal to $w_{(1,2)}(s)$ for frequencies above, say 0.1 rad/sec. Notice that Y_W is equal to $\tilde{Y}$, see Equation (7.11), except that the subsystem corresponding to the controlled variables has been selected and that the imaginary parts have been changed into real parts. For a norm based design method like μ the latter changes nothing and real weighting matrices seem more natural. Now, if the error $e(j\omega)$ is in the direction of $\tilde{y}_3$ ($e(j\omega) = \beta\tilde{y}_3$), the weighted signal will be given by:

$$e'(j\omega) = W_e(j\omega)e(j\omega) \tag{7.31}$$

$$= W_e(j\omega)\beta\tilde{y}_3 \tag{7.32}$$

$$= \begin{bmatrix} 0 & 0 & j\beta w_3(j\omega) \end{bmatrix}^T . \tag{7.33}$$

Consequently, if the gain of $w_3(s)$ is small in the low frequency range, less weight will be put on the unreachable output direction. The following performance weightings were chosen for the aircraft design:

$$w_1(s) = \frac{0.5(s+1.25)}{s+6.25\cdot 10^{-3}} \tag{7.34}$$

$$w_2(s) = \frac{0.5(s+1.25)}{s+6.25\cdot 10^{-3}} \tag{7.35}$$

$$w_3(s) = \frac{0.5(s+1.25)(s+1.25\cdot 10^{-4})}{(s+0.03125)(s+0.0125)} . \tag{7.36}$$

In Figure 7.5, the singular value Bode plot of $W_e(s)$ is shown. Notice that because Y_W is unitary the singular values are simply the gain characteristics of the individual weightings $w_1(s) \rightarrow w_3(s)$. Notice also how $w_3(s)$ reflects the small weight we impose on the unreachable output at low frequencies.

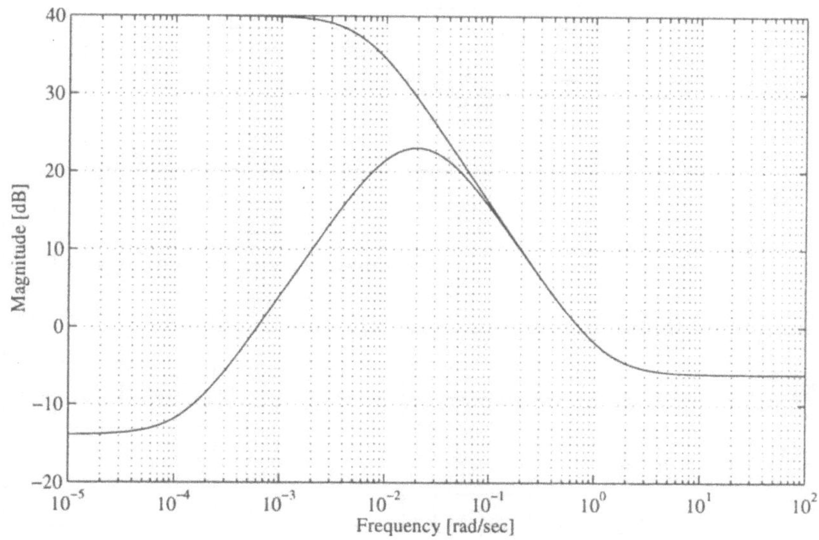

Fig. 7.5. *Singular value Bode plot of the performance weight* $W_e(s)$.

7.3 Formulation of Control Problem

Now the aircraft robust performance control problem will be formulated using the $N\Delta K$ framework introduced in Section 5.1.2. With simultaneous input and output multiplicative diagonal uncertainty and a command response performance specification as above, the control problem set-up then becomes as illustrated in Figure 7.6.

The results from Section 5.1 may then be utilized to assess the performance and robustness of the controlled system as follows. The perturbation $\Delta(s)$ belongs to the bounded subset

$$\mathbf{B\Delta} = \{\Delta(s) \in \mathbf{\Delta} \,|\bar{\sigma}(\Delta(j\omega)) < 1\} \tag{7.37}$$

where $\mathbf{\Delta}$ is given by

$$\mathbf{\Delta} = \{\text{diag}\,\{\delta_1, \delta_2, \delta_3, \delta_4, \delta_5, \delta_6, \delta_7\}\,|\delta_i \in \mathbf{C}\}\ . \tag{7.38}$$

We then have:

- The closed loop system will be robustly stable iff:

$$\sup_{\omega} \mu_{\mathbf{\Delta}}\left(P_{11}(j\omega)\right) \leq 1 \tag{7.39}$$

where $P(s) = F_\ell(N(s), K(s))$. $P_{11}(s)$ is thus the transfer matrix from the perturbations $w(s) = [w_1(s), w_2(s)]^T$ to $z(s) = [z_1(s), z_2(s)]^T$.

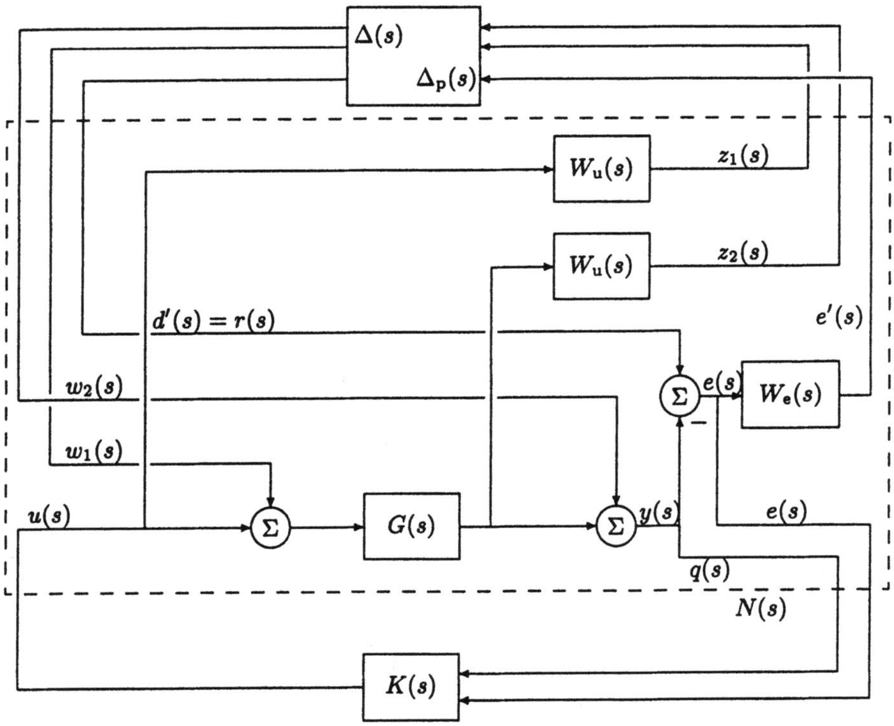

Fig. 7.6. *NΔK formulation of aircraft robust performance control problem.*

- The system will have nominal performance if:

$$\sup_{\omega} \bar{\sigma}\left(P_{22}(j\omega)\right) \leq 1 \,. \tag{7.40}$$

  $P_{22}(s)$ is the transfer matrix from the disturbances $d'(s)$ to the control errors $e'(s)$.
- The system will have robust performance, i.e it will be robustly stable and:

$$\sup_{\omega} \bar{\sigma}\left(F_{\mathrm{u}}(P(j\omega), \Delta(j\omega))\right) \leq 1 \tag{7.41}$$

  iff

$$\sup_{\omega} \mu_{\tilde{\mathbf{\Delta}}}\left(P(j\omega)\right) \leq 1 \,. \tag{7.42}$$

The augmented uncertainty block $\tilde{\mathbf{\Delta}}$ is given by:

$$\tilde{\mathbf{\Delta}} = \left\{\mathrm{diag}(\Delta, \Delta_{\mathrm{p}}) \,\middle|\, \Delta \in \mathbf{\Delta}, \Delta_{\mathrm{p}} \in \mathbf{C}^{3\times 3}\right\} \,. \tag{7.43}$$

Since $\tilde{\mathbf{\Delta}}$ consists of purely complex non-repeated blocks we may apply standard *D-K* iteration to find the optimal μ controller for the system. We will thus iteratively solve the minimization problem:

$$\inf_{K(s)\in\mathcal{K}_S} \sup_{\omega} \inf_{D(\omega)\in\mathbf{D}} \left\{\bar{\sigma}\left(D(\omega)F_\ell(N(j\omega),K(j\omega))D^{-1}(\omega)\right)\right\} . \quad (7.44)$$

The algorithms provided in the MATLAB *μ-Analysis and Synthesis Toolbox* may be readily applied to do so.

7.4 Evaluation of Classical Control Design

Before we proceed with the μ design let us evaluate the existing classical controller design for the aircraft. The existing classical PI controller configuration is displayed in Figure 7.7. The aircraft dynamics were first stabilized using pitch rate feedback through a PI controller $K_q(s)$ with lead filter $L_f(s)$:

$$K_q(s) = \frac{13(s+0.5)}{s} \quad (7.45)$$

$$L_f(s) = \frac{1+0.125s}{1+0.0208s} . \quad (7.46)$$

The remaining loops were closed in the order pitch attitude $K_\theta(s)$, height rate $K_{\dot{h}}(s)$ and longitudinal acceleration $K_{a_x}(s)$:

$$K_\theta(s) = 0.95 \quad (7.47)$$

$$K_{\dot{h}}(s) = \frac{-15(s+0.175)}{s} \quad (7.48)$$

$$K_{a_x}(s) = \frac{20(s+10.0)}{s} . \quad (7.49)$$

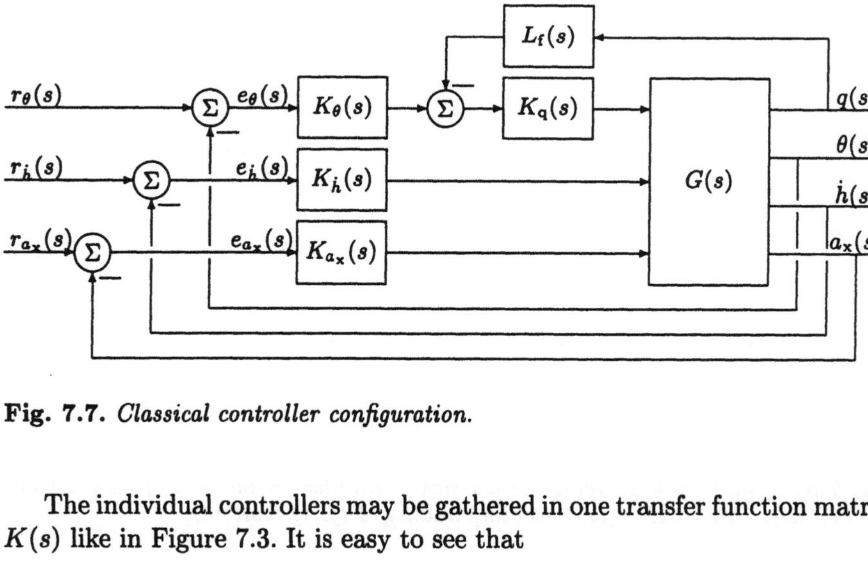

Fig. 7.7. *Classical controller configuration.*

The individual controllers may be gathered in one transfer function matrix $K(s)$ like in Figure 7.3. It is easy to see that

$$K(s) = \begin{bmatrix} K_q(s)L_f(s) & K_q(s)K_\theta(s) & 0 & 0 \\ 0 & 0 & K_{\dot{h}}(s) & 0 \\ 0 & 0 & 0 & K_{a_x}(s) \end{bmatrix}. \tag{7.50}$$

The above controllers are for the true unscaled plant. In order to evaluate the closed loop system against the robustness and performance demand introduced previously, we will use the scaled plant and scale the classical controller:

$$K_s(s) = N_u K(s) N_y^{-1} \tag{7.51}$$

where $K_s(s)$ denotes the scaled version of the classical controller $K(s)$.

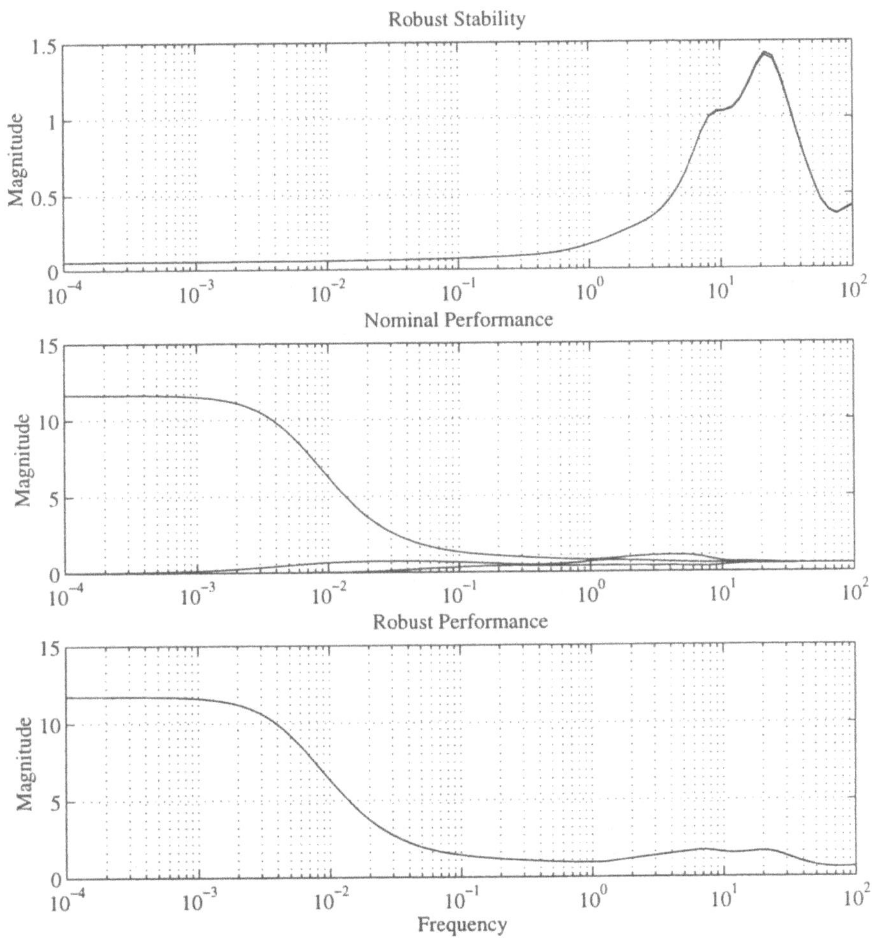

Fig. 7.8. ***Results from applying the classical controller on the aircraft. Shown are*** $\mu_\Delta(P_{11}(j\omega))$ ***(top),*** $\sigma(P_{22}(j\omega))$ ***(center) and*** $\mu_{\hat{\Delta}}(P(j\omega))$ ***(bottom).***

In order to check whether the classical controller comply with the stated demands, we check the nominal performance and robust stability conditions given in Section 7.3. It can easily be verified that the closed loop system in nominally stable since all poles of $F_\ell(N(s), K_s(s))$ are situated in the left-half s-plane. Robust stability, nominal performance and robust performance will then be checked.

In Figure 7.8 $\mu_\Delta(P_{11}(j\omega))$, the singular values of $P_{22}(j\omega)$ and $\mu_{\tilde{\Delta}}(P(j\omega))$ are shown across frequency ω.

Notice from Figure 7.8 that the closed loop system is *not* robustly stable since $\mu_\Delta(P_{11}(j\omega))$ peaks at approximately 1.4. Thus, with the applied perturbation model there exists a perturbation with

$$\bar{\sigma}(\Delta(j\omega)) = \frac{1}{1.4} = 0.71 \tag{7.52}$$

for which the closed loop system will become unstable. Furthermore, since $\sup_\omega \bar{\sigma}(P_{22}(j\omega)) > 10$ the nominal performance specification is far from being met and as a consequence we of course cannot obtain robust performance. $\mu_{\tilde{\Delta}}(P(j\omega))$ peaks at approximately 11.8. This means that there exists a perturbation $\Delta(s)$ with

$$\bar{\sigma}(\Delta(j\omega)) = \frac{1}{11.8} = 0.085 \tag{7.53}$$

for which the perturbed weighted sensitivity specification $F_u(\Delta(s), P(s))$ gets large

$$\bar{\sigma}(F_u(\Delta(j\omega), P(j\omega))) = 11.8\ . \tag{7.54}$$

Whether or not the performance specification is met for the perturbed closed loop system is of course of minor importance when closed loop stability cannot be guaranteed.

Table 7.2. *Maximum absolute and relative-to-maximum step-size deviation on pitch rate $q(t)$, pitch attitude $\theta(t)$, height rate $\dot{h}(t)$ and longitudinal acceleration $a_x(t)$ for each reference step – classical design.*

		Maximum absolute deviation on			
Step on	step size	pitch rate $\dot{\theta}$ [deg/sec]	pitch θ [deg]	height rate $\dot{h}$ [ft/sec]	acc. a_x [g]
pitch θ	30 deg	26.5	–	5.3	0.08
height rate $\dot{h}$	100 ft/sec	36.7	8.2	–	1.54
long. acc. a_x	2 g	1.24	1.07	11.3	–
		Maximum relative deviation on			
Step on	step size	$\dot{\theta}$ [%]	θ [%]	$\dot{h}$ [%]	a_x [%]
pitch θ	30 deg	–	–	5.3 %	4 %
height rate $\dot{h}$	100 ft/sec	–	27.3 %	–	77 %
long. acc. a_x	2 g	–	3.6 %	11.3 %	–

Notice how the robustness problems are concentrated at high frequencies whereas the performance problems occur at low frequencies. This should indicate that it will be possible to improve the design.

In Figure 7.9 the (nominal) step responses of the system are shown for maximum size steps applied on pitch attitude θ, height rate $\dot{h}$ and longitudinal acceleration $a_{\mathbf{x}}$. The results are further investigated in Table 7.2, where the maximum cross coupling effects are given. Both absolute and relative-to-maximum-step-size values are given. The cross coupling from height rate to pitch (27.3%) and longitudinal acceleration (77%), see Table 7.2, are unacceptable.

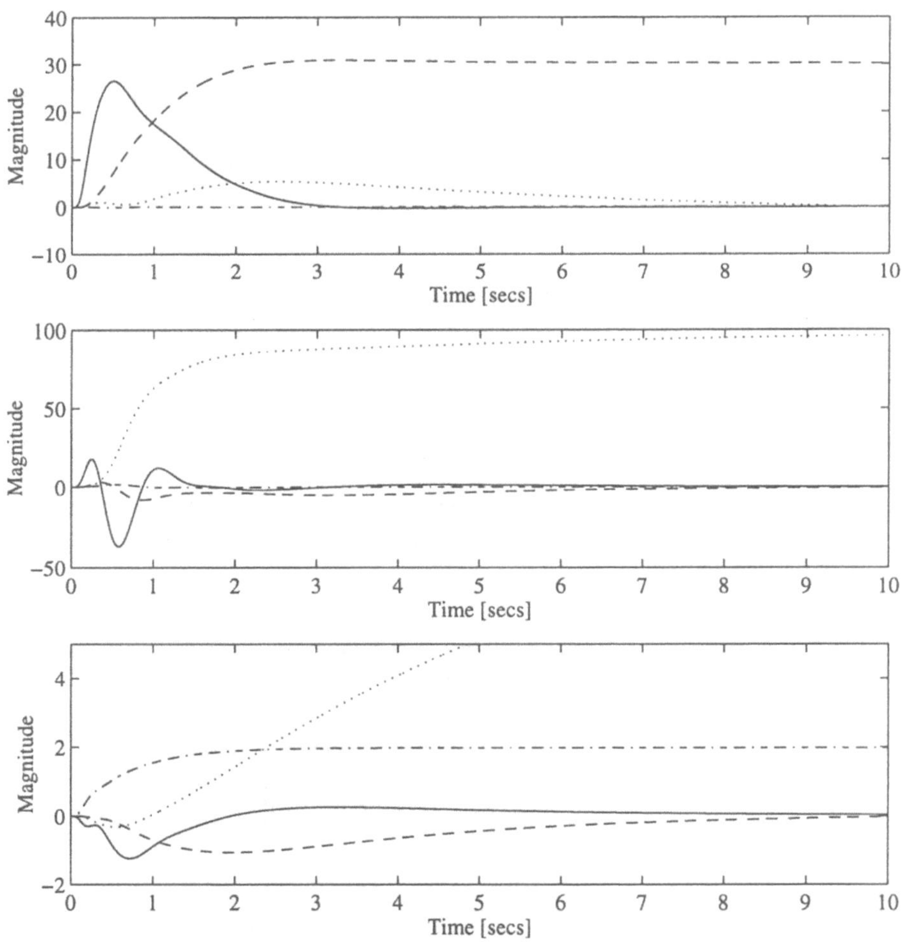

Fig. 7.9. ***Transient responses for classical control configuration. Shown are step responses from step on pitch (top), height rate (center) and acceleration (bottom). Shown signals are pitch rate (solid), pitch (dashed), height rate (dotted) and longitudinal acceleration (dash-dotted).***

7.5 Controller Design using μ

Now we will apply μ-synthesis in order to improve the classical design. The control problem is thus formulated as in Figure 7.6 and we will apply *D-K* iteration to solve it. In Figure 7.10, the result of the *D-K* iteration is shown. The maximum upper bound for $\mu_{\tilde{\Delta}}(P(j\omega))$ for selected iterations are also given in Table 7.3.

Table 7.3. *Maximum upper bound for $\mu_{\Delta}(P_{11}(j\omega))$, $\bar{\sigma}(P_{22}(j\omega))$ and $\mu_{\tilde{\Delta}}(P(s))$ in each step of D-K iteration.*

Iteration No.	$\sup_\omega \mu_{\Delta}(P_{11}(j\omega))$	$\sup_\omega \bar{\sigma}(P_{22}(j\omega))$	$\sup_\omega \mu_{\tilde{\Delta}}(P(j\omega))$
1 ($\mathcal{H}_\infty$ controller)	0.722	8.40	8.48
2	0.737	2.67	2.81
3	0.436	1.91	2.01
4	0.575	1.58	1.68
5	0.840	1.45	1.54
10	0.462	1.17	1.27
15	0.467	1.04	1.14
20	0.443	0.956	1.06
25	0.467	0.897	0.999
28	0.595	0.869	0.969
Reduced order K_μ	0.589	0.869	1.00

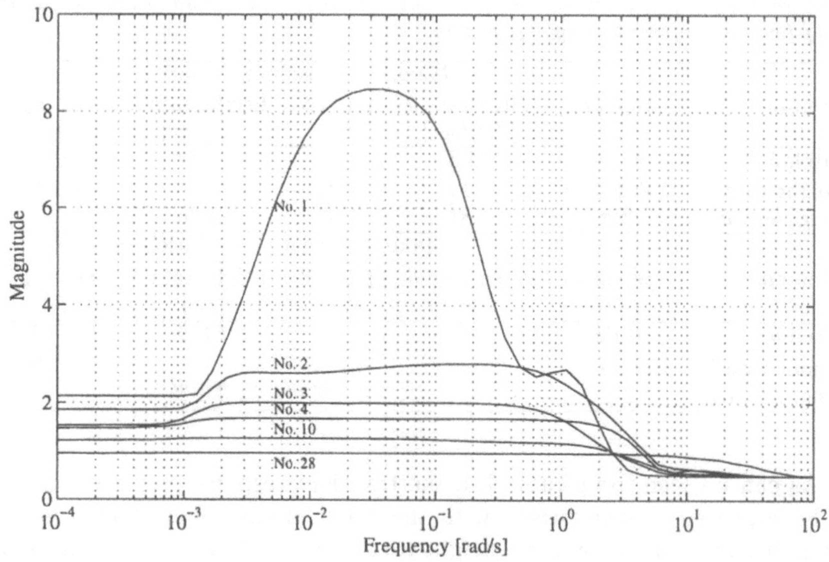

Fig. 7.10. *Upper bound on the structured singular value $\mu_{\tilde{\Delta}}(F_\ell(N(j\omega), K(j\omega)))$ in D-K iteration.*

Note that compared with the CD servo example presented in Chapter 6, the *D-K* iteration procedure here converges very slowly. However, after 25 iterations we obtain closed loop robust performance. After 28 iterations the decrease in μ per iteration became negligible. The final controller was of very high order (92 states). However, using model reduction methods it was possible to reduce the order to 30 with very little degradation in control performance, see Table 7.3. Using controller approximations with fewer states implied, however, a significant increase in μ. Since the nominal model (without weighting matrices) has 26 states, a 30th order controller does maybe not seem unreasonable. However, this is still a very high order controller and there will certainly be open questions with regards to numerics in connection with actual implementation. It is, however, beyond the scope of this book to treat this in any detail. Fixed order $\mathcal{H}_\infty$ algorithms may be a solution to the very high order controllers produced by *D-K*, *D,G-K* and μ-*K* iteration.

In Figure 7.11 the closed loop system obtained with the reduced order controller is analyzed using μ. From Figure 7.11 notice that robust stability and nominal performance are obtained with some margin to allow for robust performance.

In Figure 7.12 the transient response of the closed loop system is evaluated. Here maximum size steps has been applied on pitch attitude θ, height rate $\dot{h}$ and longitudinal acceleration a_x. The results are further investigated in Table 7.4, where the maximum cross coupling effects are given.

Table 7.4. *Maximum absolute and relative-to-maximum-step-size deviation on pitch rate $q(t)$, pitch attitude $\theta(t)$, height rate $\dot{h}(t)$ and longitudinal acceleration $a_x(t)$ for each reference step – μ design.*

		Maximum absolute deviation on			
Step on	step size	pitch rate $\dot{\theta}$ [deg/sec]	pitch θ [deg]	height rate $\dot{h}$ [ft/sec]	acc. a_x [g]
pitch θ	30 deg	25.3	–	1.82	0.073
height rate $\dot{h}$	100 ft/sec	5.14	1.24	–	0.20
long. acc. a_x	2 g	1.81	1.42	3.25	–
		Maximum relative deviation on			
Step on	step size	pitch rate $\dot{\theta}$ [%]	pitch θ [%]	height rate $\dot{h}$ [%]	acc. a_x [%]
pitch θ	30 deg	–	–	1.8 %	3.67 %
height rate $\dot{h}$	100 ft/sec	–	4.1 %	–	10.2 %
long. acc. a_x	2 g	–	4.7 %	3.3 %	–

As can be seen from Table 7.4, the transient response has been dramatically improved in comparison with the classical controller. The cross couplings between pitch attitude, height rate and longitudinal acceleration are now acceptable. Furthermore, the μ analysis guarantee that these properties are more or less maintained under the uncertainty allowed by the applied perturbation model. The steady-state errors for pitch attitude, height rate

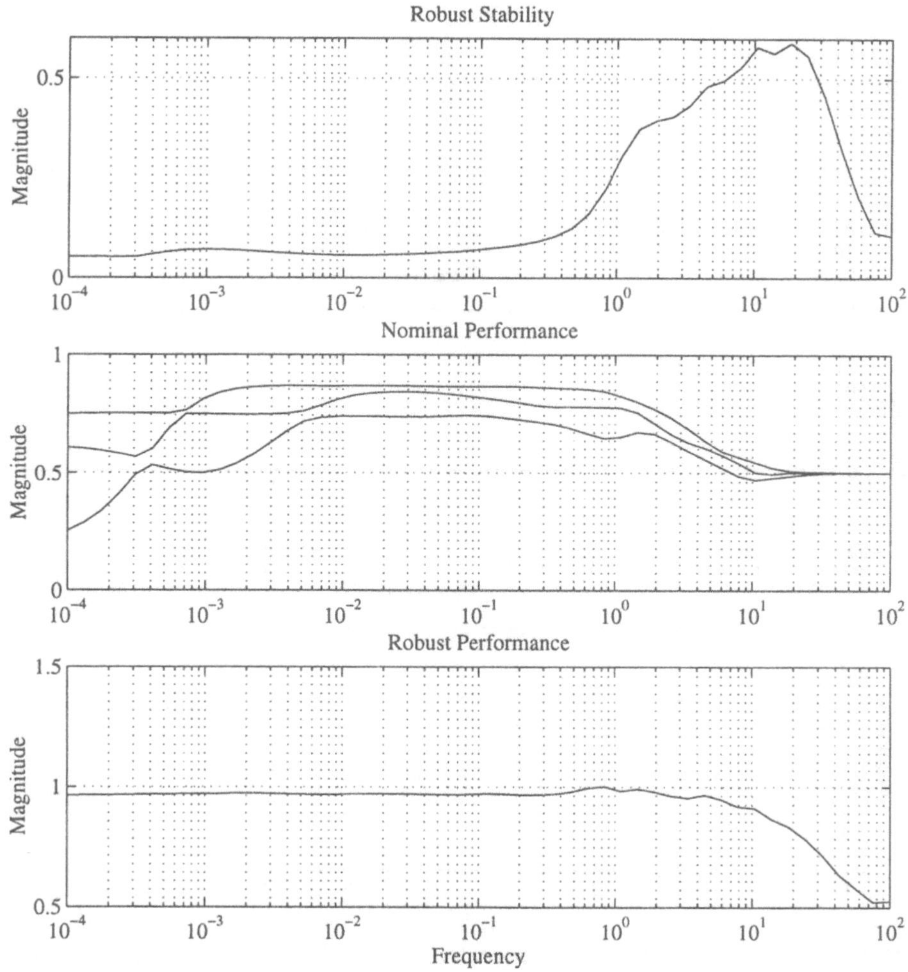

Fig. 7.11. *Results from applying the μ controller on the aircraft. Shown are $\mu_{\mathbf{\Delta}}(P_{11}(j\omega))$ (top), $\sigma(P_{22}(j\omega))$ (center) and $\mu_{\tilde{\mathbf{\Delta}}}(P(j\omega))$ (bottom).*

and longitudinal acceleration are 0.75%, 0.63% and 6.3%. The steady-state error for the acceleration may seen quite high. Remember, however, that the singular value analysis in Section 7.1.2 revealed that low frequency control of the acceleration $a_{\mathbf{x}}$ is infeasible. In fact, if the simulation in Figure 7.12 is continued for $t \to \infty$ it can be seen that $a_{\mathbf{x}}(t)$ slowly rolls off towards zero. Thus the term steady-state above for the acceleration is somewhat misused. However, these phenomena occur far below the frequency range of interest.

Finally, let us investigate the nominal output sensitivity function. As discussed previously in Section 7.2.2 the sensitivity function for the aircraft de-

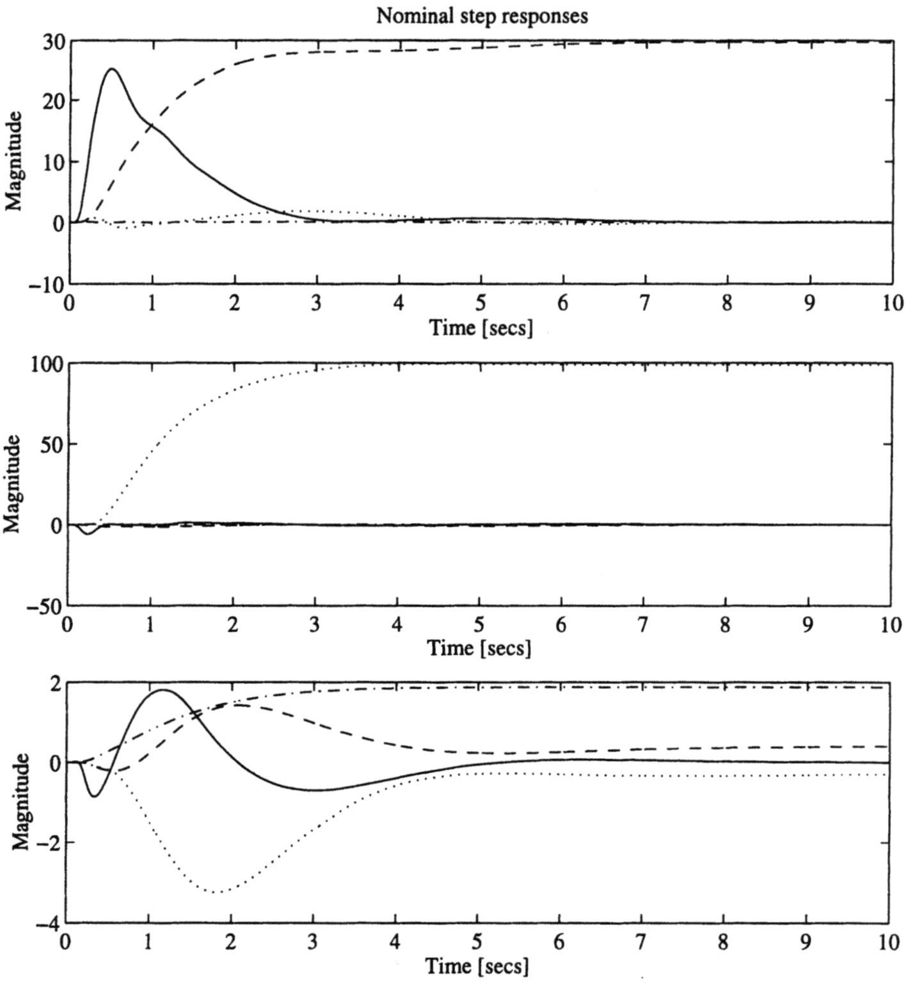

Fig. 7.12. *Transient responses for μ control system. Shown are step responses from step on pitch (top), height rate (center) and acceleration (bottom). Shown signals are pitch rate (solid), pitch (dashed), height rate (dotted) and longitudinal acceleration (dash-dotted).*

sign does not equal the standard output sensitivity $S_o(s)$ because the pitch rate feedback signal is not included in the control error $e(s)$. Let $\tilde{S}(s)$ denote the transfer function from the reference $r(s)$ to the control error $e(s)$. In Figure 7.13, the singular values of $\tilde{S}(j\omega)$ are shown together with the singular values of the performance specification $W_e^{-1}(s)$. Note how smoothly the μ controller has shaped $\tilde{S}(s)$ to comply with the specifications.

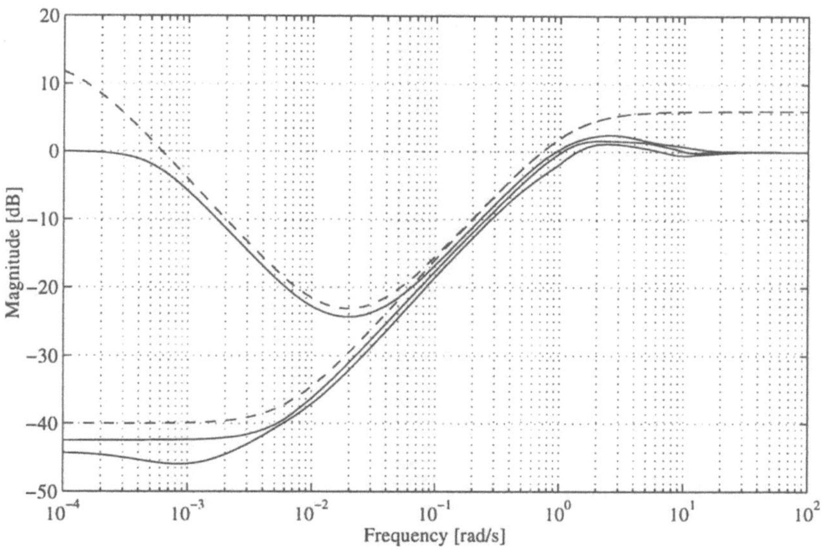

Fig. 7.13. *Nominal sensitivity and performance bounds for the aircraft design. Shown are* $\bar{\sigma}(F_\ell(N(j\omega), K(j\omega)))$ *(solid) for the reduced order controller and* $\bar{\sigma}(W_e^{-1}(j\omega))$ *(dashed).*

7.6 Summary

The rigid body dynamics of the aircraft was investigated using singular values. It was demonstrated that the system is very ill-conditioned. In fact, the condition number increases to infinity at steady-state. It then becomes impossible to use uniform weightings for controller design. One possible method of choosing performance weights in this case was suggested. A diagonal multiplicative perturbation structure was applied at both plant input and output. A classical control configuration was investigated and it was demonstrated that the cross coupling attenuation of this controller was poor. An optimal μ controller was designed using *D-K* iteration. μ analysis of the closed loop system revealed that the controller gave significant improvement in both robust stability and performance compared with the classical controller. The transient response for the controller was investigated using step responses. These proved quite satisfactory for the μ design with very good decoupling of pitch, height rate and longitudinal acceleration.

PART II

SYSTEM IDENTIFICATION AND ESTIMATION OF MODEL ERROR BOUNDS

CHAPTER 8
INTRODUCTION TO ESTIMATION THEORY

The use of system identification in control engineering has been a popular alternative to physical modeling for obtaining model descriptions of a given physical system. Application of the fundamental laws from mechanics, thermodynamics, chemistry etc. are often quite complex and time consuming tasks especially if large scale engineering systems are considered. In such cases, when plant input/output measurements are available, system identification provides an alternative for generating a model for use in control design. Usually, system identification involves linear discrete-time models. However, the general ideas are not restricted to such systems, but apply for non-linear continuous-time models as well. In this book, however, we will consider linear discrete-time models exclusively. Considering linear systems the obtained models will be readily applicable in connection with linear control systems design.

The construction of a parametric model from input/output measurements involves four basic entities:

- The data set.
- The model set (candidate models).
- The objective function for the estimation (performance function).
- Model validation.

The data-set used for the identification procedure must be "informative" of the actual physical plant during normal operation. If the system is not excited properly, we will not obtain a reasonable accurate model of the system. In more precise terms, we must require the input(s) to be ***persistently exciting***, see [Lju87, Sec. 14.2]. Sometimes, the input/output data are recorded during a specifically designed identification experiment where the user may choose the input signals. ***Experiment design*** then is concerned with choosing the input signals such that the data set becomes maximally informative subject to possible constraints at hand. Unfortunately, in other cases the user may not have the possibility to affect the experiment but must use data sets from normal operation of the system. Generally, we will only obtain knowledge of the system in the particular point of operation from which the data set was obtained. If the plant model is to be used outside this operation point considerations must be made to capture any deviations from the nominal model.

A set of candidate models is obtained by specifying within which collection of models we will look for a suitable one. This is the most important and, unfortunately, the most difficult choice of the system identification procedure. Much engineering intuition and insight is needed to choose the appropriate set of models. Usually we refer to two types of model sets, *black box* and *gray box* models. If the candidate model set is a set of standard linear models employed with no reference to the physical background it is a black box model set. Then the model parameters have no physical interpretation but are merely knobs for adjusting the fit to the data. If, on the other hand, the candidate model is obtained by physical modeling with some unknown physical parameters, it is a grey box model set. Then the model parameters are physical quantities, eg masses or moments of inertia. In this book, we will deal with black box modeling only.

Having determined our candidate model set we need to determine the "best" model within this set guided by the available measurements. This is the *identification method.* Typically the assessment of model quality is based on how the models perform when they attempt to reproduce the measured data. Three parameter estimation methods have gained particular interest; *prediction error methods* (PEM), *maximum likelihood methods* and *instrument variable methods.* Here we will consider mainly the PEM approach. However, in connection with estimation of uncertainty bounds we will use maximum likelihood methods as well.

Finally, having determined the best model in our candidate model set we need to assess the quality of this model. This is especially important in connection with robust control since this is precisely the information needed for constructing an appropriate perturbation model. In classical identification theory, model quality has been assessed under the assumption that the only source of uncertainty is noisy measurements. It is thus assumed that within our candidate model set there exists a unique model which is a perfect description of the true plant. Of course, such a strict condition is never met in practice. However, if the prediction errors, i.e the deviation between the measured and predicted outputs are almost white noise, the classical results hold with reasonable accuracy.

Unfortunately, in connection with control systems design it is often desired to have a low order approximation of the plant together with a simple quantification of the implied model imperfections in order to avoid very high order controllers. For restricted complexity models, the classical approach is inadequate since then, very often, the main source of uncertainty is due to the effect of incomplete model structures. Thus the results from classical system identification theory cannot be readily applied in estimation of model uncertainty for modern robust control design.

This gap between control theory and identification theory was recently realized by a number of researchers, see eg [GS89b, WL90a, Gev91]. The development of formal techniques for estimation of model uncertainty have since been the focus of active research and numerous results have been published, see eg [Bai92] and references therein.

8.1 Soft versus Hard Uncertainty Bounds

Considering finite dimensional linear time invariant (FDLTI) systems, estimation of model uncertainty in the case of restricted complexity model structures and a finite set of noisy data becomes a very difficult problem. On one hand, the structural model error (bias error) should be considered a deterministic quantity since the true system is hopefully deterministic. Consequently, the part of the model residuals associated with the undermodeling is deterministic. On the other hand, the model error due to noise (variance error) will be a stochastic quantity since it is caused by the noise in the data. The part of the model residuals associated with the noise is thus stochastic. Consequently, the two components of the total model error are fundamentally different quantities and require in principle fundamentally different treatment. However, in order to make coherent estimation procedures of the total model error most of the reported results in this field have relied upon treating the undermodeling and the noise as similar quantities. Generally speaking two main paths have been taken:

- The deterministic approach. The model residuals are assumed to be deterministic and bounded in magnitude. These methods will be referred to as *hard bound methods.*
- The stochastic approach. The model residuals are assumed to be stochastic, but non-stationary and correlated with the input. These methods will be referred to as *soft bound methods.*

The research on the deterministic approach has evolved along two different paths. The residuals may be assumed to be deterministic and bounded either in the frequency domain or in the time domain. The frequency domain approach is usually known as $\mathcal{H}_\infty$ *identification.* Here noisy frequency response measurements[1] are used to obtain models with $\|\cdot\|_{\mathcal{H}_\infty}$ norm error bounds that apply with probability one. In order to compute the bounds prior knowledge of the relative stability of the plant and of the noise bound on the frequency domain measurements is required. This approach is originally due to Parker & Bitmead [PB87], but has been further developed by Helmicki & co-workers [HJN91] and Gu & Khargonekar [GK92]. However, as pointed out in [NG94] the error bounds obtained through $\mathcal{H}_\infty$ identification can be very conservative. Furthermore, assessing the prior information needed to compute the bounds is a non-trivial task.

The time domain deterministic approach usually relies on *set-member-ship* identification assuming unknown, but bounded noise on the time domain data

[1] Usually obtained through sine-sweep tests.

points. In the fine paper by Wahlberg & Ljung [WL92] a thorough discussion of hard bound estimation from a least-squares perspective is presented. Hard total model error bounds are derived using prior information of the smoothness of the undermodeling impulse response, the shape of the undermodeling frequency response and a hard bound on the measurement noise. Again, the uncertainty estimates tend to be rather conservative and the necessary prior information may be difficult to obtain.

The research on the stochastic approach is mainly due to Goodwin & co-workers. In a series of papers, see eg [GS89b, GSM89, GGN91, GGN92] a methodology for estimation of total model error for restricted complexity models has been developed through stochastic embedding of the model bias. As were the case for the hard bound approaches some prior knowledge regarding the true plant and the noise are needed in order to compute the error bounds. Specifically the probability distributions for both the noise and the bias must be known. Also the second order properties – the covariances – of the noise and bias distributions must be known. However, in the crucial paper [GGN92], it is demonstrated that if the covariances for the noise and the bias are parameterized, then the parameters themselves may be estimated from the available data set through maximum likelihood methods. Thus the characterization of the necessary prior knowledge becomes qualitatively rather than quantitatively.

For a thorough discussion of the different methods for assessing the model uncertainty refer to the excellent survey paper by Ninness & Goodwin [NG94].

We feel that the possibility of estimating the necessary prior knowledge from the available measurements makes the stochastic embedding approach advantageous to the hard bound approaches. Whereas the necessary prior knowledge of the true system is qualitatively similar for both the soft and the hard bound approaches, the stochastic embedding approach provides the possibility of estimating, from data, the quantitative part of the necessary prior knowledge. Furthermore, the measurement noise may then be treated as a stochastic process which is most appropriate.

8.2 An Overview

The remainder of this part of the book is organized as follows. In Chapter 9, classical prediction error methods are reviewed for a general set of discrete-time linear models. This includes, for example, the classical approach to estimation of frequency domain model uncertainty. Next, in Chapter 10, some special cases are considered, in particular fixed denominator models with basis functions which are orthogonal in $\mathcal{H}_2$. Then, in Chapter 11, following the path outlined by Goodwin and co-workers in [GGN92], we will show how the stochastic embedding approach may result in non-conservative transfer function error bounds.

8.3 Remarks

Our treatment will be limited to scalar systems. The classical estimation methods can be extended to multivariable systems without great difficulty, but the extension is not so straightforward for the stochastic embedding approach.

Recently, De Vries and Van Den Hof [VH95] have published results on a *mixed soft and hard bound* approach to the estimation of model error bounds. In here, periodic input signals are used to distinguish bias and variance contributions to the total model error. Their results show promise, but we have not included any treatment of this new approach here.

CHAPTER 9
CLASSICAL SYSTEM IDENTIFICATION

The purpose of this chapter is to provide an introduction to system identification using prediction error methods (PEM). For a general treatment of system identification refer to any good textbook on the subject, like the two classic books [Lju87, SS89]. Here we will treat only the PEM approach. Furthermore we shall concentrate on results which are relevant for our purpose, namely estimation of model uncertainty. Specifically, the asymptotic distribution of the parameter estimates will be investigated under different assumptions. The representation and notation generally follows [Lju87, Chap. 9]. The general black-box model structure

$$A(q)y(k) = \frac{B(q)}{F(q)}u(k) + \frac{C(q)}{D(q)}e(k) \tag{9.1}$$

will be assumed with

$$A(q) = 1 + a_1q^{-1} + a_2q^{-2} + \cdots + a_{n_a}q^{-n_a} \tag{9.2}$$
$$B(q) = b_1q^{-1} + b_2q^{-2} + \cdots + b_{n_b}q^{-n_b} \tag{9.3}$$
$$F(q) = 1 + f_1q^{-1} + f_2q^{-2} + \cdots + f_{n_f}q^{-n_f} \tag{9.4}$$
$$C(q) = 1 + c_1q^{-1} + c_2q^{-2} + \cdots + c_{n_c}q^{-n_c} \tag{9.5}$$
$$D(q) = 1 + d_1q^{-1} + d_2q^{-2} + \cdots + d_{n_d}q^{-n_d}\ . \tag{9.6}$$

$e(k) \in \mathcal{N}(0, \lambda)$ denotes normal distributed zero mean noise with variance λ (white noise). q is the shift operator ($qu(k) = u(k+1)$). By $u(k)$ we mean the k'th element of the sequence $\{u(k)\}$. Let eg $\{u(k)\}$ be a sequence obtained by sampling the continuous-time signal $u_c(t)$ with sampling time T_s. Then $\{u(k)\}$ is given by

$$\{u(k)\} = u_c(kT_s) \qquad k = 1, 2, \cdots, N. \tag{9.7}$$

Frequently, we will not distinguish between the sequence $\{u(k)\}$ and the value $u(k)$ if the meaning is clear from the context. The optimal one-step-ahead predictor for (9.1) is given by

$$\hat{y}(k|\theta) = \frac{D(q)B(q)}{C(q)F(q)}u(k) + \frac{C(q) - D(q)A(q)}{C(q)}y(k) \tag{9.8}$$

where θ is the parameter vector:

$$\theta = [a_1, \cdots, a_{n_a}, b_1, \cdots, b_{n_b}, f_1, \cdots, f_{n_f}, c_1, \cdots, c_{n_c}, d_1, \cdots, d_{n_d}]^T \ . \tag{9.9}$$

A quadratic performance measure:

$$V_N(\theta, Z^N) = \frac{1}{N} \sum_{k=1}^{N} \frac{1}{2} \epsilon^2(k, \theta) \tag{9.10}$$

will be used for the PEM approach. Here $\epsilon(k,\theta) = y(k) - \hat{y}(k|\theta)$ denote the prediction errors and Z^N emphasize the measurement dependency. The parameter vector estimate $\hat{\theta}_N$ given N measurements is found as

$$\hat{\theta}_N = \arg\min_{\theta} \ V_N(\theta, Z^N) \ . \tag{9.11}$$

In Appendix E an easy implementable (in eg MATLAB) search algorithm for computing $\hat{\theta}_N$ is given. We then have

$$V'_N(\hat{\theta}_N, Z^N) = 0 \tag{9.12}$$

with prime denoting differentiation with respect to θ. Furthermore we have:

$$V'_N(\theta, Z^N) = -\frac{1}{N} \sum_{k=1}^{N} \frac{\partial \hat{y}(k|\theta)}{\partial \theta} (y(k) - \hat{y}(k|\theta)) \tag{9.13}$$

$$= -\frac{1}{N} \sum_{k=1}^{N} \psi(k,\theta)\epsilon(k,\theta) \tag{9.14}$$

$$V''_N(\theta, Z^N) = \frac{-1}{N} \sum_{k=1}^{N} \left\{ \frac{\partial^2 \hat{y}(k|\theta)}{\partial \theta^2} y(k) - \frac{\partial \hat{y}(k|\theta)}{\partial \theta} \frac{\partial \hat{y}(k|\theta)}{\partial \theta} - \frac{\partial^2 \hat{y}(k|\theta)}{\partial \theta^2} \hat{y}(k|\theta) \right\} \tag{9.15}$$

$$= \frac{1}{N} \sum_{k=1}^{N} \psi(k,\theta)\psi^T(k,\theta) - \frac{1}{N} \sum_{k=1}^{N} \frac{\partial \psi(k,\theta)}{\partial \theta} \epsilon(k,\theta) \ . \tag{9.16}$$

Here $\psi(k,\theta) = \partial \hat{y}(k|\theta)/\partial \theta$ is denoted the *model gradient.*

9.1 The Cramér-Rao Inequality for any Unbiased Estimator

Suppose that the true system is given by:

$$y(k) = G_0(q,\theta_0)u(k) + H_0(q,\theta_0)e_0(k) \tag{9.17}$$

where $e_0(k) \in \mathcal{N}(0, \lambda_0)$. Then it is a well-known result that for *any unbiased estimate* $\hat{\theta}_N$ *of* θ_0 (i.e, estimators such that $E\{\hat{\theta}_N\} = \theta_0$ regardless of the true value θ_0) the parameter covariance will obey the *Cramér-Rao Inequality:*

$$\operatorname{Cov}\left(\sqrt{N}\hat{\theta}_N\right) = NE\left\{\left(\hat{\theta}_N - E\left\{\hat{\theta}_N\right\}\right)\left(\hat{\theta}_N - E\left\{\hat{\theta}_N\right\}\right)^T\right\} \tag{9.18}$$

$$= NE\left\{\left(\hat{\theta}_N - \theta_0\right)\left(\hat{\theta}_N - \theta_0\right)^T\right\} \tag{9.19}$$

$$\geq \lambda_0 \left[\frac{1}{N}\sum_{k=1}^{N} E\left\{\psi(k,\theta_0)\psi^T(k,\theta_0)\right\}\right]^{-1} . \tag{9.20}$$

Notice that this bound applies for any N and all parameter estimation methods! Unfortunately, (9.20) is not suitable for computation of $\operatorname{Cov}(\hat{\theta}_N)$ since the true parameter θ_0 is unknown. In the next section we shall develop asymptotic expression for the parameter variance.

9.2 Time Domain Asymptotic Variance Expressions

Given the problem set-up described above it can be shown, see [Lju87, Lemma 8.2], that under weak conditions:

$$\sup_{\theta} \left|V_N(\theta, Z^N) - \bar{V}(\theta)\right| \to 0 , \qquad \text{for } N \to \infty \tag{9.21}$$

where

$$\bar{V}(\theta) = \bar{E}\left\{\frac{1}{2}\epsilon^2(k,\theta)\right\} \tag{9.22}$$

and $\bar{E}$ denotes "statistical expectation":

$$\bar{E}\left\{\frac{1}{2}\epsilon^2(k,\theta)\right\} = \lim_{N\to\infty}\frac{1}{N}\sum_{k=1}^{N} E\left\{\frac{1}{2}\epsilon^2(k,\theta)\right\} . \tag{9.23}$$

Thus the performance function $V_N(\theta, Z^N)$ converge to the limit function $\bar{V}(\theta)$. Furthermore the minimizing argument $\hat{\theta}_N$ of $V_N(\theta, Z^N)$ converges to the minimizing argument θ^* of $\bar{V}(\theta)$:

$$\left|\hat{\theta}_N - \theta^*\right| \to 0 , \qquad \text{for } N \to \infty. \tag{9.24}$$

If $V_N'(\hat{\theta}_N, Z^N)$ (for given N) is a differentiable function, then the *mean value theorem* guarantees the existence of a value $\tilde{\theta}$ "between" $\hat{\theta}_N$ and θ^* such that

$$V_N'(\hat{\theta}_N, Z^N) = V_N'(\theta^*, Z^N) + V_N''(\tilde{\theta}, Z^N)\left(\hat{\theta}_N - \theta^*\right) = 0 . \tag{9.25}$$

Under weak assumptions we then have [Lju87, pp 240]:

$$V_N''(\tilde{\theta}, Z^N) \to \bar{V}''(\theta^*)\,, \qquad \text{for } N \to \infty. \tag{9.26}$$

Thus for $N \to \infty$ we have

$$V_N'(\theta^*, Z^N) + \bar{V}''(\theta^*)\left(\hat{\theta}_N - \theta^*\right) = 0 \tag{9.27}$$

$$\Leftrightarrow \qquad \left(\hat{\theta}_N - \theta^*\right) = -\left[\bar{V}''(\theta^*)\right]^{-1} V_N'(\theta^*, Z^N)\,. \tag{9.28}$$

The second factor on the right-hand side is given by:

$$-V_N'(\theta^*, Z^N) = \frac{1}{N}\sum_{k=1}^{N} \psi(k, \theta^*)\epsilon(k, \theta^*)\,. \tag{9.29}$$

For $N \to \infty$ this is a sum of random variables $\psi(k, \theta^*)\epsilon(k, \theta^*)$ with zero mean values. Assuming they had been independent we would have for $N \to \infty$:

$$-\sqrt{N}V_N'(\theta^*, Z^N) = \frac{1}{\sqrt{N}}\sum_{k=1}^{N} \psi(k, \theta^*)\epsilon(k, \theta^*) \in \mathcal{N}(0, Q) \tag{9.30}$$

with:

$$Q = \lim_{N\to\infty} E\left\{\left[\sqrt{N}V_N'(\theta^*, Z^N)\right]\left[\sqrt{N}V_N'(\theta^*, Z^N)\right]^T\right\} \tag{9.31}$$

$$= \lim_{N\to\infty} N \cdot E\left\{\left[V_N'(\theta^*, Z^N)\right]\left[V_N'(\theta^*, Z^N)\right]^T\right\}\,. \tag{9.32}$$

However, $\psi(k, \theta^*)$ and $\epsilon(k, \theta^*)$ are not independent but under weak assumptions the dependency between distant terms will decrease and (9.30) will hold approximately. If so we have from (9.28):

$$\sqrt{N}\left(\hat{\theta}_N - \theta^*\right) \in \mathcal{N}(0, P_\theta)\,, \qquad \text{for } N \to \infty \tag{9.33}$$

with

$$P_\theta = \lim_{N\to\infty} E\left\{\sqrt{N}\left(\hat{\theta}_N - \theta^*\right)\left(\hat{\theta}_N - \theta^*\right)^T\sqrt{N}\right\} \tag{9.34}$$

$$= \lim_{N\to\infty} NE\left\{\left[\bar{V}''(\theta^*)^{-1}V_N'(\theta^*, Z^N)\right]\left[\bar{V}''(\theta^*)^{-1}V_N'(\theta^*, Z^N)\right]^T\right\} \tag{9.35}$$

$$= \lim_{N\to\infty} NE\left\{\bar{V}''(\theta^*)^{-1}\left[V_N'(\theta^*, Z^N)\right]\left[V_N'(\theta^*, Z^N)\right]^T \bar{V}''(\theta^*)^{-T}\right\} \tag{9.36}$$

$$= \bar{V}''(\theta^*)^{-1} \lim_{N\to\infty} NE\left\{\left[V_N'(\theta^*, Z^N)\right]\left[V_N'(\theta^*, Z^N)\right]^T\right\}\bar{V}''(\theta^*)^{-1} \tag{9.37}$$

$$= \left[\bar{V}''(\theta^*)\right]^{-1} Q \left[\bar{V}''(\theta^*)\right]^{-1}\,. \tag{9.38}$$

For a more rigorously justification of (9.33) and (9.38), see [Lju87, Appendix 9A] where it is shown that the above results hold under weak assumptions.

Let us take a closer look at the parameter estimate covariance matrix P_θ. It is easily seen that $\bar{V}''(\theta^*)$ is given by

$$\bar{V}''(\theta^*) = \lim_{N\to\infty} \frac{1}{N} \sum_{k=1}^{N} E\left\{\psi(k,\theta^*)\psi^T(k,\theta^*)\right\} - \lim_{N\to\infty} \frac{1}{N} \sum_{k=1}^{N} E\left\{\frac{\partial\psi(k,\theta^*)}{\partial\theta}\epsilon(k,\theta^*)\right\} \quad (9.39)$$

$$= \bar{E}\left\{\psi(k,\theta^*)\psi^T(k,\theta^*)\right\} - \bar{E}\left\{\frac{\partial\psi(k,\theta^*)}{\partial\theta}\epsilon(k,\theta^*)\right\} . \quad (9.40)$$

Furthermore Q can be written as:

$$Q = \lim_{N\to\infty} NE\left\{\left(\frac{1}{N}\sum_{k=1}^{N}\psi(k,\theta^*)\epsilon(k,\theta^*)\right)\cdot \left(\frac{1}{N}\sum_{\tau=1}^{N}\psi(\tau,\theta^*)\epsilon(\tau,\theta^*)\right)^T\right\} \quad (9.41)$$

$$= \lim_{N\to\infty} \frac{1}{N}E\left\{\sum_{k=1}^{N}\psi(k,\theta^*)\epsilon(k,\theta^*)\sum_{\tau=1}^{N}\epsilon(\tau,\theta^*)\psi^T(\tau,\theta^*)\right\} \quad (9.42)$$

$$= \lim_{N\to\infty} \frac{1}{N}\sum_{k=1}^{N}\sum_{\tau=1}^{N} E\left\{\psi(k,\theta^*)\epsilon(k,\theta^*)\epsilon(\tau,\theta^*)\psi^T(\tau,\theta^*)\right\} . \quad (9.43)$$

It is now tempting to believe that $\bar{V}''(\theta^*)$ and Q may be estimated from data simply by substitution of θ^* with $\hat{\theta}_N$. However, replacing θ^* with $\hat{\theta}_N$ in the expression for Q, see (9.32), gives, trivially, $Q = 0$ which is a useless estimate. It is consequently difficult to estimate the covariance P_θ under the given weak assumptions.

The expressions for $\bar{V}''(\theta^*)$ and Q may, however, be significantly reduced in complexity by strengthening the assumptions. Specifically, assume that the model structure admits an exact description of the true system:

$$y(k) = G_0(q,\theta_0)u(k) + H_0(q,\theta_0)e_0(k) \quad (9.44)$$

where θ_0 is the true parameter value and $e_0(k) \in N(0,\lambda_0)$. Then under (otherwise) weak assumptions

$$\left|\hat{\theta}_N - \theta_0\right| \to 0\,, \qquad \text{for } N \to \infty. \quad (9.45)$$

Furthermore it is easy to show that:

$$\begin{aligned}\epsilon(k,\theta_0) &= y(k) - \hat{y}(k|\theta_0) &(9.46)\\ &= y(k) - H^{-1}(q,\theta_0)G(q,\theta_0)u(k) - \left(1 - H^{-1}(q,\theta_0)\right) y(k) &(9.47)\\ &= H^{-1}(q,\theta_0)y(k) - H^{-1}(q,\theta_0)G(q,\theta_0)u(k) &(9.48)\\ &= e_0(k)\ . &(9.49)\end{aligned}$$

Equation (9.49) may be used to simplify the expression for P_θ considerably. We now have:

$$\begin{aligned}\bar{V}''(\theta^*) &= \bar{V}''(\theta_0) &(9.50)\\ &= \bar{E}\left\{\psi(k,\theta_0)\psi^T(k,\theta_0)\right\} - \bar{E}\left\{\frac{\partial\psi(k,\theta_0)}{\partial\theta}\epsilon(k,\theta_0)\right\} &(9.51)\\ &= \bar{E}\left\{\psi(k,\theta_0)\psi^T(k,\theta_0)\right\} - \bar{E}\left\{\frac{\partial\psi(k,\theta_0)}{\partial\theta}e_0(k)\right\} &(9.52)\\ &= \bar{E}\left\{\psi(k,\theta_0)\psi^T(k,\theta_0)\right\}\ . &(9.53)\end{aligned}$$

The last equality follows because $\partial\psi(k,\theta_0)/\partial\theta$ is formed from Z^{t-1} (past data) only and hence is uncorrelated with $e_0(k)$. Q is now given by:

$$\begin{aligned}Q &= \lim_{N\to\infty}\frac{1}{N}\sum_{k=1}^{N}\sum_{\tau=1}^{N} E\left\{\psi(k,\theta_0)\epsilon(k,\theta_0)\epsilon(\tau,\theta_0)\psi^T(\tau,\theta_0)\right\} &(9.54)\\ &= \lim_{N\to\infty}\frac{1}{N}\sum_{k=1}^{N}\sum_{\tau=1}^{N} E\left\{\psi(k,\theta_0)e_0(k)e_0(\tau)\psi^T(\tau,\theta_0)\right\} &(9.55)\\ &= \lim_{N\to\infty}\frac{1}{N}\sum_{k=1}^{N}\lambda_0 E\left\{\psi(k,\theta_0)\psi^T(k,\theta_0)\right\} &(9.56)\\ &= \lambda_0\bar{E}\left\{\psi(k,\theta_0)\psi^T(k,\theta_0)\right\} &(9.57)\end{aligned}$$

since $e_0(k)$ is white noise with variance λ_0. The expression for P_θ then reduces to:

$$\begin{aligned}P_\theta &= \lambda_0\left[\bar{E}\left\{\psi(k,\theta_0)\psi^T(k,\theta_0)\right\}\right]^{-1} &(9.58)\\ &= \lambda_0\left[\lim_{N\to\infty}\frac{1}{N}\sum_{k=1}^{N} E\left\{\psi(k,\theta_0)\psi^T(k,\theta_0)\right\}\right]^{-1}\ . &(9.59)\end{aligned}$$

Now, notice that the asymptotic covariance matrix P_θ equals the limit (as $N \to \infty$) of the Cramér-Rao bound in (9.20). Consequently, in this sense the prediction-error estimate has the best possible asymptotic properties one can hope for.

Another very important aspect of the expression for P_θ in (9.59) is that it can be estimated from data. Having processed N data points and computed $\hat{\theta}_N$ a consistent[1] estimate of P_θ is given by:

[1] A consistent estimate is an estimate that converges to the true value for $N \to \infty$.

$$\hat{P}_N = \hat{\lambda}_N \left[\frac{1}{N} \sum_{k=1}^{N} \psi(k, \hat{\theta}_N) \psi^T(k, \hat{\theta}_N) \right]^{-1} \tag{9.60}$$

$$\hat{\lambda}_N = \frac{1}{N} \sum_{k=1}^{N} \epsilon^2(k, \hat{\theta}_N) \, . \tag{9.61}$$

The estimates (9.60) and (9.61) are the classical result on the covariance of the parameter estimate. Notice that the covariance $\mathrm{Cov}(\hat{\theta}_N)$ is given by:

$$\mathrm{Cov}(\hat{\theta}_N) = \frac{1}{N} P_\theta \, . \tag{9.62}$$

9.3 Frequency Domain Asymptotic Variance Expressions

Frequency domain expression for the covariance matrix P_θ in (9.59) may also be derived. To ease notation introduce the following definitions:

$$T(q,\theta) = \left[\; G(q,\theta) \quad H(q,\theta) \; \right] \tag{9.63}$$

$$\chi(k) = \left[\; u(k) \quad e(k) \; \right]^T \tag{9.64}$$

$$W_\mathrm{u}(q,\theta) = H^{-1}(q,\theta) G(q,\theta) \tag{9.65}$$

$$W_\mathrm{y}(q,\theta) = 1 - H^{-1}(q,\theta) \tag{9.66}$$

$$W(q,\theta) = \left[\; W_\mathrm{u}(q,\theta) \quad W_\mathrm{y}(q,\theta) \; \right] \tag{9.67}$$

$$z(k) = \left[\; u(k) \quad y(k) \; \right]^T \tag{9.68}$$

$$\Psi(q,\theta) = \frac{\partial W(q,\theta)}{\partial \theta} \, . \tag{9.69}$$

The above notation corresponds to the one used in [Lju87]. Now the general model structure may be written

$$y(k) = G(q,\theta) u(k) + H(q,\theta) e(k) \tag{9.70}$$

$$= T(q,\theta) \chi(k) \, . \tag{9.71}$$

Provided $H(q,\theta)$ is minimum phase, the optimal predictor for (9.71) is

$$\hat{y}(k|\theta) = H^{-1}(q,\theta) G(q,\theta) u(k) + \left(1 - H^{-1}(q,\theta)\right) y(k) \tag{9.72}$$

which may be written

$$\hat{y}(k|\theta) = W_\mathrm{u}(q,\theta) u(k) + W_\mathrm{y}(q,\theta) y(k) = W(q,\theta) z(k) \, . \tag{9.73}$$

Furthermore, the model gradient $\psi(k,\theta)$ becomes

$$\psi(k,\theta) = \frac{\partial \hat{y}(k|\theta)}{\partial \theta} = \frac{\partial W(q,\theta)}{\partial \theta} z(k) = \Psi(q,\theta) z(k) \,. \tag{9.74}$$

We may now derive an expression for $\Psi(q,\theta)$:

$$\Psi(q,\theta) = \frac{\partial}{\partial \theta} \left[\begin{array}{cc} W_{\mathrm{u}}(q,\theta) & W_{\mathrm{y}}(q,\theta) \end{array} \right] \tag{9.75}$$

$$= \left[\begin{array}{cc} \frac{\partial}{\partial \theta} \left(H^{-1}(q,\theta) G(q,\theta) \right) & \frac{\partial}{\partial \theta} \left(1 - H^{-1}(q,\theta) \right) \end{array} \right] \tag{9.76}$$

$$= \left[\begin{array}{cc} \dfrac{\frac{\partial G(q,\theta)}{\partial \theta} H(q,\theta) - \frac{\partial H(q,\theta)}{\partial \theta} G(q,\theta)}{(H(q,\theta))^2} & \dfrac{\frac{\partial H(q,\theta)}{\partial \theta}}{(H(q,\theta))^2} \end{array} \right] \tag{9.77}$$

$$= \frac{1}{(H(q,\theta))^2} \left[\begin{array}{cc} \frac{\partial G(q,\theta)}{\partial \theta} & \frac{\partial H(q,\theta)}{\partial \theta} \end{array} \right] \left[\begin{array}{cc} H(q,\theta) & 0 \\ -G(q,\theta) & 1 \end{array} \right] \tag{9.78}$$

$$= \frac{1}{H(q,\theta)} \frac{\partial T(q,\theta)}{\partial \theta} \left[\begin{array}{cc} 1 & 0 \\ -H^{-1}(q,\theta) G(q,\theta) & H^{-1}(q,\theta) \end{array} \right] . \tag{9.79}$$

The model gradient hence becomes:

$$\psi(k,\theta) = \frac{\frac{\partial T(q,\theta)}{\partial \theta}}{H(q,\theta)} \left[\begin{array}{cc} 1 & 0 \\ -H^{-1}(q,\theta) G(q,\theta) & H^{-1}(q,\theta) \end{array} \right] \left[\begin{array}{c} u(k) \\ y(k) \end{array} \right] \tag{9.80}$$

$$= \frac{\frac{\partial T(q,\theta)}{\partial \theta}}{H(q,\theta)} \left[\begin{array}{cc} u(k) & -H^{-1}(q,\theta) G(q,\theta) u(k) + H^{-1}(q,\theta) y(k) \end{array} \right]^T \tag{9.81}$$

$$= \frac{1}{H(q,\theta)} \frac{\partial T(q,\theta)}{\partial \theta} \left[\begin{array}{c} u(k) \\ \epsilon(k,\theta) \end{array} \right] . \tag{9.82}$$

The last equality follows since

$$\epsilon(k,\theta) = y(k) - \hat{y}(k|\theta) \tag{9.83}$$

$$= y(k) - H^{-1}(q,\theta) G(q,\theta) u(k) - \left(1 - H^{-1}(q,\theta)\right) y(k) \tag{9.84}$$

$$= -H^{-1}(q,\theta) G(q,\theta) u(k) + H^{-1}(q,\theta) y(k) \,. \tag{9.85}$$

If the true system can be represented within the model set, we have $\epsilon(k,\theta_0) = e_0(k)$ and:

$$\psi(k,\theta_0) = \frac{1}{H(q,\theta_0)} \frac{\partial T(q,\theta_0)}{\partial \theta} \left[\begin{array}{c} u(k) \\ e_0(k) \end{array} \right] = \frac{1}{H(q,\theta_0)} \frac{\partial T(q,\theta_0)}{\partial \theta} \chi_0(k) \tag{9.86}$$

where $\chi_0(k) = [u(k) \;\; e_0(k)]^T$. Using Parceval's relation and Equation (9.59) we obtain:

$$P_\theta = \lambda_0 \left[\bar{E}\left\{\psi(k,\theta_0)\psi^T(k,\theta_0)\right\}\right]^{-1} \tag{9.87}$$

$$= \lambda_0 \left[\lim_{N\to\infty} \frac{1}{N}\sum_{k=1}^{N} E\left\{\psi(k,\theta_0)\psi^T(k,\theta_0)\right\}\right]^{-1} \tag{9.88}$$

$$= \lambda_0 \left[\frac{1}{2\pi}\int_{-\pi}^{\pi} \left|H\left(e^{j\omega T_s},\theta_0\right)\right|^{-2} \cdot \frac{\partial T\left(e^{j\omega T_s},\theta_0\right)}{\partial\theta}\Phi_{\chi_0}\frac{\partial T^T\left(e^{j\omega T_s},\theta_0\right)}{\partial\theta}d\omega\right]^{-1} \tag{9.89}$$

$$= \left[\frac{1}{2\pi}\int_{-\pi}^{\pi} \frac{1}{\Phi_\nu(\omega)}\frac{\partial T\left(e^{j\omega T_s},\theta_0\right)}{\partial\theta}\Phi_{\chi_0}(\omega)\frac{\partial T^T\left(e^{j\omega T_s},\theta_0\right)}{\partial\theta}d\omega\right]^{-1} \tag{9.90}$$

since

$$\Phi_\nu(\omega) = \lambda_0 \left|H\left(e^{j\omega T_s},\theta_0\right)\right|^{-2} . \tag{9.91}$$

$\Phi_{\chi_0}(\omega)$ is the spectrum of $\chi_0(k)$:

$$\Phi_{\chi_0}(\omega) = \begin{bmatrix} \Phi_{\mathrm{u}}(\omega) & \Phi_{\mathrm{ue}}(\omega) \\ \Phi_{\mathrm{eu}}(\omega) & \lambda_0 \end{bmatrix} . \tag{9.92}$$

$\Phi_{\mathrm{ue}}(\omega) = \Phi_{\mathrm{eu}}(\omega)$ is the cross-spectrum between the input $u(k)$ and the innovations $e_0(k)$. When the system operates in open loop the cross-spectrum is zero and $\Phi_{\chi_0}(\omega)$ becomes a 2-by-2 diagonal matrix. Clearly, approximate expressions for P_θ may be obtained from data by substituting θ_0 with $\hat{\theta}_N$ in (9.90).

9.4 Confidence Intervals for $\hat{\theta}_N$

If a random vector η has a Gaussian distribution:

$$\eta \in \mathcal{N}(0, P) \tag{9.93}$$

it is well-known that the scalar:

$$z = \eta^T P^{-1}\eta \tag{9.94}$$

has a χ^2 distribution with $\dim(\eta) = d$ degrees of freedom:

$$z \in \chi^2(d) . \tag{9.95}$$

Thus from (9.33) we draw the conclusion that the scalar

$$\eta_N = N \cdot \left(\hat{\theta}_N - \theta_0\right)^T P_\theta^{-1}\left(\hat{\theta}_N - \theta_0\right) \tag{9.96}$$

converges to a χ^2 distribution with $\dim\theta = n$ degrees of freedom as $N \to \infty$. Equation (9.96) may be used to "draw" n-dimensional confidence ellipsoids for $\hat{\theta}_N$.

9.5 Frequency Domain Uncertainty Bounds

Previously, we have concentrated on parameter covariance. However, in connection with control design frequency domain expression for the uncertainty is usually required. Hence, we need to translate the estimated parameter covariance $\hat{P}_N$ into bounds on the frequency response of the estimated model $G(q, \hat{\theta}_N)$. However, we will face the difficulty that the estimated frequency response $G(e^{j\omega T_s}, \hat{\theta}_N)$ generally will be non-linear in $\hat{\theta}_N$. Instead we may approximate $G(e^{j\omega T_s}, \theta^*)$ with a Taylor series expansion around $\theta = \hat{\theta}_N$:

$$G(e^{j\omega T_s}, \theta^*) \approx G(e^{j\omega T_s}, \hat{\theta}_N) + \frac{\partial G(e^{j\omega T_s}, \hat{\theta}_N)}{\partial \theta} \left(\theta^* - \hat{\theta}_N\right) . \tag{9.97}$$

Now define:

$$\tilde{g}(e^{j\omega T_s}) = \begin{bmatrix} \Re e \left\{G(e^{j\omega T_s}, \theta^*) - G(e^{j\omega T_s}, \hat{\theta}_N)\right\} \\ \Im m \left\{G(e^{j\omega T_s}, \theta^*) - G(e^{j\omega T_s}, \hat{\theta}_N)\right\} \end{bmatrix} \tag{9.98}$$

$$\Gamma(e^{j\omega T_s}) = \begin{bmatrix} \Re e \left\{\frac{\partial G(e^{j\omega T_s}, \hat{\theta}_N)}{\partial \theta}\right\} & \Im m \left\{\frac{\partial G(e^{j\omega T_s}, \hat{\theta}_N)}{\partial \theta}\right\} \end{bmatrix}^T \tag{9.99}$$

so that

$$\tilde{g}\left(e^{j\omega T_s}\right) \approx \Gamma\left(e^{j\omega T_s}\right) \left(\theta^* - \hat{\theta}_N\right) . \tag{9.100}$$

Now $\sqrt{N}\tilde{g}(e^{j\omega T_s})$ will have asymptotic normal distribution for $N \to \infty$ with covariance $P_g(\omega)$:

$$\sqrt{N}\tilde{g}(e^{j\omega T_s}) \in \mathcal{N}(0, P_g(\omega)) , \qquad \text{for } N \to \infty \tag{9.101}$$

where the covariance matrix $P_g(\omega)$ is given by

$$P_g(\omega) = \lim_{N \to \infty} E\left\{\sqrt{N}\tilde{g}(e^{j\omega T_s})\tilde{g}^T(e^{j\omega T_s})\sqrt{N}\right\} \tag{9.102}$$

$$= \lim_{N \to \infty} E\left\{\Gamma(e^{j\omega T_s})\sqrt{N}(\theta^* - \hat{\theta}_N)(\theta^* - \hat{\theta}_N)^T\sqrt{N}\Gamma^T(e^{j\omega T_s})\right\} \tag{9.103}$$

$$= \Gamma(e^{j\omega T_s})P_\theta\Gamma^T(e^{j\omega T_s}) . \tag{9.104}$$

Using arguments similar to Section 9.4 the scalar

$$z(\omega) = N\tilde{g}^T(e^{j\omega T_s})P_g^{-1}(\omega)\tilde{g}(e^{j\omega T_s}) \tag{9.105}$$

$$= \tilde{g}^T(e^{j\omega T_s}) \left[\Gamma(e^{j\omega T_s})\frac{1}{N}P_\theta\Gamma^T(e^{j\omega T_s})\right]^{-1} \tilde{g}(e^{j\omega T_s}) \tag{9.106}$$

will have χ^2 distribution with dim $\tilde{g}(e^{j\omega T_s}) = 2$ degrees of freedom. Equation (9.106) may then be used to draw confidence ellipsis in the complex plane for the frequency response estimate $G(e^{j\omega T_s}, \hat{\theta}_N)$.

Completely analogue, confidence ellipses may be computed for the frequency response estimate $H(e^{j\omega T_s}, \hat{\theta}_N)$.

9.6 A Numerical Example

To illustrate the classical approach to system identification and estimation of frequency domain error bounds we will give a practical example. We will consider identification of a discrete-time model for a wind turbine. The model should be suitable for subsequent controller synthesis.

The wind turbine considered is a horizontal axis 400 kW variable pitch machine with generator connected directly to the grid such that the velocity of the main shaft is forced to be constant. The control system consists of 3 independent controllers:

- A speed controller which during start-up procedure synchronize the angular velocity of the generator shaft with the grid frequency 50 Hz before the generator is connected to the grid. This controller is used only during start-up of the wind turbine.
- A pitch servo loop which controls the hydraulic pitch actuators. The purpose is to make the pitch follow a given pitch reference from either the speed controller above or the power controller below.
- A power controller which becomes active after the generator has been connected to the grid. The operation of the power controller is twofold. If wind speed is less than rated, the control output (pitch reference) will simply follow some pre-specified "optimal" curve empirically determined from wind-tunnel experiments. If wind speed exceeds rated value, the power controller shall maintain 400 kW output.

Usually the power controller for wind speeds above rated value is the most difficult part of the control system design. It seems reasonable that modern robust control design methods here should be superior to classical designs. We will thus assume that the control problem at hand is one of maintaining rated power output (400 kW) (attenuating wind speed fluctuations) by varying the pitch of the turbine blades. Thus a nominal model and an uncertainty estimate will be required in rated operation mode, that is for average wind speeds above rated value (approximately 13 m/s). For this purpose 14.400 samples were collected with sampling frequency 8 Hz. The input signal was a fundamental square wave with period 40 seconds applied as pitch reference. 45 periods were thus collected. As outlined above the control system includes also a pitch servo mechanism. During the experiment it was equipped with a non-linear proportional controller. However, the intention is not to include the pitch servo into the model of the turbine itself. Thus the actual pitch ϑ [deg] was also recorded and used as input for the identification procedure. The first period of the pitch ϑ is shown in Figure 9.1 together with measured power output P [100kW] and wind speed v [m/sec] measured at the top of the nacelle. Note how the actual pitch is not purely square due to the pitch servo loop. Furthermore, it is clear that the wind speed fluctuations introduce considerable noise on the power output. It is tempting, since a wind speed measurement is available, to include this measurement as an additional input to the estimated model. Unfortunately, because the wind speed measurement is taken at the top of the nacelle, it will be affected by changes in the pitch angle. Also, every time a blade passes the tower disturbances in the wind field around the turbine will be introduced. This will further distort the wind measurement. It is therefore most common to treat the wind speed as a non-measurable disturbance when designing controllers for wind turbines. Thus we will consider it in the same manner for our identification experiment.

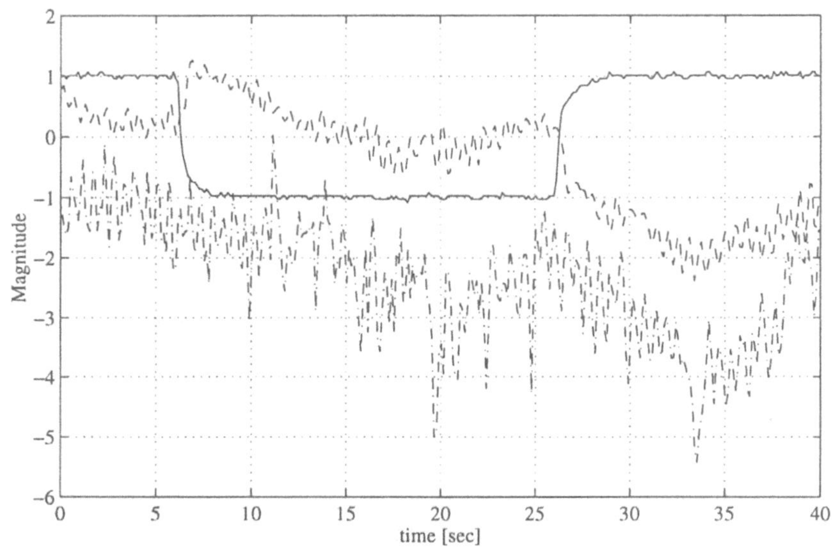

Fig. 9.1. *First period of the signals used for identification of the wind turbine. Shown are small signal values for pitch [deg] (solid), power [100kW] (dashed) and wind speed [m/sec] (dash-dotted, off-set 2 m/sec).*

We have subtracted mean values from the signals in Figure 9.1. The means are given in Table 9.1. Note that the working point for the power output is slightly below nominal value. However, during an open loop experiment it will be very difficult to maintain an average power output of 400 kW without considerable risk of overloading the generator. Closed loop experiments could have been considered instead. However, this has not been treated here.

Table 9.1. *Working point values for the signals displayed in Figure 9.1.*

	Pitch ϑ [deg]	Power P [kW]	Wind speed v [m/sec]
Working point	11	346	14.2

9.6.1 Choosing the Model Structure

As discussed in Chapter 8, choosing a set of candidate models for the identification procedure is by no means a trivial task. Much engineering ingenuity is often necessary to make this choice and some iterations with different model structures is almost always necessary to obtain useful results. The detailed physics for the wind turbine are very complicated, see eg [Lei89]. However, a useful model for control need not be so involved. In Figure 9.2 a block diagram representation of the main components in a dynamic model of the wind turbine is shown.

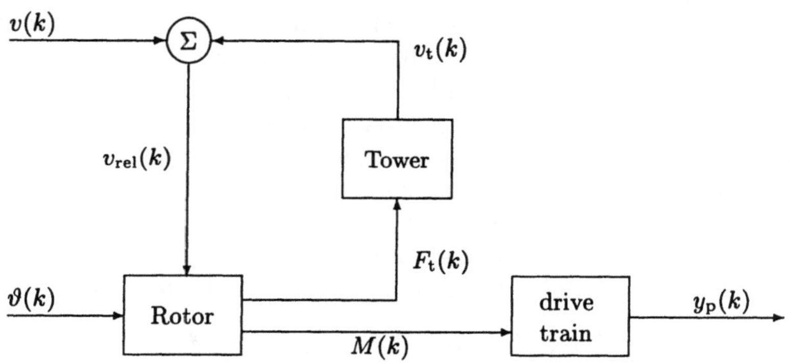

Fig. 9.2. *Simple model of a pitch controlled wind turbine.*

The wind flow around the turbine blades results in an axial force F_t which bends the tower and a force in the rotational direction which creates the driving torque M on the main shaft. The torque is then transmitted to the generator via the drive train. A reasonable physical model of the drive train (including the generator) is two rotating masses, one on each side of the gear box, connected with a spring and damper; the generator also acting like a damper. The three necessary states to describe this system is the velocity of the two masses and the difference of their angle.

The tower can also be modeled with a spring and damper. This will require two additional states (position and velocity of the hub). Furthermore, recent results indicate that dynamical effects in the wind flow around the blades also exists. The explanation is that when the pitch changes the wind distribution around the turbine will also change creating a new wind "tail". A first order system with time constant from 1–5 secs is a reasonable approximation for single turbines. However, if the turbine is placed in connection with several other turbines, then for certain wind directions the time constant could be much larger.

Furthermore, notice how not only the power output but also the tower motion is affected when controlling the pitch. Tower movements then affect the relative wind speed v_{rel}. Thus we have closed loops in the system.

The spectrum for the power output will generally have two high frequency peaks. The first peak is located at the rotational frequency of the rotor. This frequency is usually denoted 1P. A non-symmetric rotor with, for example, one blade having a slightly different pitch than the others cause this peak. The second peak which is normally more significant occurs at the 3P frequency and is caused by the three blades passing through the inhomogeneous wind field around the tower. A possible model for the 3P (1P) effect could be white noise filtered through a second order system with a sharp resonance.

Finally, the dynamics of the wind speed are also important. These dynamics are very complex. Low-pass filtered white noise can be used to model wind speed variations around the 10 mins average wind speed. However, both dominating frequencies and variance grow with increasing wind speeds. Consequently, if the wind speed is fully described by filtered white noise we will expect the necessary order to be quite large.

Since the wind speed affects the driving torque and thus passes through the same dynamics as the pitch angle a reasonable discrete-time linear model structure for the wind turbine, neglecting the 1P effect, could be

$$y_{\mathrm{p}}(k) = G_{\mathrm{t}}(q)\left(G_{\mathrm{f}}(q)\vartheta(k) + k_{\mathrm{v}}v(k)\right) + d(k) \tag{9.107}$$

where y_{p}, ϑ, v and d denote power output, pitch angle, wind speed and 3P disturbance respectively. $G_{\mathrm{t}}(q)$ is a transfer function describing the drive train and $G_{\mathrm{f}}(q)$ describes the pitch rotor dynamics. A possible model for the wind speed could be filtered white noise

$$v(k) = G_{\mathrm{w}}(q)e_{\mathrm{v}}(k)\ , \qquad e_{\mathrm{v}}(k) \in \mathcal{N}(0, \lambda_{\mathrm{v}})\ . \tag{9.108}$$

Similarly filtered white noise could be used for $d(k)$:

$$d(k) = G_{\mathrm{d}}(q)e_{\mathrm{d}}(k)\ , \qquad e_{\mathrm{d}}(k) \in \mathcal{N}(0, \lambda_{\mathrm{d}})\ . \tag{9.109}$$

Let $G_{\mathrm{x}}(q)$ be given by

$$G_{\mathrm{x}}(q) = \frac{N_{\mathrm{x}}(q)}{D_{\mathrm{x}}(q)} \tag{9.110}$$

with

$$N_{\mathrm{x}}(q) = n_1q^{-1} + n_2q^{-2} + \cdots + n_{n_{\mathrm{x}}}q^{-n_{\mathrm{x}}} \tag{9.111}$$

$$D_{\mathrm{x}}(q) = 1 + d_1q^{-1} + d_2q^{-2} + \cdots + d_{d_{\mathrm{x}}}q^{-d_{\mathrm{x}}}\ . \tag{9.112}$$

Then $y_{\mathrm{p}}(k)$ can be written

$$
\begin{aligned}
y_{\mathrm{p}}(k) &= G_{\mathrm{t}}(q)G_{\mathrm{f}}(q)\vartheta(k) + k_{\mathrm{v}}G_{\mathrm{t}}(q)G_{\mathrm{v}}(q)e_{\mathrm{v}}(k) + G_{\mathrm{d}}(q)e_{\mathrm{d}}(k) && (9.113)\\
&= \frac{N_{\mathrm{t}}(q)N_{\mathrm{f}}(q)}{D_{\mathrm{t}}(q)D_{\mathrm{f}}(q)}\vartheta(k) + k_{\mathrm{v}}\frac{N_{\mathrm{t}}(q)}{D_{\mathrm{t}}(q)}\frac{N_{\mathrm{v}}(q)}{D_{\mathrm{v}}(q)}e_{\mathrm{v}}(k) + \frac{N_{\mathrm{d}}(q)}{D_{\mathrm{d}}(q)}e_{\mathrm{d}}(k) && (9.114)\\
&= \frac{N_{\mathrm{t}}(q)N_{\mathrm{f}}(q)}{D_{\mathrm{t}}(q)D_{\mathrm{f}}(q)}\vartheta(k) + \\
&\quad \frac{k_{\mathrm{v}}N_{\mathrm{t}}(q)N_{\mathrm{v}}(q)D_{\mathrm{d}}(q)e_{\mathrm{v}}(k) + N_{\mathrm{d}}(q)D_{\mathrm{t}}(q)D_{\mathrm{v}}(q)e_{\mathrm{d}}(k)}{D_{\mathrm{t}}(q)D_{\mathrm{v}}(q)D_{\mathrm{d}}(q)} . && (9.115)
\end{aligned}
$$

Note that since we have two white noise sources, the general PEM model (9.1) does not include (9.115). However, the nominator in the last term of (9.115) can be modeled by a moving average (MA) process with a single noise source. Then the wind turbine model fits the PEM structure with

$$
\begin{aligned}
A(q) &= D_{\mathrm{t}}(q) && (9.116)\\
B(q) &= N_{\mathrm{t}}(q)N_{\mathrm{f}}(q) && (9.117)\\
F(q) &= D_{\mathrm{f}}(q) && (9.118)\\
D(q) &= D_{\mathrm{v}}(q)D_{\mathrm{d}}(q) && (9.119)
\end{aligned}
$$

and $C(q)$ some polynomial modeling the single noise source MA process.

Neglecting the tower dynamics, the degrees for each polynomial were chosen as follows

$$
\begin{aligned}
\deg(A) &= 3 && (9.120)\\
\deg(B) &= 2 && (9.121)\\
\deg(F) &= 1 && (9.122)\\
\deg(C) &= \deg(D) = k && (9.123)
\end{aligned}
$$

where k is some positive natural number. The identification procedure was then repeated for increasing values of k until the residuals were close to white noise.

9.6.2 Estimation and Model Validation

A Marquardts search procedure as outlined in Appendix E was implemented in MATLAB in order to find the least squares estimate of the parameter vector θ. 320 samples were used to get rid of initial conditions effects. The data set was split into two parts, each consisting of 7200 samples. The identification procedure was then used on the first data set and the other set was used for model validation. Model validation was performed by analysis of the residuals. The standard two residual tests were performed:

- Are the residuals white noise?
- Are the residuals and past inputs independent?

The typical whiteness test is to determine the covariance estimate

$$\hat{R}_N^{\epsilon}(\kappa) = \frac{1}{N}\sum_{k=1}^{N-\kappa} \epsilon(k,\hat{\theta}_N)\epsilon(k+\kappa,\hat{\theta}_N) . \tag{9.124}$$

If $\{\epsilon(k,\hat{\theta}_N)\}$ indeed is a white noise sequence with variance λ then for $\kappa \ll N$ $\hat{R}_N^{\epsilon}(\kappa)$ will be approximately normal distributed with zero mean and variance $\lambda^2 N^{-1}$ (as usual N denotes the number of measurements):

$$E\left\{\hat{R}_N^{\epsilon}(\kappa)\right\} = 0 , \qquad \kappa \geq 1 \tag{9.125}$$

$$E\left\{\hat{R}_N^{\epsilon}(\kappa)\hat{R}_N^{\epsilon}(\kappa)\right\} = \lambda^2 N^{-1} . \tag{9.126}$$

Since $\hat{R}_N^{\epsilon}(0)$ is an estimate of λ we may reject the hypothesis that the prediction errors are white noise on an α level if

$$\frac{\hat{R}_N^{\epsilon}(\kappa)}{\hat{R}_N^{\epsilon}(0)} > \frac{\alpha}{\sqrt{N}} . \tag{9.127}$$

For example, if $\alpha = 2.58$ we obtain 99% confidence levels. We may thus plot $\hat{R}_N^{\epsilon}(\kappa)/\hat{R}_N^{\epsilon}(0)$ for $0 < \kappa \ll N$ and if the maximum numerical value exceeds $2.58/\sqrt{N}$ for any value of κ the white noise hypothesis must be rejected.

Note that white residuals will require both the deterministic part $G(q,\theta)$ and the stochastic part $H(q,\theta)$ to be an adequate description of the true system. If we are mainly interested in the deterministic system $G_{\mathrm{T}}(q)$ the question is whether the residuals $\epsilon(k,\hat{\theta}_N)$ are independent of the inputs. If not, there is more in the output that originates from the input than explained by the current model $G(q,\hat{\theta}_N)$. Independence between inputs and residuals is usually tested using the cross-covariance estimate:

$$\hat{R}_N^{\epsilon,\mathrm{u}}(\kappa) = \frac{1}{N}\sum_{k=\kappa}^{N} \epsilon(k)u(k-\kappa) . \tag{9.128}$$

If $\{\epsilon(k,\hat{\theta}_N)\}$ and $\{u(k)\}$ are independent then $\sqrt{N}\hat{R}_N^{\epsilon,\mathrm{u}}(\kappa)$ will be asymptotically normal distributed with zero mean and variance P [Lju87, pp 429]:

$$\sqrt{N}\hat{R}_N^{\epsilon,\mathrm{u}}(\kappa) \in \mathcal{N}(0,P) , \qquad \text{for } N \rightarrow \infty \tag{9.129}$$

with P given by

$$P = \sum_{\kappa=-\infty}^{\infty} R^{\epsilon}(\kappa)R^{\mathrm{u}}(\kappa) \tag{9.130}$$

and

$$R^{\epsilon}(\kappa) = \lim_{N\to\infty} \frac{1}{N} \sum_{k=1}^{N} \epsilon(\kappa, \hat{\theta}_N)\epsilon(\kappa - k, \hat{\theta}_N) \tag{9.131}$$

$$R^{\mathrm{u}}(\kappa) = \lim_{N\to\infty} \frac{1}{N} \sum_{k=1}^{N} u(\kappa)u(\kappa - k) \,. \tag{9.132}$$

Using a data estimate $\hat{P}_N$ for the covariance we may reject the hypothesis that the inputs and prediction errors are independent on an α level if

$$\left|\hat{R}_N^{\epsilon,\mathrm{u}}(\kappa)\right| > \alpha\sqrt{\frac{\hat{P}_N}{N}} \,. \tag{9.133}$$

Residual tests should always be performed on a separate test set different from the data set used for the identification procedure.

9.6.3 Results

As described above the degree of the C and D polynomials was increased until the residual tests were both passed. The required number of states for the noise filter was 18. Thus 42 parameters were estimated. Despite the high number of parameters the identification algorithm converged without difficulty. Using numerical search algorithms only convergence to local minima can be obtained. For large number of parameters it can be very difficult to judge whether a given local minima is in fact the global minima. Experiments with different initial values indicated, however, that the obtained minimum was global.

The corresponding residual tests are shown in Figure 9.3. Here 99% confidence levels has been used. The pole-zero configuration for the deterministic and stochastic part of the system is shown in Figure 9.4 and the Bode plots are given in Figure 9.5.

The deterministic part $G(q, \hat{\theta}_N)$ has one pair of complex poles with relative damping 0.3 and natural frequency 1.1 Hz (6.91 rad/s). This corresponds well to values obtained from the turbine manufacturer *Vestas Wind Systems A/S* for a second order approximation to the drive train. We then have an additional pole in -0.15 from $A(q)$ and the pole 0.18 from $F(q)$. The latter corresponds to a time constant of 0.07 secs. This is much faster than reported results on the rotor dynamics. There is also two zeros at $z = 0$ since $\deg A + \deg F - \deg B = n_{n_a} + n_{n_f} - n_{n_b} = 2$. Finally we have a zero outside the unit circle at approximately -2. It is then customary to term the system non-minimum phase. However, the interpretation of non-minimum phase zeros for discrete-time systems is not straightforward. The step-response of $G(q, \hat{\theta}_N)$ does not show traditional minimum-phase behavior, see Figure 9.6.

Notice how the poles and zeros for the stochastic part $H(q, \hat{\theta}_N)$ mostly comes in pole-zero pairs. Traveling up the unit circle from the point $(1, 0)$, the

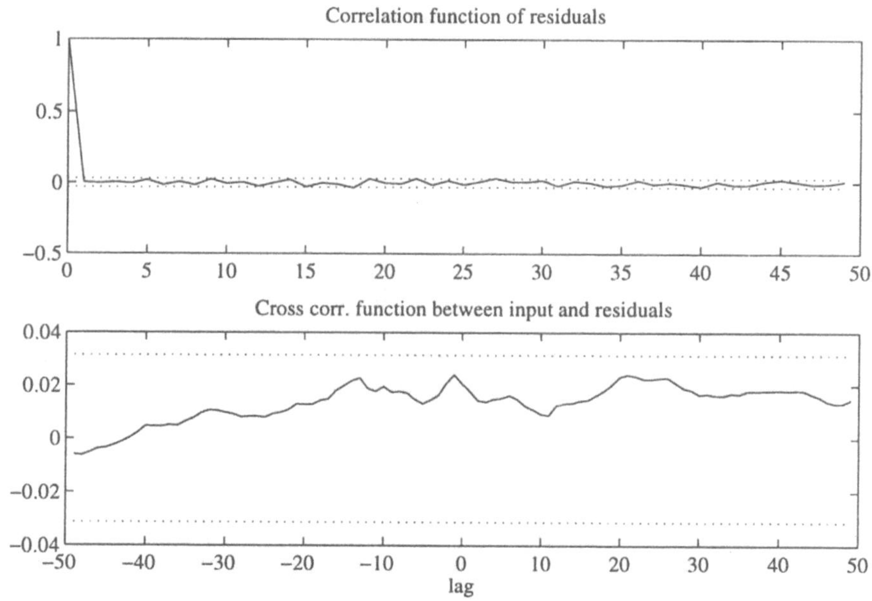

Fig. 9.3. *Residual tests for the wind turbine identification. 99% confidence limits are indicated in each plot.*

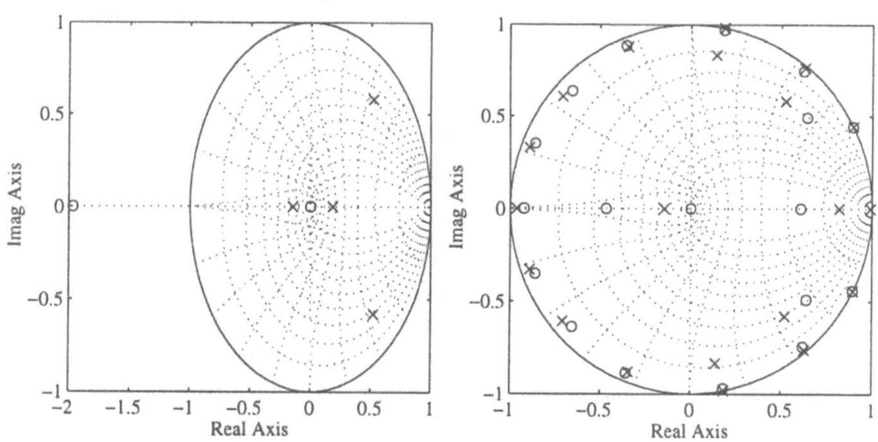

Fig. 9.4. *Pole-zero configuration for the deterministic part $G(q, \hat{\theta}_N)$ (left) and stochastic part $H(q, \hat{\theta}_N)$ (right).*

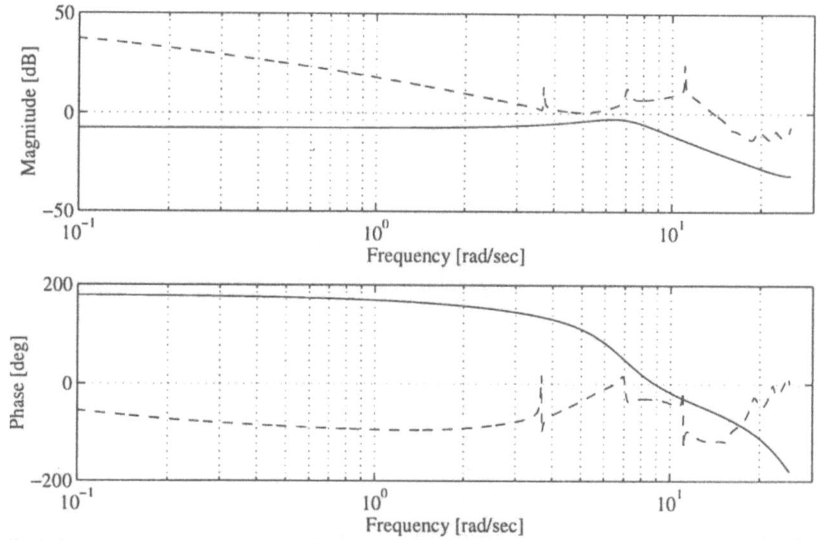

Fig. 9.5. *Bode plot for the deterministic (solid) and stochastic (dashed) part of the estimated wind turbine model.*

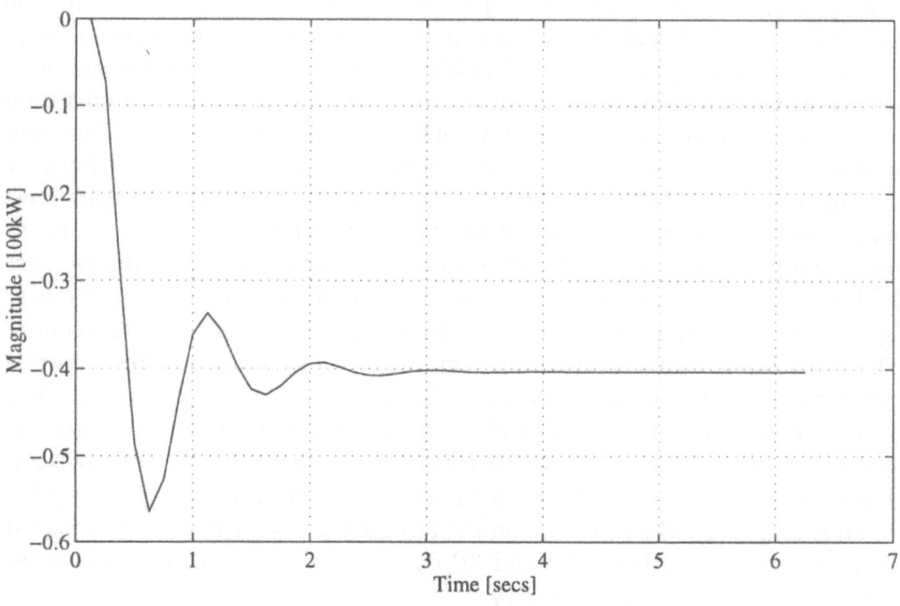

Fig. 9.6. *Step response of deterministic part of estimated wind turbine model.*

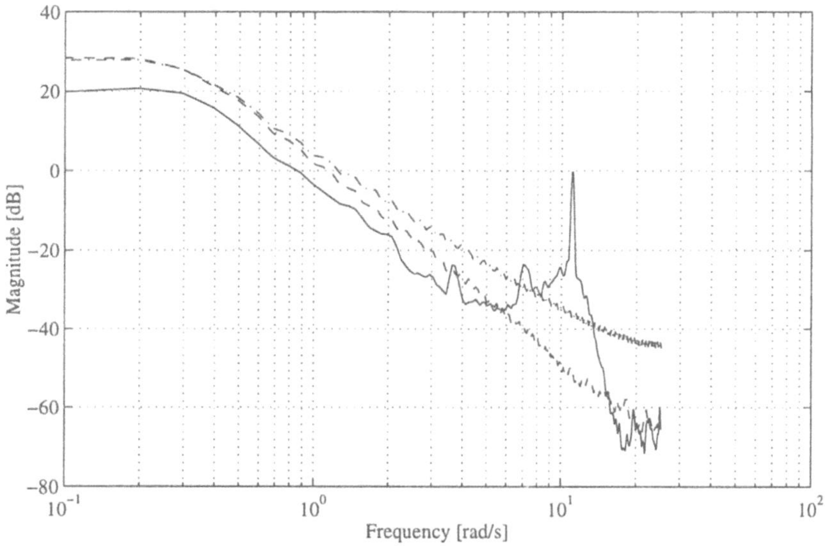

Fig. 9.7. *Estimated power spectrum for the pitch input signal (dashed), a pure square wave with the same amplitude and frequency (dash-dotted) and the power output (solid).*

first three pole-zero pairs are very lightly damped. The effects of these pole-zero pairs can be seen in the Bode plot for $H(q, \hat{\theta}_N)$ as sharp resonances; one placed at the 3P frequency and one at the 1P frequency. Furthermore there is a resonance at approximately 7 rad/sec corresponding to the second pole-zero pair encountered traveling up the unit circle. This probably stems from the tower resonance which has been included in the noise model. The tower resonance clearly belongs in the $A(q)$ polynomial since it is struck both by the pitch input and the wind speed. Several identification experiments have been conducted with additional parameters in $A(q)$ and $B(q)$. However, it has not been possible to represent the tower resonance in the deterministic part of the model. An explanation of this can be given from the estimated power spectra of both input and output, see Figure 9.7. Note that the spectrum of the pitch input signal in the frequency range where the tower resonance is located has decayed with more than 60 dB compared with the DC value. Thus in our model the output power P in this frequency range must be almost completely characterized by white noise filtered through $C(q)/(D(q)A(q))$. However, then it is not possible to distinguish between high frequency poles in $A(q)$ and $D(q)$. For some reason the resonant poles for the tower is placed in $D(q)$ together with the 1P and 3P dynamics whereas the complex poles for the drive train is placed in $A(q)$.

In other words the frequency content in the input signal (pitch angle) is too small in the high frequency range. The main problem is that the pitch servo loop is much too slow so that the square wave input in fact is filtered through a low pass filter with small cut-off frequency, see Figure 9.7. However even a pure square wave input with period 40 secs will have little relative energy around 10 rad/sec as shown in Figure 9.7. In order to avoid this problem the pitch servo loop must be retuned and preferably another pitch input signal applied with more energy in the high frequency area.

In Figure 9.7 note the resonant peak for the 1P, tower and 3P dynamics can be easily identified in the output power spectrum. Compare with the Bode plot in Figure 9.5. Also notice that the drive train complex pole pair are located at almost exactly the same frequency as the tower resonance. If they were included in the same filter it would be very difficult for the identification procedure to distinguish between them. It will furthermore be very difficult for the controller to compensate for both. Finally note that the noise filter also contains a pole at approximately 0.99 corresponding to a time constant of 13.5 secs. This pole is probably due to the rather slow rotor dynamics. The time constant is somewhat larger than usually reported in the literature, but this is probably due to neighboring wind turbines. Again the corresponding pole actually belongs in $F(q)$ or $A(q)$ rather than the noise filter denominator $D(q)$. However, the phenomena could be described by filtered white noise which is probably the case here due to the high number of parameters in $D(q)$.

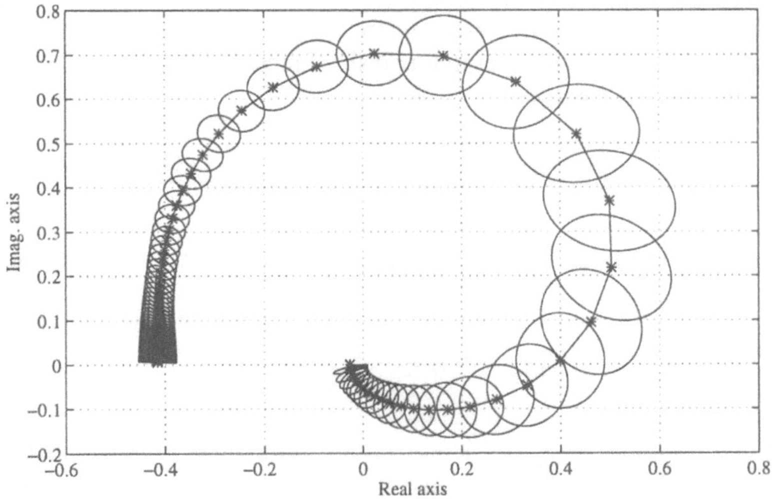

Fig. 9.8. *Nominal Nyquist curve with 2 standard deviations uncertainty ellipses.*

The frequency domain model uncertainty was estimated using the approach outlined in Section 9.5. In Figure 9.8 the nominal Nyquist for the deterministic part $G(q, \hat{\theta}_N)$ is shown superimposed by 2 standard deviation (95.5%) uncertainty ellipses. These uncertainty ellipses may then be used

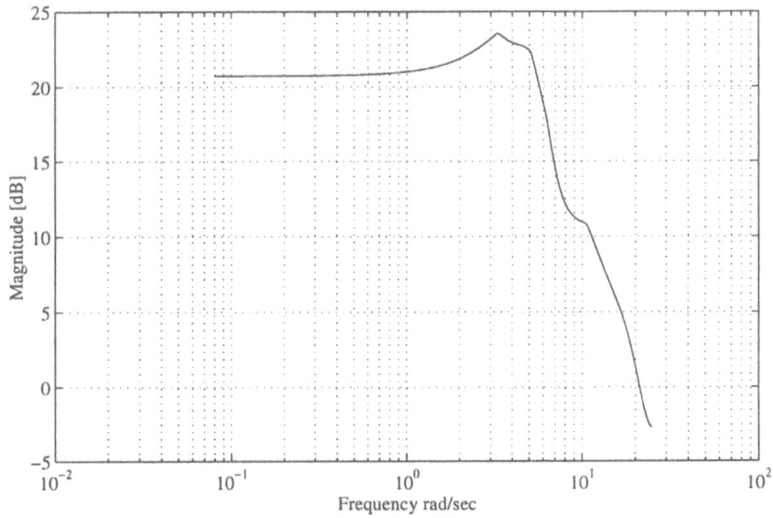

Fig. 9.9. *Robustness bound on the complementary sensitivity. (closed loop transfer function).*

in a robust control design. In standard $\mathcal{H}_\infty$ control only frequency domain uncertainty circles can be represented. Thus an $\mathcal{H}_\infty$ design approach could be taken by approximating the ellipses by circumscribed circles. This will provide a set of additive uncertainty circles at each frequency ω. Fitting the radii of these circles with a stable transfer function $G_{\mathrm{A}}(z)$ will provide the uncertain system description

$$G_{\mathrm{T}}(z) = G(z,\hat{\theta}_N) + G_{\mathrm{A}}(z) = G(z,\hat{\theta}_N)\left(1 + \frac{G_{\mathrm{A}}(z)}{G(z,\hat{\theta}_N)}\right) . \tag{9.134}$$

Given the results in Table 4.1 a controller will robustly stabilize the plant iff

$$\left|T(e^{j\omega T_s})\right| < \left|\frac{G_{\mathrm{A}}(e^{j\omega T_s})}{G(e^{j\omega T_s},\hat{\theta}_N)}\right| \tag{9.135}$$

where $T(z)$ denotes the closed loop transfer function as usual. Thus by plotting $G_{\mathrm{A}}(e^{j\omega T_s})/G(e^{j\omega T_s},\hat{\theta}_N)$ against frequency ω we may obtain an upper bound on the achievable closed loop bandwidth. This is done in Figure 9.9. The maximum bandwidth is approximately 20 rad/sec which is quite close to the Nyquist frequency $\omega_n = \pi/T_s = 25$ rad/sec. Thus a fast control design should be possible with the derived model.

However, because the tower resonance was not included into the deterministic part of the system, the high frequency uncertainty estimates will probably be misleading even though the residual tests were passed. Since the true system will have a resonance in the high frequency area, if the pitch input ϑ has large energy in this area the resonance will be struck from the input. This will surely deteriorate the performance properties in the high frequency area and may even result in instability.

The above example illustrates that care must be taken in connection with high order noise models since some of the deterministic behavior of the true system may in fact be explained by filtered white noise.

9.7 Summary

Classical system identification using the prediction error method (PEM) was reviewed for a general class of black-box models. In particular, we considered asymptotic expressions for the parameter variance and showed how these parameter uncertainties can be translated to frequency domain uncertainty ellipses. For the general model structure we must use a first order Taylor approximation since the model is not linear in the parameter vector θ. Unfortunately, we may only obtain consistent data estimates of the parameter covariance in the case where the model admits an exact description of the true system. This severely hampers the usefulness of the classical approach in connection with robust control design. Finally, a numerical example was presented, namely identification of a model for a 400 kW constant speed, variable pitch wind turbine from a set of input/output measurements. The input applied was a 0.025 Hz fundamental square wave. The model structure was chosen from physical considerations. The order of the deterministic part was fixed and the order of the noise filter was then increased until the prediction errors for a separate test-set became white. The result was a third order model for the deterministic part and a 18th order model for the noise filter. Frequency domain uncertainty ellipses were then estimated for the deterministic part. However, it was argued that some of the truly deterministic dynamics were in fact explained by filtered white noise and thus included in the noise filter. The main reason for this was inadequate excitation in the high frequency area. This caused the frequency domain uncertainty estimate for the deterministic part to be overly optimistic in the high frequency area.

CHAPTER 10

ORTHONORMAL FILTERS IN SYSTEM IDENTIFICATION

In this chapter we will provide an introduction to the use of orthonormal basis functions in system identification. First let us, however, consider some special cases of the general model structure (9.1) introduced in the previous chapter. Throughout this chapter we assume that the true system is finite-dimensional, linear and time-invariant (FDLTI) and given by

$$y(k) = G_{\mathrm{T}}(q)u(k) + H_{\mathrm{T}}(q)e_0(k) \tag{10.1}$$

where $e_0(k) \in \mathcal{N}(0, \lambda_0)$ is white noise.

10.1 ARX Models

A very popular set of models is the autoregressive structure with exogenous input, abbreviated ARX models. The ARX model structure can be derived as follows. Assume that our model set is given by

$$y(k) = G(q,\theta)u(k) + H(q,\theta)e(k) \tag{10.2}$$

where $e(k)$ is white noise with variance λ. If the noise filter $H(q,\theta)$ is minimum phase the optimal predictor for $y(k)$ is

$$\hat{y}(k|\theta) = H^{-1}(q,\theta)G(q,\theta)u(k) + \left(1 - H^{-1}(q,\theta)\right) y(k)\ . \tag{10.3}$$

Then, if $H^{-1}(q,\theta)G(q,\theta)$ and $H^{-1}(q,\theta)$ are both stable[1], we may write:

$$H^{-1}(q,\theta)G(q,\theta) = \sum_{k=1}^{\infty} b_k q^{-k} \tag{10.4}$$

$$H^{-1}(q,\theta) - 1 = \sum_{k=1}^{\infty} a_k q^{-k}\ . \tag{10.5}$$

Truncating these expansions at $k = n$ we may rewrite (10.2) as

[1] This is assured, of course, if $H(q,\theta)$ is minimum-phase and if $H(q,\theta)$ and $G(q,\theta)$ have the same unstable poles.

$$\left(1+\sum_{k=1}^{n_a} a_k q^{-k}\right) y(k) = \sum_{k=1}^{n_b} b_k q^{-k} u(k) + e(k) \tag{10.6}$$

$$\Leftrightarrow \quad A(q)y(k) = B(q)u(k) + e(k) \tag{10.7}$$

with $A(q)$ and $B(q)$ as in (9.2) and (9.3). This is the ARX model structure. For $n_a = 0$ a finite impulse response (FIR) model is derived and for $n_b = 0$ we obtain an autoregressive (AR) model structure. For the ARX model the optimal predictor (10.3) becomes

$$\hat{y}(k|\theta) = (1 - A(q))y(k) + B(q)u(k) . \tag{10.8}$$

Thus, if we take

$$\theta = [a_1, \cdots, a_{n_a}, b_1, \cdots, b_{n_b}]^T \tag{10.9}$$

and

$$\phi(k) = [-y(k-1), \cdots, -y(k-n_a), u(k-1), \cdots, u(k-n_b)]^T \tag{10.10}$$

we may write

$$\hat{y}(k|\theta) = \phi^T(k)\theta . \tag{10.11}$$

The predictor for the ARX model structure is consequently linear in θ. The prediction errors $\epsilon(k, \theta)$ are given by

$$\begin{aligned} \epsilon(k,\theta) &= y(t) - (1 - A(q))y(t) - B(q)u(t) && (10.12) \\ &= [A(q)G_T(q) - B(q)]\, u(t) + A(q)H_T(q)e(k) . && (10.13) \end{aligned}$$

In Appendix E it is shown that the parameter estimate

$$\hat{\theta}_N = \arg\min_{\theta} \; V_N(\theta, Z^N) \tag{10.14}$$

with $V_N(\theta, Z^N)$ as in (9.10) is given by the analytic expression

$$\hat{\theta}_N = \left[\sum_{k=1}^{N} \phi(k)\phi^T(k)\right]^{-1} \sum_{k=1}^{N} \phi(k)y(k) . \tag{10.15}$$

If we define

$$\begin{aligned} Y &\triangleq [y(1), y(2), \cdots, y(N)]^T && (10.16) \\ \Phi &\triangleq \left[\phi^T(1), \phi^T(2), \cdots, \phi^T(N)\right]^T && (10.17) \end{aligned}$$

it is easily shown that an equivalent expression for (10.15) is

$$\hat{\theta}_N = \left(\Phi^T \Phi\right)^{-1} \Phi^T Y . \tag{10.18}$$

10.1.1 Variance of ARX Parameter Estimate

Let us then investigate the quality of the estimate. Under weak assumptions we then have from (9.33) that:

$$\sqrt{N}\left(\hat{\theta}_N - \theta^*\right) \in \mathcal{N}(0, P_\theta)\,, \qquad \text{for } N \to \infty \tag{10.19}$$

where θ^* is defined by

$$\theta^* = \arg\min_{\theta} \left\{ \lim_{N\to\infty} \frac{1}{N} \sum_{k=1}^{N} E\left\{\frac{1}{2}\epsilon^2(t,\theta)\right\}\right\}. \tag{10.20}$$

The covariance matrix P_θ is given by (9.38):

$$P_\theta = \left[\bar{V}''(\theta^*)\right]^{-1} Q \left[\bar{V}''(\theta^*)\right]^{-1} \tag{10.21}$$

with $\bar{V}''(\theta^*)$ and Q as in (9.40) and (9.43) respectively. For the ARX structure, the model gradient $\psi(k,\theta)$ is given by

$$\psi(k,\theta) = \frac{\partial \hat{y}(k|\theta)}{\partial \theta} = \phi(k)\,. \tag{10.22}$$

Thus $\partial\psi(k,\theta)/\partial\theta = 0$ and the expressions for $\bar{V}''(\theta^*)$ and Q reduce to

$$\bar{V}''(\theta^*) = \lim_{N\to\infty} \frac{1}{N} \sum_{k=1}^{N} E\left\{\phi(k)\phi^T(k)\right\} \tag{10.23}$$

$$Q = \lim_{N\to\infty} \frac{1}{N} \sum_{k=1}^{N}\sum_{l=1}^{N} E\left\{\phi(k)\phi^T(l)\epsilon(k,\theta^*)\epsilon(l,\theta^*)\right\} \tag{10.24}$$

where the prediction errors are given by (10.13). The expression for the parameter covariance P_θ is, however, still quite complicated. The data estimates (9.60) and (9.61) will *not* be a consistent estimate, i.e converge for $N \to \infty$, unless in the special case where $G_{\mathrm{T}}(q) = G(q,\theta_0)$ and $H_{\mathrm{T}}(q) = H(q,\theta_0)$ (the true system can be represented within the model set). In this case $\theta^* = \theta_0$, $\epsilon(k,\theta^*) = e_0(k)$ and the expression for P_θ reduce to

$$P_\theta = \lambda_0 \left[\lim_{N\to\infty} \frac{1}{N} \sum_{k=1}^{N} E\left\{\phi(k)\phi^T(k)\right\}\right]. \tag{10.25}$$

We thus need to require that both the deterministic and stochastic part of the true system can be exactly modeled by the ARX structure in order for the data estimate

$$\hat{P}_N = \hat{\lambda}_N \left[\frac{1}{N} \sum_{k=1}^{N} \phi(k)\phi^T(k) \right]^{-1} \tag{10.26}$$

$$\hat{\lambda}_N = \frac{1}{N} \sum_{k=1}^{N} \epsilon^2(k, \hat{\theta}_N) \tag{10.27}$$

to be a consistent estimate of the parameter covariance matrix P_θ. A well-known related problem with ARX models is that even if the true system $G_{\mathrm{T}}(q)$ can be described by $G(q, \theta_0)$ then if the true noise filter $H_{\mathrm{T}}(q)$ cannot be represented by $H(q, \theta) = A(q)$, the parameter estimate $\hat{\theta}_N$ will *not* converge to the true value θ_0 for $N \to \infty$ because $G(q, \theta)$ and $H(q, \theta)$ are not *independently* parameterized.

10.2 Output Error Models

The output error (OE) model structure is given by

$$y(k) = G(q, \theta)u(k) + e(k) = \frac{B(q)}{F(q)}u(k) + e(k) \tag{10.28}$$

with $B(q)$ and $F(q)$ defined by (9.3) and (9.4). The optimal predictor for the OE model structure is given by

$$\begin{aligned} \hat{y}(k|\theta) &= H^{-1}(q, \theta)G(q, \theta)u(k) + \left(1 - H^{-1}(q, \theta)\right) y(k) & (10.29) \\ &= G(q, \theta)u(k) \; . & (10.30) \end{aligned}$$

Note that the OE predictor is *not* linear in θ. This means that we have no analytic expression for the parameter estimate $\hat{\theta}_N$, but must resort to numerical methods to determine $\hat{\theta}_N$ as in the general PEM case. The model gradient $\psi(k, \theta)$ for the OE model structure is given by

$$\psi(k, \theta) = \frac{\partial G(q, \theta)}{\partial \theta} u(k) \; . \tag{10.31}$$

Thus $\psi(k, \theta)$ is a deterministic sequence; a crucial property which we will utilize in computing the parameter covariance P_θ. The prediction errors $\epsilon(k, \theta)$ are given by

$$\begin{aligned} \epsilon(k, \theta) &= y(k) - \hat{y}(k|\theta) & (10.32) \\ &= \left(G_{\mathrm{T}}(q) - G(q, \theta)\right) u(k) + H_{\mathrm{T}}(q)e_0(k) \; . & (10.33) \end{aligned}$$

We will then consider the case where $G_{\mathrm{T}}(q) = G(q, \theta_0)$, i.e where $G_{\mathrm{T}}(q)$ is a member of the model set $G(q, \theta)$. We will, however, *not* assume that the true noise filter $H_{\mathrm{T}}(q)$ is given by $H(q, \theta) = 1$. Since $G(q, \theta)$ and $H(q, \theta)$ are independently parameterized, then according to [Lju87, Theorem 8.4]:

$$\theta^* = \arg\min_{\theta} \left\{ \lim_{N\to\infty} \frac{1}{N} \sum_{k=1}^{N} E\left\{ \frac{1}{2}\epsilon^2(t,\theta) \right\} \right\} = \theta_0 \tag{10.34}$$

and the prediction errors becomes

$$\epsilon(k,\theta^*) = H_{\mathrm{T}}(q)e_0(k) \ . \tag{10.35}$$

10.2.1 Variance of OE Parameter Estimate

Now the parameter covariance is given by

$$P_\theta = \lim_{N\to\infty} \frac{1}{N} E\left\{ \left(\hat{\theta}_N - \theta_0\right)\left(\hat{\theta}_N - \theta_0\right)^T \right\} \tag{10.36}$$

$$= \left[\bar{V}''(\theta_0)\right]^{-1} Q \left[\bar{V}''(\theta_0)\right]^{-1} \ . \tag{10.37}$$

$\bar{V}''(\theta_0)$ is given by

$$\bar{V}''(\theta_0) = \lim_{N\to\infty} \frac{1}{N} \sum_{k=1}^{N} E\left\{ \psi(k,\theta_0)\psi^T(k,\theta_0) \right\} - \lim_{N\to\infty} \frac{1}{N} \sum_{k=1}^{N} E\left\{ \frac{\partial\psi(k,\theta_0)}{\partial\theta} H_{\mathrm{T}}(q)e_0(k) \right\} \tag{10.38}$$

$$= \lim_{N\to\infty} \frac{1}{N} \sum_{k=1}^{N} \psi(k,\theta_0)\psi^T(k,\theta_0) - 0 \tag{10.39}$$

$$= \lim_{N\to\infty} \frac{1}{N} \Psi^T(\theta_0, Z^N)\Psi(\theta_0, Z^N) \tag{10.40}$$

since $\psi(k,\theta)$ is deterministic. $\Psi(\theta, Z^N)$ is given by

$$\Psi(\theta, Z^N) = \left[\psi(1,\theta), \psi(2,\theta), \cdots, \psi(N,\theta)\right]^T \ . \tag{10.41}$$

Q is given by

$$Q = \lim_{N\to\infty} \frac{1}{N} \sum_{k=1}^{N}\sum_{h=1}^{N} E\left\{ \psi(k,\theta_0)\psi^T(h,\theta_0)\epsilon(k,\theta_0)\epsilon(h,\theta_0) \right\} \tag{10.42}$$

$$= \lim_{N\to\infty} \frac{1}{N} \sum_{k=1}^{N}\sum_{h=1}^{N} \psi(k,\theta_0)\psi^T(h,\theta_0) E\left\{ H_{\mathrm{T}}(q)e_0(k)H_{\mathrm{T}}(q)e_0(h) \right\} \tag{10.43}$$

$$= \lim_{N\to\infty} \frac{1}{N} \Psi^T(\theta_0, Z^N) C_\nu \Psi(\theta_0, Z^N) \tag{10.44}$$

where C_ν is given by

$$C_\nu = E\left\{VV^T\right\} \ , \qquad V = \left[\nu(1), \nu(2), \cdots, \nu(N)\right] \tag{10.45}$$

with $\nu(k) = H_{\mathrm{T}}(q)e_0(k)$. Thus the parameter covariance P_θ is given by

$$P_\theta = \lim_{N\to\infty} N \left(\Psi_{\mathrm{o}}^T\Psi_{\mathrm{o}}\right)^{-1} \Psi_{\mathrm{o}}^T C_\nu \Psi_{\mathrm{o}} \left(\Psi_{\mathrm{o}}^T\Psi_{\mathrm{o}}\right)^{-1} \tag{10.46}$$

with $\Psi_{\mathrm{o}} = \Psi(\theta_0, Z^N)$. Then

$$\hat{P}_N = N \left(\Psi_N^T\Psi_N\right)^{-1} \Psi_N^T C_\nu \Psi_N \left(\Psi_N^T\Psi_N\right)^{-1} \tag{10.47}$$

with $\Psi_N = \Psi(\hat{\theta}_N, Z^N)$ will be a consistent estimate of the covariance P_θ. Notice that this is true even if the true noise filter $H_{\mathrm{T}}(q)$ cannot be captured within the model set $H(q,\theta) = 1$. Thus we can obtain a consistent estimate of the parameter covariance provided the true system $G_{\mathrm{T}}(q)$ can be described by our model set $G(q,\theta)$. This is an important property for OE models. Of course, in order to evaluate (10.47) the covariance matrix for the true noise $\nu(k)$ must be known. However, provided $G(q,\theta_0) = G_{\mathrm{T}}(q)$ we have that $\epsilon(k,\theta_0) = \nu(k)$, see (10.35). If we assume that $\nu(k)$ is a stationary stochastic process such that

$$E\{\nu(k)\nu(h)\} = E\{\nu(m)\nu(n)\}\ , \qquad \text{for } k-h = m-n. \tag{10.48}$$

C_ν will be constant along the diagonal and all off-diagonals as well. Furthermore, we may estimate these diagonals from the prediction errors by averaging over all $\nu(i)\nu(j)$ for which $i - j = M$ with M constant. $M = 0$ then corresponds to the main diagonal, $M = 1$ to the first off-diagonal etc. The covariance estimate $R_N^\epsilon(\kappa)$ in (9.124) used for model validation in Section 9.6.2 is precisely what we need. Note, of course that C_ν is symmetric.

We may then replace C_ν by its estimate in (10.47) to obtain an estimate of the parameter covariance.

10.3 Fixed Denominator Models

Now we consider models which are linear in θ (fixed denominator models):

$$G(q,\theta) = \Lambda(q)\theta \tag{10.49}$$

with

$$\Lambda(q) = [\Lambda_1(q), \cdots, \Lambda_n(q)] \tag{10.50}$$

$$\theta = [\theta_1, \cdots, \theta_n]^T\ . \tag{10.51}$$

$\Lambda_i(q)$ is known as the basis functions for $G(q,\theta)$. The optimal predictor in this case is

$$\hat{y}(k|\theta) = \Lambda(q)\theta u(k) = \phi(k)\theta\ . \tag{10.52}$$

Clearly,

$$\phi(k) = \Lambda(q)u(k) \tag{10.53}$$

is a deterministic sequence. Thus for fixed denominator models, the predictor is linear in θ and the regressors $\phi(k)$ is deterministic. With Y and Φ as before, see (10.16) and (10.17), the least squares estimate of θ is given by

$$\hat{\theta}_N = \left(\Phi^T\Phi\right)^{-1}\Phi^T Y\,. \tag{10.54}$$

The output vector Y may be written

$$Y = \begin{bmatrix} y(1) \\ \vdots \\ y(N) \end{bmatrix} = \begin{bmatrix} G_{\mathrm{T}}(q)u(1) \\ \vdots \\ G_{\mathrm{T}}(q)u(N) \end{bmatrix} + \begin{bmatrix} H_{\mathrm{T}}(q)e_0(1) \\ \vdots \\ H_{\mathrm{T}}(q)e_0(N) \end{bmatrix} \tag{10.55}$$

$$= G_{\mathrm{T}}(q)\begin{bmatrix} u(1) \\ \vdots \\ u(N) \end{bmatrix} + \begin{bmatrix} \nu(1) \\ \vdots \\ \nu(N) \end{bmatrix} \tag{10.56}$$

$$= G_{\mathrm{T}}(q)U + V\,. \tag{10.57}$$

Thus

$$\hat{\theta}_N = \left(\Phi^T\Phi\right)^{-1}\Phi^T\left(G_{\mathrm{T}}(q)U + V\right)\,. \tag{10.58}$$

10.3.1 Variance of Fixed Denominator Parameter Estimate

Since Φ and $G_{\mathrm{T}}(q)U$ are deterministic we may easily compute the expectation of $\hat{\theta}_N$:

$$E\left\{\hat{\theta}_N\right\} = \theta^* = \left(\Phi^T\Phi\right)^{-1}\Phi^T G_{\mathrm{T}}(q)U\,. \tag{10.59}$$

Then

$$\tilde{\theta}_N = \hat{\theta}_N - \theta^* = \left(\Phi^T\Phi\right)^{-1}\Phi^T V \tag{10.60}$$

which gives us the covariance matrix for the parameter estimate

$$P_\theta = NE\left\{\tilde{\theta}_N\tilde{\theta}_N^T\right\} = N\left(\Phi^T\Phi\right)^{-1}\Phi^T C_\nu \Phi\left(\Phi^T\Phi\right)^{-1} \tag{10.61}$$

where $C_\nu = E\left\{VV^T\right\}$. Note that (10.61) is *not* an asymptotic expression like (10.46). Thus the data estimate will be equal to the true covariance matrix for finite data. Furthermore, this holds *regardless of the form of the true system* $G_{\mathrm{T}}(q)$. Consequently, for fixed denominator models an finite-data unbiased estimate of the parameter variance can be obtained *even in the case of undermodeling.* Then of course the parameter θ^* will not reflect the "true" properties of $G_{\mathrm{T}}(q)$. Thus, when our main objective is to estimate the total model error it is of little use that we can estimate P_θ. This will only provide the "variance" part of the total model error, see Section 13.1.1.

If $G_{\mathrm{T}}(q)$ can be represented within the model set ($G_{\mathrm{T}}(q) = \Lambda(q)\theta_0$) then $\theta^* \to \theta_0$ for $N \to \infty$ and we may use (10.61) to map the parametric uncertainty into the frequency domain. Here we furthermore have the advantage that $G(q, \theta)$ is linear in θ so we need not to resort to Taylor approximations. Let

$$\tilde{g}(e^{j\omega T_s}) = \begin{bmatrix} \Re e \left\{ G(e^{j\omega T_s}, \theta_0) - G(e^{j\omega T_s}, \hat{\theta}_N) \right\} \\ \Im m \left\{ G(e^{j\omega T_s}, \theta_0) - G(e^{j\omega T_s}, \hat{\theta}_N) \right\} \end{bmatrix} \tag{10.62}$$

$$\Gamma(e^{j\omega T_s}) = \begin{bmatrix} \Re e \ \{\Lambda(e^{j\omega T_s})\} & \Im m \{\Lambda(e^{j\omega T_s})\} \end{bmatrix}^T \tag{10.63}$$

so that

$$\tilde{g}(e^{j\omega T_s}) = \Gamma(e^{j\omega T_s})(\theta_0 - \hat{\theta}_N) \ . \tag{10.64}$$

Now $\sqrt{N}\tilde{g}(e^{j\omega T_s})$ will have asymptotic normal distribution for $N \to \infty$ with covariance $P_g(\omega)$:

$$\sqrt{N}\tilde{g}(e^{j\omega T_s}) \in \mathcal{N}(0, P_g(\omega)) \ , \qquad \text{for } N \to \infty \tag{10.65}$$

where the covariance matrix $P_g(\omega)$ is given by

$$P_g(\omega) = \lim_{N\to\infty} E\left\{ \sqrt{N}\tilde{g}(e^{j\omega T_s})\tilde{g}^T(e^{j\omega T_s})\sqrt{N} \right\} \tag{10.66}$$

$$= \lim_{N\to\infty} E\left\{ \Gamma(e^{j\omega T_s})\sqrt{N}(\theta_0 - \hat{\theta}_N)(\theta_0 - \hat{\theta}_N)^T\sqrt{N}\Gamma^T(e^{j\omega T_s}) \right\} \tag{10.67}$$

$$= \Gamma(e^{j\omega T_s})P_\theta\Gamma^T(e^{j\omega T_s}) \ . \tag{10.68}$$

Thus the scalar

$$z(\omega) = N\tilde{g}^T(e^{j\omega T_s})P_g^{-1}(\omega)\tilde{g}(e^{j\omega T_s}) \tag{10.69}$$

$$= \tilde{g}^T(e^{j\omega T_s}) \left[\Gamma(e^{j\omega T_s})\frac{1}{N}P_\theta\Gamma^T(e^{j\omega T_s}) \right]^{-1} \tilde{g}(e^{j\omega T_s}) \tag{10.70}$$

will have χ^2 distribution with 2 degrees of freedom. Equation (10.70) may then be used to draw confidence ellipsis in the complex plane for the frequency response estimate $G(e^{j\omega T_s}, \hat{\theta}_N)$.

10.3.2 FIR Models

The most well-known fixed denominator model structure is probably the Finite Impulse Response (FIR) model:

$$G(q,\theta) = \sum_{k=1}^{n} g_k q^{-k} . \tag{10.71}$$

The basis functions for the FIR model is q^{-k}, $k = 1, \cdots, n$. Let $\mathcal{H}_2^*(\mathbf{C},\mathbf{C})$ denote the Hardy space of scalar functions $G(z)$ which are analytical in $|z| > 1$ (no unstable poles) and continuous in $|z| \geq 1$ (strictly proper). The scalar product on $\mathcal{H}_2^*(\mathbf{C},\mathbf{C})$ is given as

$$< G, F > = \frac{1}{2\pi j} \int_{-\pi}^{\pi} G^*(e^{-j\omega}) F(e^{-j\omega}) d\omega \tag{10.72}$$

$$= \frac{1}{2\pi j} \oint_{|z|=1} G^*(z) F(z) \frac{dz}{z} \tag{10.73}$$

where $G^*(z) = G(z^{-1})$. $\mathcal{H}_2^*(\mathbf{C},\mathbf{C})$ is the space of all real rational stable strictly proper scalar discrete-time transfer functions and is the discrete-time counterpart of $\mathcal{H}_2(\mathbf{C},\mathbf{C})$ introduced in Section 3.5.2. Before we proceed, let us review the concept of orthonormal basis.

Lemma 10.1 (Orthonormal Basis). *Let* H *be a Hilbert space. Then the set of elements* $M = \{e_\alpha : \alpha \in A\}$ *from* H *is called an orthonormal system if*

$$< e_\alpha, e_\beta > = \delta_{\alpha\beta} \qquad \textit{for } \alpha, \beta \in A \tag{10.74}$$

where $\delta_{\alpha\beta}$ *denotes the Kronecker delta, i.e.* $\delta_{\alpha\alpha} = 1, \forall \alpha \in A$ *and* $\delta_{\alpha\beta} = 0$ *for* $\alpha \neq \beta$. *An orthonormal system* M *is called an orthonormal basis of* H *if* M *is total in* H *which means that the smallest subspace of* H *that contains* M *is dense.*

Example 10.1 (Orthonormal Basis). The set of unit vectors $\{e_1, \cdots, e_n\}$ is an orthonormal basis in $\mathbf{R}^n$. e_j is the vector with 1 at the j'th place and zero otherwise. ∎

An important property of orthonormal basis functions is that if M is an orthonormal basis in H, then every element of H may be given as a linear combination of M.

It is now easy to show that the basis functions $\Lambda_k = e^{-j\omega T_s k}$ are orthonormal in $\mathcal{H}_2^*(\mathbf{C},\mathbf{C})$;

$$< \Lambda_k, \Lambda_l > = \frac{1}{2\pi j} \int_{-\pi}^{\pi} e^{j\omega T_s k} e^{-j\omega T_s l} d\omega \tag{10.75}$$

$$= \frac{1}{2\pi j} \int_{-\pi}^{\pi} e^{j\omega T_s (k-l)} d\omega \tag{10.76}$$

$$= \delta_{kl} . \tag{10.77}$$

Since $e^{-j\omega T_s k}$, $k = 1, 2, \cdots$ form an orthonormal basis for functions in $\mathcal{H}_2^*(\mathbf{C},\mathbf{C})$, then for $G(z) \in \mathcal{H}_2^*(\mathbf{C},\mathbf{C})$ there exists a sequence such that

$$G(e^{j\omega T_s}) = \sum_{k=1}^{\infty} g_k e^{j\omega T_s k} \tag{10.78}$$

or equivalently

$$G(z) = \sum_{k=1}^{\infty} g_k z^{-k} . \tag{10.79}$$

(10.79) is the well known relation between $G(z)$ and its impulse response g_k.

The orthonormality properties of the basis functions $\Lambda_k(q)$ ensure that the corresponding asymptotic covariance matrix for the regression vector $\phi(k)$:

$$R = \lim_{N\to\infty} \frac{1}{N} \sum_{k=1}^{N} \phi(k,\theta)\phi^T(k,\theta) \tag{10.80}$$

$$= \lim_{N\to\infty} \frac{1}{N} \Phi^T \Phi \tag{10.81}$$

has a certain (Toeplitz) structure, see [Wah94], which can be shown to improve the numerics in the least squares estimate of θ.

The above properties for FIR models are all quite well-known. It is probably less well-known that the FIR model structure is in some sense an optimal fixed denominator model structure *if the only a priori information we have about the system is that it is exponentially stable.* To see this we must introduce the so-called *n-width* of a set of transfer functions. The following definition is taken from [Wah94]:

Definition 10.1 (n-Width Measures). *Assume that we know that the transfer function $G(z)$ belongs to an a priori given bounded set $G(z) \in \mathcal{S}$. The n-width then measures the smallest approximation error for the worst possible system $G(z) \in \mathcal{S}$ using the best possible n-dimensional fixed denominator model set:*

$$d_n(\mathcal{S};\mathcal{B}) = \inf_{\Phi_n \in M_n(\mathcal{B})} \sup_{G(z)\in\mathcal{S}} \inf_{G_n(z)\in\Phi_n} \|G(z) - G_n(z)\|_B \tag{10.82}$$

where $\mathcal{B}$ denotes some Banach space with norm B, e.g. $\mathcal{H}_2$ or $\mathcal{H}_\infty$. Φ_n is the n-dimensional linear subspace spanned by the basis functions $\{\Lambda_k\}$, $k = 1, \cdots, n$. $M_n(\mathcal{B})$ denotes the collection of all such subspaces. If

$$d_n(\mathcal{S};\mathcal{B}) = \sup_{G(z)\in\mathcal{S}} \inf_{G_n(z)\in\Phi_n^*} \|G(z) - G_n(z)\|_B \tag{10.83}$$

then Φ_n^* is called an **optimal** *subspace for $d_n(\mathcal{S};\mathcal{B})$.*

Using the above definition we may now address the *optimal model set problem* for fixed denominator model structures. Now assume that the only a priori information about $G(z)$ is exponential stability. This means that $G(z)$ belongs to $\mathcal{S} = \mathcal{H}^*_{2,\mathrm{R}}(\mathbf{C},\mathbf{C})$, the Hardy space of real rational strictly proper stable scalar discrete-time transfer functions which are analytic outside the disc $|z| \leq R$, $R < 1$. Thus we know that the poles of $G(z)$ are all located within the disc $|z| \leq R$. The impulse response of $G(z)$ thus decreases at least as R^k. Then we have the following theorem from [Wah94]

Theorem 10.1 (n-Width of FIR Models). *The space Φ^*_n spanned by $\{z^{-1},\cdots,z^{-n}\}$ is an optimal n-dimensional subspace in the n-width sense for $\mathcal{H}^*_{2,\mathrm{R}}(\mathbf{C},\mathbf{C})$.*

Thus the FIR model structure is "optimal" if the a priori information is just exponential stability. The main drawback of the FIR model structure is that it may converge quite slowly if the true system has poles located closed to the unit circle; thus if R above is close to one such that the a priori information is "weak". Then very high order models must be used. However, implementation, information and sensitivity aspects limit the allowable size of the model order n that can be tolerated in practice.

10.3.3 Laguerre Models

Using more appropriate basis functions than the delay operator q^{-1} we may dramatically reduce the number of parameter necessary for obtaining a useful estimate of $G_{\mathrm{T}}(q)$. The problem is that the delay operator is not a suitable operator if the true system has poles closed to the unit circle because it has too short memory. Using operators with longer memory, the number of parameters necessary to describe useful approximations can be reduced. We will, however, for numerical purposes like to preserve the orthonormality properties of the basis functions and to maintain the asymptotic properties of the parameter estimate.

The Laguerre functions are a popular alternative to FIR models which comply with the above. The discrete-time Laguerre models are defined by:

$$G(z,\theta) = \sum_{i=1}^{n} \Lambda_i(z)\theta_i \tag{10.84}$$

$$\Lambda_i(z) = L_i(z,a) = \frac{\sqrt{1-a^2}}{z-a}\left[\frac{1-az}{z-a}\right]^{i-1}, \quad \begin{array}{l} a \in \mathbf{R},\ |a|<1, \\ i = 1,\cdots,n. \end{array} \tag{10.85}$$

It can be shown, see eg [NG94], that the Laguerre functions form an orthonormal basis for functions in $\mathcal{H}^*_2(\mathbf{C},\mathbf{C})$. This means that if $G(z) \in \mathcal{H}^*_2(\mathbf{C},\mathbf{C})$ then there exists a sequence $\{g_k\}$ such that

$$G(z) = \sum_{k=1}^{\infty} g_k L_k(z,a) , \qquad a \in \mathbf{R}, |a| < 1 \tag{10.86}$$

$$= \sum_{k=1}^{\infty} g_k \frac{\sqrt{1-a^2}}{z-a} \left[\frac{1-az}{z-a}\right]^{k-1} , \qquad a \in \mathbf{R}, |a| < 1 . \tag{10.87}$$

Notice the similarity with the FIR representation. For Laguerre models, however, the necessary number of basis functions $L_k(z,a)$ to obtain a useful estimate of $G(z)$ is usually much smaller than for the FIR models.

It can be shown that the Laguerre model structure is optimal in the n-width sense if a priori information about the dominating time constant of the system is available. This is a very common situation. We may obtain this knowledge from e.g. physical insight or step response experiments. Here we will first present the continuous-time results since they are more transparent.

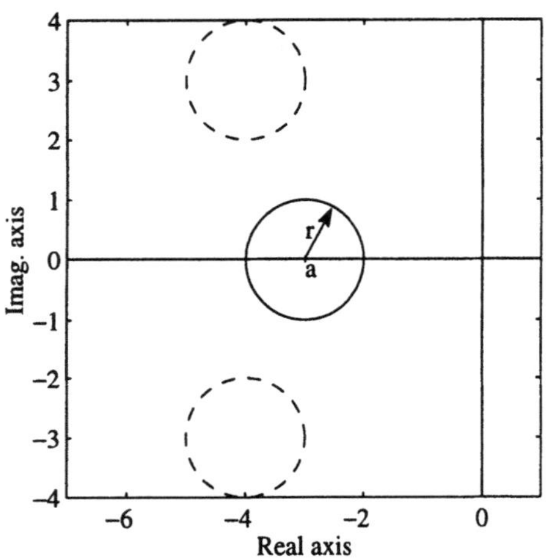

Fig. 10.1. *A priori information for Laguerre (solid) and Kautz (dashed) models.*

Assume that the true system $G_T(s)$ is analytical outside the disc $|s+\alpha| \leq r$ in the left half s-plane with $0 < r < \alpha$. This corresponds to the situation where we know that the (dominating) poles of the system are located within $|s+\alpha| \leq r$, see Figure 10.1. It can then be shown that the continuous-time Laguerre functions are the optimal fixed denominator model structure in the n-width sense. Again the following theorem is from [Wah94].

Theorem 10.2 (n-Width of Continuous-time Laguerre Models). *The space Φ_n^* spanned by the continuous-time Laguerre basis functions $\{L_1(s,a), \cdots, L_n(s,a)\}$ with*

$$L_j(s,a) = \frac{\sqrt{2a}}{s+a} \left(\frac{s-a}{s+a}\right)^{j-1} , \qquad j = 1, 2, \cdots, n \tag{10.88}$$

and $a = \sqrt{\alpha^2 - r^2}$ is an optimal n-dimensional subspace in the n-width sense for functions $G(s) \in \mathcal{H}_2(\mathbf{C}, \mathbf{C})$ which are analytical outside the domain $|s + \alpha| \leq r$, $0 < r < \alpha$.

Note that the optimal Laguerre pole a is smaller than the center pole α of the set $|s + \alpha| \leq r$. Thus the Laguerre pole should generally be chosen slightly faster than the dominant pole of the system to be approximated. The corresponding discrete-time result is as follows [Wah94].

Theorem 10.3 (n-Width of Discrete-Time Laguerre Models). *The space Φ_n^* spanned by the discrete-time Laguerre basis function $\{L_1(z, a), \cdots, L_n(z, a)\}$ with*

$$a = \frac{1}{2\alpha}\left(1 + \alpha^2 - r^2 + \sqrt{(1 + \alpha^2 - r^2)^2 - 4\alpha^2}\right) \tag{10.89}$$

is an optimal n-dimensional subspace in the n-width sense for functions $G(z) \in \mathcal{H}_2^*(\mathbf{C}, \mathbf{C})$ which are analytic outside the domain $|z - \alpha| \leq r$, with $r > |\alpha| + 1$.

The Laguerre model (10.87) can be represented by a *Laguerre network*, see Figure 10.2. The structure is the same for continuous-time and discrete-time systems. Notice that the network consists of a single low-pass filter in series with several first order all-pass filters.

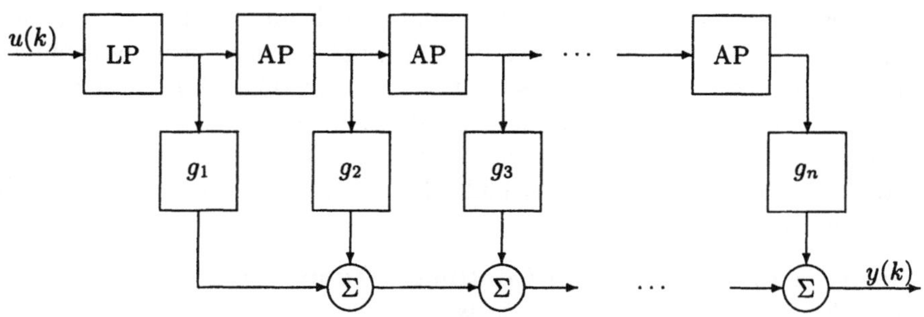

Fig. 10.2. *Laguerre network. LP denotes a first order low-pass filter and AP denotes first order all-pass filters.*

The Laguerre functions provide useful approximations for well-damped systems with one dominating time constant. However for poorly damped systems the convergence will be slow since resonant poles occur in complex conjugate pairs. Furthermore systems with scattered poles cannot be well described. To obtain useful approximations for such systems we must use other basis functions, like eg the Kautz functions introduced in the next section.

10.3.4 Kautz Models

Assume that we know that the true continuous-time system $G_T(s)$ is resonant and that we know approximately the corresponding dominant mode. This is equivalent to assuming that the system is analytic outside the two regions shown in Figure 10.1. In order to determine which basis functions should be used when approximating resonant systems the methodology is to find a mapping which maps these regions onto $|s + \alpha| \le r$ for which we know that the Laguerre functions are optimal. Then the Laguerre functions are mapped into the first domain and the result will be the desired basis functions for resonant systems. Because this mapping is two-to-one the procedure becomes somewhat technical and it will be beyond the scope of this book to treat it in detail. The interested reader may refer to [Wah94] and references therein. In [Wah94] it is shown that the continuous-time Laguerre functions map into the *Kautz functions* defined by

$$\Psi_k(s,b,c) = \begin{cases} \dfrac{\sqrt{2bs}}{s^2+bs+c}\left[\dfrac{s^2-bs+c}{s^2+bs+c}\right]^{\frac{k-1}{2}}, & k \text{ odd} \\[2ex] \dfrac{\sqrt{2bc}}{s^2+bs+c}\left[\dfrac{s^2-bs+c}{s^2+bs+c}\right]^{\frac{k-2}{2}}, & k \text{ even} \end{cases} \tag{10.90}$$

with $b > 0$, $c > 0$, $k = 1, 2, \cdots$. The Kautz functions form an orthonormal basis for functions in $\mathcal{H}_2(\mathbf{C}, \mathbf{C})$. Furthermore the Kautz functions are optimal in the n-width sense for functions $G(s) \in \mathcal{H}_2(\mathbf{C}, \mathbf{C})$ which are analytic outside the domain $|(s^2 + \alpha s + c)/s| \le r$, $r < \alpha$ since this set is mapped onto $|s + \alpha| \le r$ by the transformation in question. $|(s^2 + \alpha s + c)/s| \le r$ describes a rather complicated set; nevertheless, it can capture a priori information about complex poles as well as multiple real ones. Formally, we have the following theorem [Wah94].

Theorem 10.4 (n-Width of Continuous-time Kautz Models). *The space Φ_n^* spanned by the Kautz basis functions $\{\Psi_1(s,b,c), \cdots, \Psi_n(s,b,c)\}$ with $b = \sqrt{\alpha^2 - r^2}$ is an optimal $2n$-dimensional subspace in the n-width sense for functions $G(s) \in \mathcal{H}_2(\mathbf{C}, \mathbf{C})$, which are analytic outside the domain $|(s^2 + \alpha s + c)/s| \le r$, $r < \alpha$.*

The discrete-time Kautz functions given by

$$\Psi_k(z,b,c) = \begin{cases} \dfrac{\sqrt{1-c^2}(z-b)}{z^2+b(c-1)z-c}\left[\dfrac{-cz^2+b(c-1)z+1}{z^2+b(c-1)z-c}\right]^{\frac{k-1}{2}}, & k \text{ odd} \\[2ex] \dfrac{\sqrt{(1-c^2)(1-b^2)}}{z^2+b(c-1)z-c}\left[\dfrac{-cz^2+b(c-1)z+1}{z^2+b(c-1)z-c}\right]^{\frac{k-2}{2}}, & k \text{ even} \end{cases} \tag{10.91}$$

with $-1 < b < 1$, $-1 < c < 1$ and $k = 1, 2, \cdots$ form an orthonormal basis for functions in $\mathcal{H}_2^*(\mathbf{C},\mathbf{C})$ such that any $G(z) \in \mathcal{H}_2^*(\mathbf{C},\mathbf{C})$ may be written

$$G(z) = \sum_{k=1}^{\infty} g_k \Psi_k(z,b,c)\,, \qquad b, c \in \mathbf{R}, |b| < 1, |c| < 1\,. \tag{10.92}$$

Theorem 10.5 (n-Width of discrete-time Kautz Models). *The space Φ_n^* spanned by the discrete-time Kautz basis functions $\{\Psi_1(z,b,c), \cdots, \Psi_n(z,b,c)\}$ with*

$$c = \frac{1}{2\alpha}\left(1 + \alpha^2 - r^2 + \sqrt{(1+\alpha^2-r^2)^2 - 4\alpha^2}\right) \tag{10.93}$$

is an optimal $2n$-dimensional subspace in the n-width sense for functions $G(s) \in \mathcal{H}_2^*(\mathbf{C},\mathbf{C})$ which are analytic outside the domain

$$\left|\frac{z(z-b)}{1-bz} - \alpha\right| \leq r\,. \tag{10.94}$$

Assume we know that the dominating resonant mode in the true system is given approximately by the complex pair of poles β and β^*. Then the choice

$$b = \frac{\beta + \beta^*}{1 + \beta\beta^*}, \qquad c = -\beta\beta^* \tag{10.95}$$

places the resonant mode of the Kautz filters close to that of the true system.

10.3.5 Combined Laguerre and Kautz Structures

Due to the linear properties of the Laguerre and Kautz model structures we can approximate systems with several possibly real and complex poles by cascading a number of Laguerre and Kautz filters. Such a general fixed denominator model structure can be written

$$G(q,\theta) = \Lambda(q)\theta \tag{10.96}$$

$$= \sum_{i=1}^{m^L}\sum_{k=1}^{n^{L_i}} L_k(q,a_i)\theta_k^{L,i} + \sum_{i=1}^{m^\Psi}\sum_{k=1}^{n^{\Psi_i}} \Psi_k(q,b_i,c_i)\theta_k^{\Psi,i}\,. \tag{10.97}$$

10.4 Summary

The prediction error method was considered for some special cases. First the popular ARX model structure was investigated. An analytical expression for the parameter estimate could be obtained since the predictor for the ARX model is linear in θ. Unfortunately, the corresponding expression for the parameter covariance does not allow for a consistent data estimate unless in the case when ***both the deterministic and stochastic part of the model*** admit a perfect description of the true system. It is well known that if the ARX model admits a perfect description of the deterministic part $G(q,\theta)$ of the true system, unless the same is true for the stochastic part also, then the parameter estimate for the deterministic part will be biased.

Next output error (OE) models were considered. The OE predictor is *not* linear in θ. Thus the parameter estimate $\hat{\theta}_N$ must be found using numerical search using, for example, Marquardts algorithm. On the other hand it was shown that the expression for the parameter covariance P_θ was considerably simplified in the case where $G(q,\theta)$ admits a perfect description for the deterministic part of the true system ***regardless of the true stochastic part.*** A consistent estimate can then be obtained from the data.

Finally, fixed denominator models were considered. Models which are linear in the parameter vector have attracted much attention within the system identification community during the last 5 years. From a classical point of view it seems odd that the poles of the model is determined a priori and just the zeros estimated from the data. However, in many cases very good a priori estimates of the dominating poles of the system can be obtained from, for example, step response tests. The predictor for fixed denominator models is linear in θ. Thus analytic expressions for the parameter estimate can be derived. Furthermore the regressors $\phi(k)$ are deterministic. This enables us to derive ***non-asymptotic*** expressions for the parameter covariance P_θ in the general case. Thus an unbiased estimate of the parameter estimate can be obtained ***even in the case of undermodeling.*** Of course, for "true" frequency domain error bounds we must require that the true system can be represented within the model set. Otherwise we will only obtain the variance part of the model error. Finite impulse response (FIR), Laguerre and Kautz model structures were then presented. Using n-width measures for optimality it was shown that the FIR model structure is an optimal fixed denominator model set if the only a priori information is exponential stability. However, usually we know more than that. If we know that the (dominating) poles of the system is located within a disc with radius r located at $(-a,0)$ in the complex plane, then the Laguerre model structure becomes the optimal one in the n-width sense. Thus Laguerre models are useful in modeling well damped systems with one dominating time constant. For lightly damped systems or for systems with scattered poles, Kautz models provides useful descriptions.

CHAPTER 11
THE STOCHASTIC EMBEDDING APPROACH

In this chapter we will give an introduction to the stochastic embedding approach for estimation of frequency domain model error bounds. It is shown how the approach may be developed to construct a flexible framework for estimation of error bounds under different noise and undermodeling conditions. Furthermore quantitative properties of prior assumptions on the noise and the undermodeling may be estimated from current data. The main new contribution in this book is to investigate and improve the qualitative prior assumptions regarding the covariance structures of the noise and undermodeling.

11.1 The Methodology

As discussed in the introduction on page 123 estimation of frequency domain model error bounds from a finite set of noisy data is a very difficult problem unless we assume that the true system can be represented within our model set. Some sort of a priori knowledge of the noise and the undermodeling will be required to obtain such bounds in the general case. The main paradigm of the stochastic embedding approach is to assume that the undermodeling is a realization of a stochastic process with known distribution. Henceforth the true transfer function of the system is assumed to be a realization of a stochastic process.

It is furthermore assumed that the transfer function for the true system may be decomposed into a parametric part $G(q,\theta_0)$ and a bias contribution $G_\Delta(q)$. Thus the frequency response of the true system is supposed given by:

$$G_{\mathrm{T}}\left(e^{j\omega T_s}\right) = G\left(e^{j\omega T_s},\theta_0\right) + G_\Delta\left(e^{j\omega T_s}\right) \ . \tag{11.1}$$

We then assume that $G_\Delta(e^{j\omega T_s})$ is a realization of a zero mean stochastic process

$$E\left\{G_\Delta\left(e^{j\omega T_s}\right)\right\} = 0 \tag{11.2}$$

$$\Rightarrow \quad E\left\{G_{\mathrm{T}}\left(e^{j\omega T_s}\right)\right\} = G(e^{j\omega T_s},\theta_0) \ . \tag{11.3}$$

The expectation $E\{\cdot\}$ in Equation (11.2) and (11.3) means averaging over different realizations of the undermodeling. Of course, for any given system we will have just one realization and the undermodeling will be deterministic. However, the stochastic embedding of the undermodeling enables us to treat it within the same framework as the noise and to derive results which can be viewed as a natural extension of the classical results.

11.1.1 Necessary Assumptions

The following assumptions will be made:

1. The observed data are assumed to be generated by the linear system $\mathcal{S}$ according to:

$$\mathcal{S} : y(k) = G_{\mathrm{T}}(q)u(k) + H_{\mathrm{T}}(q)e_0(k) \tag{11.4}$$

   where $G_{\mathrm{T}}(q)$ and $H_{\mathrm{T}}(q)$ are strictly stable rational transfer functions, $e_0(k)$ is a zero mean i.i.d. (independent identically distributed) stochastic process, $u(k)$ is a quasi-stationary sequence (see [Lju87, pp 27]). Furthermore, $u(k)$ and $e(k)$ are independent. Applying Equation (11.1) we may write (11.4) as:

$$y(k) = G\left(q, \theta_0\right) u(k) + G_{\Delta}\left(q\right) u(k) + H\left(q\right) e_0(k)\ . \tag{11.5}$$

2. The process noise $\nu(k) = H_{\mathrm{T}}(q)e_0(k)$ is independent of the undermodeling described by $\delta(k) = G_{\Delta}\left(q\right) u(k)$.
3. The probability density function f_{Δ} for the zero mean stochastic process $\delta(k)$ can be parameterized as a function of a parameter vector β.
4. The probability density function f_{ν} for the process noise $\nu(k)$ can be parameterized as a function of a parameter vector γ.

Assumption 1 is more or less standard in system identification. Note that we have not restricted the distribution function for the noise to be Gaussian. Assumption 2 is obviously reasonable. From assumption 3 we now get:

$$f_{\Delta} = f_{\Delta}(G_{\Delta}, \beta) \tag{11.6}$$

and similar from 4:

$$f_{\nu} = f_{\nu}(H, \gamma) \tag{11.7}$$

11.1.2 Model Formulation

We will use fixed denominator modeling both for the parametric part $G(q, \theta)$ and the undermodeling $G_{\Delta}(q)$. In particular we will use a FIR model of order $L \leq N$ for $G_{\Delta}(q)$:

$$G_{\Delta}(q) = \sum_{k=1}^{L} \eta(k) q^{-k} \tag{11.8}$$

where $\eta(k)$ denotes the model error impulse response. Replacing q with $e^{j\omega T_s}$ we get the frequency response:

$$G_\Delta\left(e^{j\omega T_s}\right) = \Pi\left(e^{j\omega T_s}\right)\eta \tag{11.9}$$

with

$$\Pi\left(e^{j\omega T_s}\right) = \left[e^{-j\omega T_s}, e^{-2j\omega T_s}, \cdots, e^{-Lj\omega T_s}\right] \tag{11.10}$$

$$\eta = [\eta(1), \eta(2), \cdots, \eta(L)]^T \tag{11.11}$$

where T denotes transpose. We will use a general fixed denominator model for the parametric part $G(q,\theta)$:

$$G(e^{j\omega T_s}, \theta) = \Lambda\left(e^{j\omega T_s}\right)\theta \tag{11.12}$$

with

$$\Lambda\left(e^{j\omega T_s}\right) = \left[\Lambda_1\left(e^{j\omega T_s}\right), \Lambda_2\left(e^{j\omega T_s}\right), \cdots, \Lambda_n\left(e^{j\omega T_s}\right)\right] \tag{11.13}$$

where n is the order of the model. Hence the Laguerre and Kautz model structures introduced in Section 10.3 will be valid model structures whereas the general polynomial model (9.1) cannot readily be used. We may now rewrite Equation (11.4) as:

$$\begin{aligned} y(k) &= G_{\mathrm{T}}(q)u(k) + H_{\mathrm{T}}(q)e_0(k) && (11.14)\\ &= G(q,\theta_0)u(k) + G_\Delta(q)u(k) + \nu(k) && (11.15)\\ &= \Lambda(q)\theta_0 u(k) + \Pi(q)\eta u(k) + \nu(k) && (11.16)\\ &= \phi^T(k)\theta_0 + \xi^T(k)\eta + \nu(k) && (11.17)\end{aligned}$$

where

$$\phi^T(k) = \Lambda(q)u(k) = [\phi_1(k), \phi_2(k), \cdots, \phi_n(k)] \tag{11.18}$$

$$\xi^T(k) = \Pi(q)u(k) = [u(k-1), u(k-2), \cdots, u(k-L)]\ . \tag{11.19}$$

Introducing:

$$Y = [y(1), y(2), \cdots, y(N)]^T \tag{11.20}$$

$$V = [\nu(1), \nu(2), \cdots, \nu(N)]^T \tag{11.21}$$

Equation (11.17) may be rewritten as:

$$Y = \begin{bmatrix} y(1) \\ y(2) \\ \vdots \\ y(N) \end{bmatrix} = \begin{bmatrix} \phi^T(1)\theta_0 \\ \phi^T(2)\theta_0 \\ \vdots \\ \phi^T(N)\theta_0 \end{bmatrix} + \begin{bmatrix} \xi^T(1)\eta \\ \xi^T(2)\eta \\ \vdots \\ \xi^T(N)\eta \end{bmatrix} + \begin{bmatrix} \nu(1) \\ \nu(2) \\ \vdots \\ \nu(N) \end{bmatrix} \tag{11.22}$$

$$= \Phi\theta_0 + X\eta + V \tag{11.23}$$

with Φ and X defined by:

$$\Phi = \begin{bmatrix} \phi_1(1) & \phi_2(1) & \cdots & \phi_n(1) \\ \phi_1(2) & \phi_2(2) & \cdots & \phi_n(2) \\ \vdots & \vdots & \ddots & \vdots \\ \phi_1(N) & \phi_2(N) & \cdots & \phi_n(N) \end{bmatrix} \tag{11.24}$$

$$X = \begin{bmatrix} \xi_1(1) & \xi_2(1) & \cdots & \xi_L(1) \\ \xi_1(2) & \xi_2(2) & \cdots & \xi_L(2) \\ \vdots & \vdots & \ddots & \vdots \\ \xi_1(N) & \xi_2(N) & \cdots & \xi_L(N) \end{bmatrix} \tag{11.25}$$

$$= \begin{bmatrix} u(1-1) & u(1-2) & \cdots & u(1-L) \\ u(2-1) & u(2-2) & \cdots & u(2-L) \\ \vdots & \vdots & \ddots & \vdots \\ u(N-1) & u(N-2) & \cdots & u(N-L) \end{bmatrix} . \tag{11.26}$$

11.1.3 Computing the Parameter Estimate

Adopting a quadratic performance function for the estimation problem gives:

$$V_N(\theta, Z^N) = \sum_{k=1}^{N} (y(k) - \hat{y}(k|\theta))^2 \tag{11.27}$$

$$= \sum_{k=1}^{N} \left(y(k) - \phi^T(k)\theta\right)^2 \tag{11.28}$$

$$= (Y - \Phi\theta)^T (Y - \Phi\theta) \tag{11.29}$$

$$= Y^T Y - Y^T \Phi\theta - \theta^T \Phi^T Y + \theta^T \Phi^T \Phi\theta . \tag{11.30}$$

The derivative of $V_N(\theta, Z^N)$ with respect to θ is henceforth given by:

$$V_N'(\theta, Z^N) = 0 - \frac{\partial}{\partial\theta}\left\{Y^T\Phi\theta\right\} - \frac{\partial}{\partial\theta}\left\{\theta^T\Phi^T Y\right\} + \frac{\partial}{\partial\theta}\left\{\theta^T\Phi^T\Phi\theta\right\} \tag{11.31}$$

$$= -\Phi^T Y - \Phi^T Y + \left(\Phi^T\Phi + \left(\Phi^T\Phi\right)^T\right)\theta \tag{11.32}$$

$$= -2\Phi^T Y + 2\left(\Phi^T\Phi\right)\theta . \tag{11.33}$$

Minimizing the performance function gives the desired estimate of the parameter vector θ as:

$$\hat{\theta}_N = \arg\min_{\theta} \ \left\{V_N(\theta, Z^N)\right\} \tag{11.34}$$

$$\Rightarrow \quad \hat{\theta}_N = \arg_{\theta} \ \left\{V_N'(\theta, Z^N) = 0\right\} \tag{11.35}$$

$$\Leftrightarrow \quad \Phi^T Y = \left(\Phi^T \Phi\right) \hat{\theta}_N \tag{11.36}$$

$$\Leftrightarrow \quad \hat{\theta}_N = \left(\Phi^T \Phi\right)^{-1} \Phi^T Y . \tag{11.37}$$

Equation (11.37) is the well known solution to the standard linear regression least-squares estimate of θ.

If the matrix $\Phi^T \Phi$ is ill-conditioned numerically more robust algorithms exists for computing the parameter estimate, eg through QR factorization, see Appendix G. However, as we noted in Section 10.3.2, see page 162, if we use orthonormal basis functions for $G(q, \theta)$ then $\Phi^T \Phi$ will asymptotically (for $N \to \infty$) have a Toeplitz structure which guarantees that the parameter estimate (11.37) will be well conditioned.

11.1.4 Variance of Parameter Estimate

In order to evaluate the covariance of the parameter estimate first introduce:

$$\Omega = \left(\Phi^T \Phi\right)^{-1} \Phi^T \tag{11.38}$$

to achieve:

$$\begin{aligned}
\hat{\theta}_N &= \Omega Y && (11.39)\\
&= \Omega \Phi \theta_0 + \Omega X \eta + \Omega V && (11.40)\\
&= \left(\Phi^T \Phi\right)^{-1} \Phi^T \Phi \theta_0 + \Omega X \eta + \Omega V && (11.41)\\
&= \theta_0 + \Omega X \eta + \Omega V && (11.42)
\end{aligned}$$

equivalent to:

$$\hat{\theta}_N - \theta_0 = \Omega X \eta + \Omega V . \tag{11.43}$$

The covariance of the parameter vector $\hat{\theta}_N$ with respect to the nominal value θ_0 is consequently given by:

$$\begin{aligned}
\mathrm{Cov}\left(\hat{\theta}_N - \theta_0\right) &= E\left\{\left(\hat{\theta}_N - \theta_0\right)\left(\hat{\theta}_N - \theta_0\right)^T\right\} && (11.44)\\
&= E\left\{\left(\Omega X \eta + \Omega V\right)\left(\Omega X \eta + \Omega V\right)^T\right\} && (11.45)\\
&= E\left\{\Omega \left(X\eta + V\right)\left(\eta^T X^T + V^T\right)\Omega^T\right\} && (11.46)\\
&= E\left\{\Omega\left(X\eta\eta^T X^T + X\eta V^T + V\eta^T X^T + VV^T\right)\Omega^T\right\} && (11.47)\\
&= E\left\{\Omega\left(X\eta\eta^T X^T + VV^T\right)\Omega^T\right\} && (11.48)
\end{aligned}$$

since η and V are assumed uncorrelated. Defining the covariance matrices for the noise C_ν and the undermodeling C_η as:

$$E\left\{\eta\eta^T\right\} = C_\eta \tag{11.49}$$

$$E\left\{VV^T\right\} = C_\nu . \tag{11.50}$$

Equation (11.48) may be written as:

$$\operatorname{Cov}\left(\hat{\theta}_N - \theta_0\right) = \Omega \left[X C_\eta X^T + C_\nu\right] \Omega^T \tag{11.51}$$

$$= \left(\Phi\Phi^T\right)^{-1} \Phi^T \left[X C_\eta X^T + C_\nu\right] \Phi \left(\Phi^T\Phi\right)^{-1} . \tag{11.52}$$

Equation (11.52) states that under the assumptions given the variance on the parameter estimate is a combination of a bias term ($\Omega X C_\eta X^T \Omega^T$) and a noise term ($\Omega C_\nu \Omega^T$). Compare with the classical expression for fixed denominator models (10.61).

11.1.5 Estimating the Model Error

Combine Equation (11.1), (11.9) and (11.12) to obtain:

$$G_{\mathrm{T}}\left(e^{j\omega T_s}\right) = \Lambda\theta_0 + \Pi\eta . \tag{11.53}$$

Furthermore substituting θ_0 with $\hat{\theta}_N$ in (11.12) gives:

$$G\left(e^{j\omega T_s}, \hat{\theta}_N\right) = \Lambda\hat{\theta}_N . \tag{11.54}$$

Combining Equation (11.53) and (11.54) we find that:

$$G_{\mathrm{T}}\left(e^{j\omega T_s}\right) - G\left(e^{j\omega T_s}, \hat{\theta}_N\right) = \Lambda\left(\theta_0 - \hat{\theta}_N\right) + \Pi\eta . \tag{11.55}$$

Combining this with Equation (11.43) we achieve the key result:

$$G_{\mathrm{T}}\left(e^{j\omega T_s}\right) - G\left(e^{j\omega T_s}, \hat{\theta}_N\right) = \Lambda\left(-\Omega X\eta - \Omega V\right) + \Pi\eta \tag{11.56}$$

$$= \left(\Pi - \Lambda\Omega X\right)\eta - \Lambda\Omega V . \tag{11.57}$$

Equation (11.57) is a central result of the stochastic embedding approach. Π and Λ are known functions of the frequency ω whereas Ω and X are known functions of the input signals. Consequently, Equation (11.57) expresses the modeling error as a known linear combination of two independent random vectors η and V.

The expression (11.57) furthermore clearly separates the bias term $(\Pi - \Lambda\Omega X)\eta$ and the noise term $\Lambda\Omega V$. The bias term is especially interesting since it separates into an a priori term $\Pi\eta$, see Equation (11.9), and a data induced correction term $\Lambda\Omega X\eta$ which compensate for the difference between θ_0 and $\hat{\theta}_N$, see also Equation (11.43).

Also notice that if $\hat{\theta}_N = \theta_0$ then:

$$G_{\mathrm{T}}\left(e^{j\omega T_s}\right) - G\left(e^{j\omega T_s}, \hat{\theta}_N\right) = \Pi\eta , \qquad \text{if } \hat{\theta}_N = \theta_0 \tag{11.58}$$

that is, our a priori estimate of $G_\Delta\left(e^{j\omega T_s}\right)$ will not be changed as we will expect from Equation (11.1) and the assumptions.

We will now investigate the second order properties of the modeling error. Introduce the following parameters:

$$\hat{\rho}_N = \begin{bmatrix} \hat{\theta}_N \\ 0 \end{bmatrix}, \qquad \rho_0 = \begin{bmatrix} \theta_0 \\ \eta \end{bmatrix} \tag{11.59}$$

$$\Gamma\left(e^{j\omega T_s}\right) = \begin{bmatrix} \Re e\ \left\{\Lambda\left(e^{j\omega T_s}\right), \Pi\left(e^{j\omega T_s}\right)\right\} \\ \Im m\left\{\Lambda\left(e^{j\omega T_s}\right), \Pi\left(e^{j\omega T_s}\right)\right\} \end{bmatrix} \tag{11.60}$$

$$\tilde{g}\left(e^{j\omega T_s}\right) = \begin{bmatrix} \Re e\ \left\{G_{\mathrm{T}}\left(e^{j\omega T_s}\right) - G\left(e^{j\omega T_s}, \hat{\theta}_N\right)\right\} \\ \Im m\left\{G_{\mathrm{T}}\left(e^{j\omega T_s}\right) - G\left(e^{j\omega T_s}, \hat{\theta}_N\right)\right\} \end{bmatrix} \tag{11.61}$$

to obtain:

$$\Gamma\left(e^{j\omega T_s}\right)\left(\rho_0 - \hat{\rho}_N\right) = \tag{11.62}$$

$$\begin{bmatrix} \Re e\ \left\{\Lambda_1 \cdots \Lambda_n, e^{-j\omega T_s} \cdots e^{-Lj\omega T_s}\right\} \\ \Im m\left\{\Lambda_1 \cdots \Lambda_n, e^{-j\omega T_s} \cdots e^{-Lj\omega T_s}\right\} \end{bmatrix} \begin{bmatrix} \theta_0(1) - \hat{\theta}_N(1) \\ \vdots \\ \theta_0(n) - \hat{\theta}_N(n) \\ \eta(1) \\ \vdots \\ \eta(L) \end{bmatrix} \tag{11.63}$$

$$= \begin{bmatrix} \Re e\ \left\{(\Lambda_1\theta_0(1) + \cdots + \Lambda_n\theta_0(n)) - \left(\Lambda_1\hat{\theta}_N(1) + \cdots + \Lambda_n\hat{\theta}_N(n)\right)\right. \\ \Im m\left\{(\Lambda_1\theta_0(1) + \cdots + \Lambda_n\theta_0(n)) - \left(\Lambda_1\hat{\theta}_N(1) - \cdots - \Lambda_n\hat{\theta}_N(n)\right)\right. \end{bmatrix}$$
$$\begin{matrix} \left.+\left(e^{-j\omega T_s}\eta(1) + \cdots + e^{-Lj\omega T_s}\eta(L)\right)\right\} \\ \left.+\left(e^{-j\omega T_s}\eta(1) + \cdots + e^{-Lj\omega T_s}\eta(L)\right)\right\} \end{matrix} \tag{11.64}$$

$$= \begin{bmatrix} \Re e\ \left\{G\left(e^{j\omega T_s}, \theta_0\right) - G\left(e^{j\omega T_s}, \hat{\theta}_N\right) + G_\Delta\left(e^{j\omega T_s}\right)\right\} \\ \Im m\left\{G\left(e^{j\omega T_s}, \theta_0\right) - G\left(e^{j\omega T_s}, \hat{\theta}_N\right) + G_\Delta\left(e^{j\omega T_s}\right)\right\} \end{bmatrix} \tag{11.65}$$

$$= \begin{bmatrix} \Re e\ \left\{G_{\mathrm{T}}\left(e^{j\omega T_s}\right) - G\left(e^{j\omega T_s}, \hat{\theta}_N\right)\right\} \\ \Im m\left\{G_{\mathrm{T}}\left(e^{j\omega T_s}\right) - G\left(e^{j\omega T_s}, \hat{\theta}_N\right)\right\} \end{bmatrix} \tag{11.66}$$

$$= \tilde{g}\left(e^{j\omega T_s}\right) . \tag{11.67}$$

From Equation (11.67) follows:

$$P_{\tilde{g}}(\omega) = E\left\{\tilde{g}\left(e^{j\omega T_s}\right)\tilde{g}\left(e^{j\omega T_s}\right)^T\right\} \tag{11.68}$$

$$= E\left\{\Gamma\left(e^{j\omega T_s}\right)\left(\rho_0 - \hat{\rho}_N\right)\left(\rho_0 - \hat{\rho}_N\right)^T\Gamma^T\left(e^{j\omega T_s}\right)\right\} \tag{11.69}$$

$$= \Gamma\left(e^{j\omega T_s}\right)E\left\{\left(\rho_0 - \hat{\rho}_N\right)\left(\rho_0 - \hat{\rho}_N\right)^T\right\}\Gamma^T\left(e^{j\omega T_s}\right) \tag{11.70}$$

$$= \Gamma\left(e^{j\omega T_s}\right)\Upsilon\Gamma^T\left(e^{j\omega T_s}\right) . \tag{11.71}$$

Remembering that η and V are assumed uncorrelated, we obtain:

$$\Upsilon = E\left\{(\rho_0 - \hat{\rho}_N)(\rho_0 - \hat{\rho}_N)^T\right\} \tag{11.72}$$

$$= E\left\{\begin{bmatrix} \theta_0 - \hat{\theta}_N \\ \eta \end{bmatrix}\begin{bmatrix} \theta_0 - \hat{\theta}_N \\ \eta \end{bmatrix}^T\right\} \tag{11.73}$$

$$= E\left\{\begin{bmatrix} -\Omega X\eta - \Omega V \\ \eta \end{bmatrix}\begin{bmatrix} (-\Omega X\eta - \Omega V)^T & \eta^T \end{bmatrix}\right\} \tag{11.74}$$

$$= E\left\{\begin{bmatrix} \Omega(-X\eta - V)(-X\eta - V)^T\Omega^T & \Omega(-X\eta - V)\eta^T \\ \eta(-X\eta - V)^T\Omega^T & \eta\eta^T \end{bmatrix}\right\} \tag{11.75}$$

$$= E\left\{\begin{bmatrix} \Omega\left(X\eta\eta^T X^T + VV^T\right)\Omega^T & -\Omega X\eta\eta^T \\ -\eta\eta^T X^T\Omega^T & \eta\eta^T \end{bmatrix}\right\} \tag{11.76}$$

$$= \begin{bmatrix} E\{\Omega\left(X\eta\eta^T X^T + VV^T\right)\Omega^T\} & E\left\{-\Omega X\eta\eta^T\right\} \\ E\{-\eta\eta^T X^T\Omega^T\} & E\{\eta\eta^T\} \end{bmatrix} \tag{11.77}$$

$$= \begin{bmatrix} \Omega\left(XC_\eta X^T + C_\nu\right)\Omega^T & -\Omega XC_\eta \\ -C_\eta X^T\Omega^T & C_\eta \end{bmatrix}. \tag{11.78}$$

The second order properties of the total model error is then described by Equation (11.71) and (11.78). Once the distributions f_Δ and f_ν are determined (and the parameterizations of $C_\eta(\beta)$ and $C_\nu(\gamma)$ have been estimated) we may use (11.71) to map the uncertainty levels into the frequency domain.

We will now show how to find the magnitude expectation of the squared undermodeling (the total model error variance in the stochastic setting). We first show that it may be expressed as the trace of the covariance matrix $P_{\tilde{g}}$:

$$E\left\{\left|G(e^{j\omega T_s}, \hat{\theta}_N) - G_{\mathrm{T}}\left(e^{j\omega T_s}\right)\right|^2\right\}$$

$$= E\left\{\left(G(e^{j\omega T_s}, \hat{\theta}_N) - G_{\mathrm{T}}\left(e^{j\omega T_s}\right)\right)\left(G(e^{j\omega T_s}, \hat{\theta}_N) - G_{\mathrm{T}}\left(e^{j\omega T_s}\right)\right)^*\right\} \tag{11.79}$$

$$= E\left\{\Re e\left\{G(e^{j\omega T_s}, \hat{\theta}_N) - G_{\mathrm{T}}\left(e^{j\omega T_s}\right)\right\}^2 + \Im m\left\{G(e^{j\omega T_s}, \hat{\theta}_N) - G_{\mathrm{T}}\left(e^{j\omega T_s}\right)\right\}^2\right\} \tag{11.80}$$

$$= E\left\{\tilde{g}_1^2\left(e^{j\omega T_s}\right) + \tilde{g}_2^2\left(e^{j\omega T_s}\right)\right\} \tag{11.81}$$

$$= \mathrm{tr}\left\{P_{\tilde{g}}(\omega)\right\}. \tag{11.82}$$

We may also express the undermodeling directly via Equation (11.57) as:

$$E\left\{\left|G(e^{j\omega T_s}, \hat{\theta}_N) - G_{\mathrm{T}}\left(e^{j\omega T_s}\right)\right|^2\right\}$$

$$= E\left\{((\Pi - \Lambda\Omega X)\eta - \Lambda\Omega V)((\Pi - \Lambda\Omega X)\eta - \Lambda\Omega V)^*\right\} \tag{11.83}$$

$$= E\left\{(\Pi - \Lambda\Omega X)\,\eta\eta^T\,(\Pi - \Lambda\Omega X)^* + (\Pi - \Lambda\Omega X)\,\eta V^T\Omega^T\Lambda^* \right.$$
$$\left. -\Lambda\Omega V\eta^T\,(\Pi - \Lambda\Omega X)^* + \Lambda\Omega V V^T\Omega^T\Lambda^*\right\} \tag{11.84}$$
$$= (\Pi - \Lambda\Omega X)\,E\left\{\eta\eta^T\right\}(\Pi - \Lambda\Omega X)^* + 0 - 0 + \Lambda\Omega E\left\{VV^T\right\}\Omega^T\Lambda^* \tag{11.85}$$
$$= (\Pi - \Lambda\Omega X)\,C_\eta\,(\Pi - \Lambda\Omega X)^* + \Lambda\Omega C_\nu\Omega^T\Lambda^* \,. \tag{11.86}$$

Notice that in Equation (11.86) we have separated the contributions from the bias error and the variance error. Notice furthermore that in (11.86) and (11.71) everything but C_η and C_ν are uniquely determined from the parametric model structure and the input. Thus the total model error is shaped by the covariance matrices for the undermodeling and the noise. Remember that in the classical approach for fixed denominator models, see Section 10.3, the model error was shaped by the noise covariance matrix in a similar manner.

The preliminary approach by Goodwin & Salgado [GS89b] – in which C_η and C_ν were assumed quantitatively given a priori – suffers from the fact that no input/output information was in fact used to estimate the model error! It consequently seems essential to obtain good estimates of C_η and C_ν from the available data.

11.1.6 Recapitulation

We have now demonstrated the main results of the stochastic embedding approach as proposed by Goodwin *et al.* Let us shortly recapitulate the key assumptions and results:

Main assumptions.

- We will describe the (deterministic) undermodeling of our process as one realization of a stochastic process. From this realization we will estimate the properties of the stochastic description.
- We assume that the true system $G_{\mathrm{T}}\left(e^{j\omega T_s}\right)$ may be decomposed as:

  $$G_{\mathrm{T}}\left(e^{j\omega T_s}\right) = G(e^{j\omega T_s}, \theta_0) + G_\Delta\left(e^{j\omega T_s}\right)$$

  where:

  $$E\left\{G_\Delta\left(e^{j\omega T_s}\right)\right\} = 0$$

  and consequently:

  $$E\left\{G_{\mathrm{T}}\left(e^{j\omega T_s}\right)\right\} = G(e^{j\omega T_s}, \theta_0)$$

  where $E\{\cdot\}$ means expectation over different realizations of the stochastic process we associate with the undermodeling.
- The model $G(e^{j\omega T_s}, \theta_0)$ shall be linear in the parameter vector θ as e.g. are the class of Laguerre/Kautz models.

Main Results.

- The difference between the true system $G_{\mathrm{T}}\left(e^{j\omega T_s}\right)$ and the estimated model $G(e^{j\omega T_s}, \hat{\theta}_N)$ is given by:

$$G_{\mathrm{T}}\left(e^{j\omega T_s}\right) - G(e^{j\omega T_s}, \hat{\theta}_N) = \left(\Pi - \Lambda\Omega X\right)\eta - \Lambda\Omega V$$

with symbols as defined above. This central result shows that under the stochastic embedding approach the modeling error may be expressed as a known linear combination of two independent random vectors η and V. Furthermore the expression separates the undermodeling from the noise contribution. The undermodeling contribution to the total model error $\left(\Pi - \Lambda\Omega X\right)\eta$ consists of two terms. $\Pi\eta$ is an a priori assumption of the undermodeling based on estimates of the impulse response η. $\Lambda\Omega X\eta$ is a data-induced correction term to this prior due to the shift from θ_0 to $\hat{\theta}_N$. Furthermore:

$$E\left\{\left|G(e^{j\omega T_s}, \hat{\theta}_N) - G_{\mathrm{T}}\left(e^{j\omega T_s}\right)\right|^2\right\} = \\ \left(\Pi - \Lambda\Omega X\right) C_\eta \left(\Pi - \Lambda\Omega X\right)^* + \Lambda\Omega C_\nu \Omega^T \Lambda^* .$$

In the above expression for the total model error variance the covariances C_ν and C_η are the only unknowns. If reasonable estimates of these matrices may be obtained and substituted in the above expression, estimates of the total model error and its contributions may be determined.
- With $\tilde{g}\left(e^{j\omega T_s}\right)$ defined as in Section 11.1.5 we have:

$$E\left\{\tilde{g}\left(e^{j\omega T_s}\right)\right\} = 0$$

$$E\left\{\tilde{g}\left(e^{j\omega T_s}\right)\tilde{g}\left(e^{j\omega T_s}\right)^T\right\} = P_{\tilde{g}} = \Gamma\left(e^{j\omega T_s}\right)\Upsilon\Gamma\left(e^{j\omega T_s}\right)^T$$

$$E\left\{\left|G(e^{j\omega T_s}, \hat{\theta}_N) - G_{\mathrm{T}}\left(e^{j\omega T_s}\right)\right|^2\right\} = \mathrm{tr}\left\{P_{\tilde{g}}\right\} .$$

Assumptions NOT made.

- No assumptions have been made on the probability density functions for the undermodeling and the process noise other than that they may be parameterized as functions of β and γ.
- Also we have not limited us to consider special structures of the parameterization of these density functions. For the undermodeling this is equivalent to specifying the structure of undermodeling impulse response variance $E\left\{\eta^2(k)\right\}$.
For the process noise, selection of parameterization structure amounts to specifying the necessary parameters for the chosen probability function.

11.2 Estimating the Parameterizations of f_Δ and f_ν

Having presented the central methodology, we now turn to specifying and estimating the parameterizations of the undermodeling and process noise.

11.2.1 Estimation Techniques

Goodwin *et al.* [GGN92] propose maximum likelihood methods in order to estimate the parameter vectors β and γ from the residuals vector ϵ defined by:

$$\begin{aligned}\epsilon &= Y - \Phi\hat{\theta}_N &(11.87)\\ &= Y - \Phi\Omega Y &(11.88)\\ &= \left[I - \Phi\left(\Phi^T\Phi\right)^{-1}\Phi^T\right]Y &(11.89)\\ &\triangleq P_\Phi Y &(11.90)\end{aligned}$$

where P_Φ is a $N \times N$ matrix. However, it is possible to show that the rank of P_Φ is only $N-p$ and hence ϵ will have a singular distribution and maximum likelihood methods will fail. However, we may transform ϵ into another coordinate system by:

$$\begin{aligned}\varpi &= R^T\epsilon &(11.91)\\ &= R^TY - R^T\Phi\hat{\theta}_N &(11.92)\\ &= R^T\Phi\theta_0 + R^TX\eta + R^TV - R^T\Phi\hat{\theta}_N &(11.93)\\ &= R^T\Phi\left(\theta_0 - \hat{\theta}_N\right) + R^TX\eta + R^TV\,. &(11.94)\end{aligned}$$

If we choose a full rank $(N \times N - p)$ transformation matrix R so that:

$$R^T\Phi = 0 \tag{11.95}$$

then

$$\varpi = R^TY = R^TX\eta + R^TV\,. \tag{11.96}$$

With R chosen as above the distribution on ϖ will be non-singular. Notice that the transformed data vector ϖ has only $N-p$ elements. Possible ways of choosing R is discussed in Section 11.2.3.

From Equation (11.96), note that ϖ is the sum of two independent random vectors, η and V, whose probability density functions we have parameterized as functions of β and γ, see Equation (11.6) and (11.7).

Now introduce the combined parameter vector ζ as:

$$\zeta = \begin{bmatrix}\beta\\ \gamma\end{bmatrix}. \tag{11.97}$$

Then it will be possible to compute the likelihood function for the data vector ϖ given ζ and the inputs U under the given assumptions on the probability density functions. The likelihood function will be denoted $\mathcal{L}(\varpi|U,\zeta)$. Maximizing the likelihood function will yield the desired estimate for the parameter vector ζ:

$$\hat{\zeta} = \arg\max_{\zeta} \ \mathcal{L}(\varpi|U,\zeta) \, . \tag{11.98}$$

Notice that in the above procedure we have not narrowed the applicable class of probability density functions to be chosen.

11.2.2 Choosing the Probability Distributions

To proceed further we now have to make a specific choice of the assumed probability structure of the undermodeling and process noise. Assume, for example, that these are zero mean Gaussian distributed:

$$\eta \sim \mathcal{N}(0, C_\eta(\beta)) \tag{11.99}$$

$$V \sim \mathcal{N}(0, C_\nu(\gamma)) \tag{11.100}$$

with $C_\eta(\beta)$ being the parameterized covariance matrix for the impulse response of the undermodeling and $C_\nu(\gamma)$ the parameterized covariance matrix for the noise.

With these prior distributions on f_Δ and f_ν the parameter estimate $\hat{\theta}_N$ and $\tilde{g}\left(e^{j\omega T_s}\right)$ will be Gaussian distributed as well:

$$\hat{\theta}_N \sim \mathcal{N}(\theta_0, P_\theta) \tag{11.101}$$

$$\tilde{g}\left(e^{j\omega T_s}\right) \sim \mathcal{N}(0, P_{\tilde{g}}) \, . \tag{11.102}$$

Remember that this is still averaging over different realizations of a class of systems, from where the true system is just one realization. Consequently we will not generally expect $\hat{\theta}_N$ to converge against θ_0 for $N \to \infty$.

Equation (11.102) describes the assumed probability characteristics of the total model error for the class of systems defined. Since the true system is one such system it will constitute one event in the probability space assumed.

Since $\tilde{g}\left(e^{j\omega T_s}\right)$ is zero mean Gaussian distributed with covariance $P_{\tilde{g}}$ the scalar $z\left(e^{j\omega T_s}\right)$ given by:

$$z\left(e^{j\omega T_s}\right) = \tilde{g}\left(e^{j\omega T_s}\right)^T P_{\tilde{g}}^{-1} \tilde{g}\left(e^{j\omega T_s}\right) \tag{11.103}$$

has χ^2 distribution with $\dim \tilde{g}\left(e^{j\omega T_s}\right) = 2$ degrees of freedom.

Equation (11.103) may thus be used to draw confidence ellipses in the complex plane for the frequency response estimate $G(e^{j\omega T_s}, \hat{\theta}_N)$ with respect to the class of true systems defined by Equation (11.1).

The tractability of the results for the Gaussian assumptions on η and V motivates us to use the corresponding results in practice.

11.2.3 Maximum Likelihood Estimation of ζ

Equation (11.98) may now be solved (numerically) under the given assumptions on the probability distributions and the structure of the covariance matrices. Unfortunately, this may be a non-convex problem with multiple local maxima so we cannot guarantee to find the global maximum. The likelihood function of the $N-p$ data vector ϖ is equivalent to the probability density of ϖ. Since f_Δ and f_ν are assumed Gaussian, the probability density of ϖ will be an $N-p$ dimensional Gaussian distribution with zero mean:

$$f(\varpi|U,\zeta) = \frac{1}{(2\pi)^{(N-p)/2}} \cdot \frac{1}{(\det\Sigma)^{1/2}} \cdot e^{-\frac{1}{2}\varpi^T\Sigma^{-1}\varpi} \tag{11.104}$$

where Σ is given by:

$$\Sigma = E\left\{\varpi\varpi^T\right\} \tag{11.105}$$
$$= E\left\{\left(R^TX\eta + R^TV\right)\left(R^TX\eta + R^TV\right)^T\right\} \tag{11.106}$$
$$= E\left\{R^TX\eta\eta^TX^TR\right\} + E\left\{R^TVV^TR\right\} \tag{11.107}$$
$$= R^TXC_\eta X^TR + R^TC_\nu R \tag{11.108}$$

with given parameterizations on C_η and C_ν. Maximizing the likelihood function is equivalent to maximizing the loglikelihood function. For computational purposes this is preferred. The loglikelihood function $\ell(\varpi|U,\zeta)$ is given by:

$$\ell(\varpi|U,\zeta) = \ln\left(f(\varpi|U,\zeta)\right) \tag{11.109}$$
$$= -\frac{N-p}{2}\ln(2\pi) - \frac{1}{2}\ln(\det\Sigma) - \frac{1}{2}\varpi^T\Sigma^{-1}\varpi \tag{11.110}$$
$$= -\frac{1}{2}\ln(\det\Sigma) - \frac{1}{2}\varpi^T\Sigma^{-1}\varpi + k \tag{11.111}$$

where $k = -\ln(2\pi)(N-p)/2$ is some constant. In Appendix H it is demonstrated how we may derive explicit expressions for the partial derivatives of $\ell(\varpi|U,\zeta)$ with respect to the elements of ζ and for the Hessian matrix. This enables us to construct numerical search algorithms for the maximum based on Marquardts method. Such search procedures usually converge quickly and efficiently at least to a local maximum.

Assuming that the global maximum is found, it will now be possible to find the parameter vector ζ that maximizes the loglikelihood function for ϖ given the parameterizations on C_η and C_ν.

Choosing the Transformation *R*. The choice of R affect the singular value ratio[1] of the ϖ covariance matrix Σ. This is important since taking

[1] The condition number.

the loglikelihood of ϖ involves inversion of Σ, see Equation (11.111). Consequently we must require that the singular value ratio of Σ is reasonably small. Furthermore, since the determinant of Σ is the product of the eigenvalues then if some of the eigenvalues are close to zero[2], then the determinant will be close to zero as well and we may have problems when taking the logarithm in Equation (11.111). Consequently we will also require that the ratio of the largest eigenvalue of Σ to the smallest should be reasonably small. We propose that R is obtained in either of the following two ways:

- as the orthonormal basis for P_Φ in (11.90).
- through QR-factorization of Φ, see Appendix G.

Both these choices result in well balanced covariances Σ both with respect to eigenvalue and singular value ratios.

11.3 Parameterizing the Covariances

Parameterizing C_ν and C_η we may now estimate the parameters themselves as described in the previous section. Several parameterizations have been investigated.

11.3.1 Parameterizing the Noise Covariance C_ν

Two assumptions on the noise $\nu(k)$ have been investigated:

- $\nu(k)$ is assumed to be white measurement noise. This corresponds to $H(q) = 1$, see Equation (11.4) on page 170.
- $\nu(k)$ is assumed to be white noise filtered through a first order ARMA filter $H(q) = (1 + cq^{-1})/(1 + aq^{-1})$, see Equation (11.4) on page 170.

White Noise Assumption. The white noise assumption on $\nu(k)$ is equivalent to specifying the covariance matrix for the noise as a diagonal matrix of the white noise variance σ_e^2:

$$C_\nu(\gamma) = \sigma_e^2 \cdot I_N \tag{11.112}$$

where N is the number of measurements. The parameterization in (11.112) constitutes the simplest possible assumption on the noise, but may prove adequate in many applications.

[2] Relatively to the remaining eigenvalues.

Colored Noise Assumption. Assume that $\nu(k)$ is given by a first order ARMA model

$$\nu(k) = \frac{1+cq^{-1}}{1+aq^{-1}} e(k) \tag{11.113}$$

where $e(k) \in \mathcal{N}(0, \sigma_e^2)$ is white noise with variance σ_e^2. It is then straightforward to show, see Appendix K that the noise impulse response $h_\nu(k)$ will be given by:

$$h_\nu(k) = \begin{cases} 1\,, & k = 0 \\ (-a)^{k-1}(-a+c)\,, & k > 0 \end{cases} . \tag{11.114}$$

The corresponding noise covariance matrix $C_\nu(\gamma)$ will furthermore be given by:

$$C_\nu(i,j) = \sigma_e^2 \sum_{k=0}^{\infty} h_\nu(k) h_\nu(k+j-i) \tag{11.115}$$

$$= \begin{cases} \sigma_e^2 \left(1 + \dfrac{(c-a)^2}{1-a^2}\right), & i = j \\[2ex] \dfrac{\sigma_e^2 (-a)^M \left(1 + c^2 - a * c - c/a\right)}{1-a^2}, & |j-i| = M \end{cases} \tag{11.116}$$

where $|\cdot|$ denotes absolute value. Other noise assumptions may be incorporated into $C_\nu(\gamma)$ if, for example, more detailed noise models are available. The computational labour for higher order noise descriptions may however be quite severe.

11.3.2 Parameterizing the Undermodeling Covariance C_η

3 different parameterizations of the undermodeling impulse response covariance have been investigated:

- A constant diagonal structure corresponding to a constant undermodeling impulse response variance.
- An exponential decaying diagonal structure corresponding to an exponentially decaying undermodeling impulse response variance.
- A non-diagonal structure corresponding to a first order undermodeling impulse response.

A Constant Diagonal Assumption. The simplest possible structure on the covariance matrix for the undermodeling impulse response would be a constant diagonal matrix:

$$C_\eta(\beta) = \alpha \cdot I_L \tag{11.117}$$

where L is the order of the undermodeling FIR description. The corresponding undermodeling impulse response variance is:

$$E\left\{\eta^2(k)\right\} = \alpha \, . \tag{11.118}$$

This corresponds for example to the assumption that 68,3% (1 standard deviation) of the undermodeling impulse responses will fall within the rectangle shown in Figure 11.1 for $\sqrt{\alpha} = 5$ and $L = 25$.

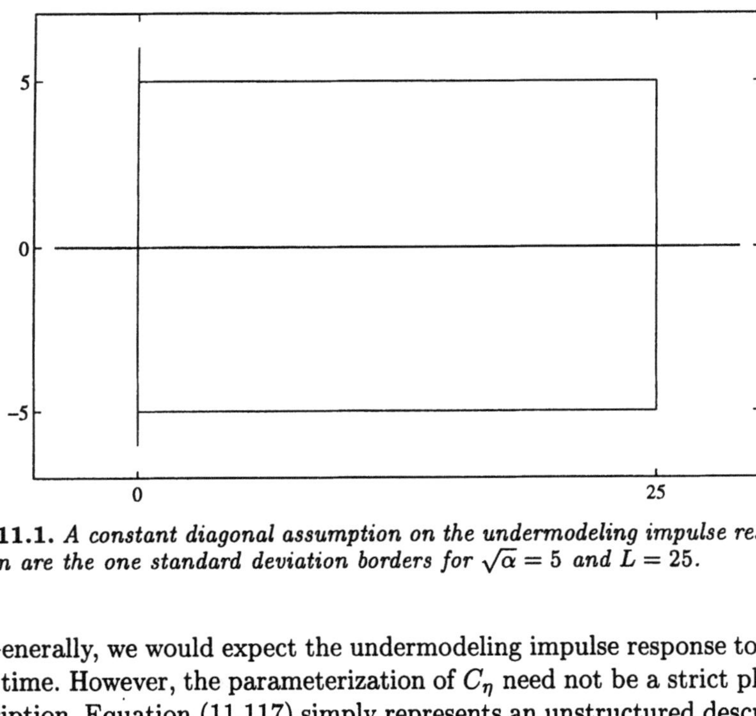

Fig. 11.1. *A constant diagonal assumption on the undermodeling impulse response. Shown are the one standard deviation borders for* $\sqrt{\alpha} = 5$ *and* $L = 25$.

Generally, we would expect the undermodeling impulse response to decay with time. However, the parameterization of C_η need not be a strict physical description. Equation (11.117) simply represents an unstructured description of the undermodeling.

Notice that the diagonal assumption on C_η causes the impulse response $\eta(k)$ to be mutually uncorrelated:

$$E\left\{\eta(k)\eta(\tau)\right\} = 0 \, , \qquad \text{for } k \neq \tau. \tag{11.119}$$

It is then easy to show that the initial estimate of the undermodeling is given by:

$$E\left\{\left|G_{\Delta}\left(e^{j\omega T_s}\right)\right|^2\right\} = \alpha \cdot L\,. \tag{11.120}$$

Consequently the initial estimate of $\left|G_{\Delta}\left(e^{j\omega T_s}\right)\right|^2$ is a frequency domain stationary process. However, if the true system is strictly proper its frequency response will decay to zero at high frequencies and as a consequence, so should the absolute modeling error.

An Exponentially Decaying Assumption. Goodwin and co-workers [GGN92] proposes a diagonal structure on the covariance matrix for the undermodeling:

$$C_\eta(\beta) = \underset{1\leq k\leq L}{\text{diag}}\ \alpha\lambda^k \tag{11.121}$$

corresponding to the undermodeling impulse response variance:

$$E\left\{\eta^2(k)\right\} = \alpha\lambda^k\,. \tag{11.122}$$

This corresponds, for example, to the assumption that 68,3% (1 standard deviation) of the undermodeling impulse responses will fall within the rectangle shown in Figure 11.2 for $\sqrt{\alpha} = 5$, $\sqrt{\lambda} = 0.9$ and $L = 25$.

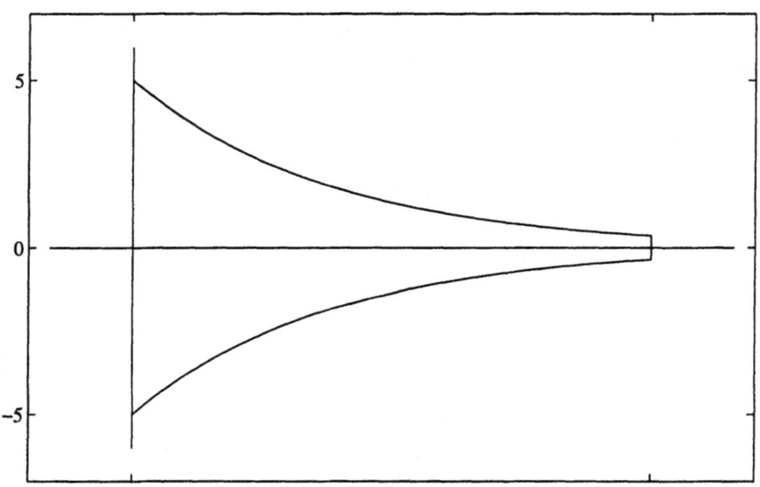

Fig. 11.2. *A exponentially decaying assumption on the undermodeling impulse response. Shown are the one standard deviation borders for* $\sqrt{\alpha} = 5$, $\sqrt{\lambda} = 0.9$ *and* $L = 25$.

Compared with the constant undermodeling impulse response variance above we now have decaying properties on the undermodeling impulse response.

Notice however that the diagonal assumption on C_η still causes the impulse response $\eta(k)$ to be mutually uncorrelated:

$$E\{\eta(k)\eta(\tau)\} = 0\,, \qquad \text{for } k \neq \tau. \tag{11.123}$$

It is then easy to show that the initial estimate of the undermodeling is given by:

$$E\left\{\left|G_\Delta\left(e^{j\omega T_s}\right)\right|^2\right\} = \frac{\alpha\lambda}{1-\lambda}\,. \tag{11.124}$$

Consequently the initial estimate of $\left|G_\Delta\left(e^{j\omega T_s}\right)\right|^2$ is still a frequency domain stationary process which is undesirable.

A Non-Diagonal Assumption. In order to obtain a frequency domain non-stationary initial estimate of $\left|G_\Delta\left(e^{j\omega T_s}\right)\right|^2$, we propose that the covariance matrix C_η may be parameterized as:

$$C_\eta(\beta) = \left[\alpha\lambda^0 \cdots \alpha\lambda^{L-1}\right]^T \left[\alpha\lambda^0 \cdots \alpha\lambda^{L-1}\right] \tag{11.125}$$

corresponding to a first order undermodeling impulse response η. Notice that the above C_η is singular. Thus the undermodeling impulse response $\{\eta(k)\}$ is a stochastic process with singular distribution and in fact of the form:

$$\eta(k) = a\lambda^{k-1} \tag{11.126}$$

where $a \in \mathcal{N}(0, \alpha^2)$ and λ some fixed number. The covariance matrix for the stochastic process (11.126) is namely given by:

$$C_\eta = E\left\{[\eta(1), \eta(2), \cdots, \eta(L)]^T [\eta(1), \eta(2), \cdots, \eta(L)]\right\} \tag{11.127}$$

$$= E\left\{\left[a\lambda^0, a\lambda^{-1}, \cdots, a\lambda^{L-1}\right]^T \left[a\lambda^0, a\lambda^{-1}, \cdots, a\lambda^{L-1}\right]\right\} \tag{11.128}$$

$$= E\left\{a^2 \left[\lambda^0, \lambda^{-1}, \cdots, \lambda^{L-1}\right]^T \left[\lambda^0, \lambda^{-1}, \cdots, \lambda^{L-1}\right]\right\} \tag{11.129}$$

$$= E\left\{a^2\right\} \left[\lambda^0, \lambda^{-1}, \cdots, \lambda^{L-1}\right]^T \left[\lambda^0, \lambda^{-1}, \cdots, \lambda^{L-1}\right] \tag{11.130}$$

$$= \alpha^2 \left[\lambda^0, \lambda^{-1}, \cdots, \lambda^{L-1}\right]^T \left[\lambda^0, \lambda^{-1}, \cdots, \lambda^{L-1}\right] \tag{11.131}$$

which is the same as (11.125). Consequently every realization of η is smooth in precisely a first order fashion. This is a highly structured description of the undermodeling compared with the diagonal assumptions on $C_\eta(\beta)$. The parameterization (11.125) corresponds for example to the assumption that 68,3% (1 standard deviation) of the undermodeling impulse responses will fall within the rectangle shown in Figure 11.3 for $\alpha = 5$, $\lambda = 0.9$ and $L = 25$.

The corresponding initial undermodeling estimate will then be given by:

$$E\left\{\left|G_\Delta\left(e^{j\omega T_s}\right)\right|^2\right\} = \Pi\left(e^{j\omega T_s}\right) C_\eta(\beta) \Pi^T\left(e^{j\omega T_s}\right) \tag{11.132}$$

which is, in general, not frequency domain stationary. Notice that we have not increased the number of parameters.

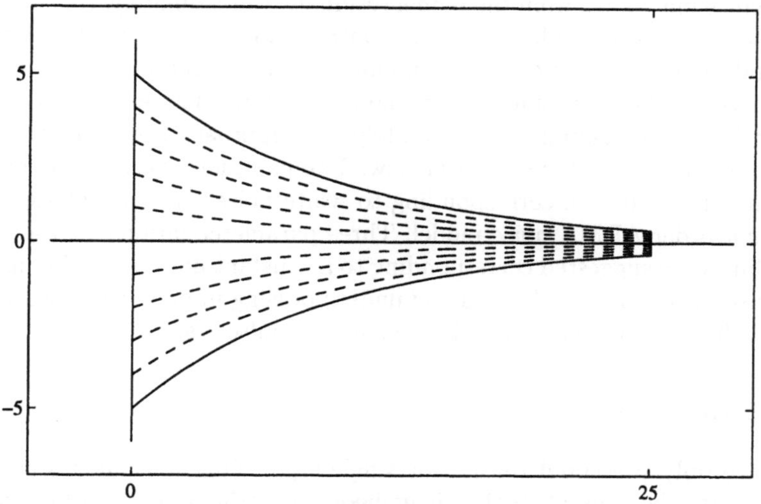

Fig. 11.3. *A first order decaying assumption on the undermodeling impulse response. Shown are the one standard deviation borders for* $\alpha = 5$, $\lambda = 0.9$ *and* $L = 25$. *The impulse response is smooth in precisely a first order fashion.*

11.3.3 Combined Covariance Structures

The different structures on the noise and undermodeling covariances may then be combined to provide 6 different parameterizations of the covariance matrices $C_\nu(\gamma)$ and $C_\eta(\beta)$.

11.4 Summary

The stochastic embedding approach for estimation of frequency domain model uncertainty was introduced. The main paradigm for the stochastic embedding approach is to assume that the undermodeling is a realization of a stochastic process. With this "technical" assumption we may extend the classical results on fixed denominator modeling to the case where no exact description of the true system exists within the model set. The expression for the parameter covariance matrix then contains two contributions: one due to the noise in the data set and one due to the undermodeling. We may furthermore derive an explicit expression for the total model error which nicely separates bias and variance contributions. In the expression for the frequency domain error bounds, two covariance terms appear; one for the noise and one for the undermodeling impulse response. An estimate of these two covariances must be obtained in order to evaluate the frequency domain expression. Remember that for the classical fixed denominator approach we could quite easily obtain a non-parametric estimate of the noise covariance matrix C_ν from the model residuals. Unfortunately, this is not the case now since the residuals contain contributions both from the noise and the undermodeling. Instead a maximum likelihood approach is taken. If we parameterize the covariances, then the parameters themselves may be estimated from the residuals. Unfortunately, this approach is only tractable for small data series, see the remarks below. Two parameterizations for the noise were investigated; one corresponding to white noise and one corresponding to a first order ARMA noise model. Three parameterizations for the undermodeling were suggested corresponding to a constant undermodeling impulse response, an exponentially decaying undermodeling impulse response and, finally, a first order decaying undermodeling impulse response.

11.4.1 Remarks

We have only presented results for single-input single-output systems. We believe that the general stochastic embedding methodology can be straightforwardly extended to multivariable systems. However, the parameterization of the noise and undermodeling covariances and the maximum likelihood estimation of the parameters do not seem to readily fit into a multivariable framework.

The maximum likelihood estimation is also a severe bottleneck for large data series. The approach outlined here is only tractable for rather small data series (less than 300 measurements) because evaluation of the $N-p$ dimensional Gaussian distribution for the transformed residuals ϖ involves inversion of $N \times N$ matrices.

CHAPTER 12

ESTIMATING UNCERTAINTY USING STOCHASTIC EMBEDDING

In this chapter we will illustrate the stochastic embedding approach through a numerical example. We will estimate a second order Laguerre model from data generated by a third order system corrupted by white noise. Different parameterizations of the noise and undermodeling will then be investigated using montecarlo experiments.

12.1 The True System

Suppose that the true system is given by:

$$G_{\mathrm{T}}(s) = \frac{458}{(s+1)(s^2+30s+229)} . \tag{12.1}$$

$G_{\mathrm{T}}(s)$ is taken from the well-known paper by Rohrs *et al.* [RVAS85]. The system was sampled with sampling frequency 10 [Hz], the input sequence was a 0.4 [Hz] fundamental square wave and the output were corrupted by the white noise sequence:

$$\nu(k) \sim \mathcal{N}(0, 0.05) . \tag{12.2}$$

200 samples were collected, 125 of which were used to get rid of initial condition effects. The last 75 samples were used for estimation. It is thus assumed that only a very limited data set is available for the identification procedure. A second order Laguerre model:

$$G(q,\theta) = \frac{\theta_1 q^{-1}}{1+\xi q^{-1}} + \frac{\theta_2 q^{-1}\left(1-\xi^{-1}q^{-1}\right)}{\left(1+\xi q^{-1}\right)^2} \tag{12.3}$$

was fitted to the data. The Laguerre pole $-\xi$ was chosen as 0.85. The order of the FIR model for the undermodeling $G_{\Delta}(q)$ was chosen as $L = 20$.

The discrete-time equivalent of the true system assuming zero order holds on the input is:

$$G_{\mathrm{T}}(q) = \frac{0.0370q^{-1} + 0.0717q^{-2} + 0.00785q^{-3}}{1 - 1.34q^{-1} + 0.446q^{-2} - 0.0450q^{-3}} . \tag{12.4}$$

The discrete-time poles of the true system are consequently:

$$p_1 = 0.905 \tag{12.5}$$
$$p_2 = 0.219 + 0.0443j \tag{12.6}$$
$$p_2 = 0.219 - 0.0443j\,. \tag{12.7}$$

The Laguerre pole is thus slightly faster than the dominant pole of the true system as usually recommended for Laguerre models, see Section 10.3.3 on page 163.

The least squares estimate of θ averaging over 100.000 realizations of the noise was found as:

$$\theta^* \approx \hat{\theta}_{N,av} = \begin{bmatrix} 0.1079 \\ 0.0138 \end{bmatrix}. \tag{12.8}$$

The estimated covariance matrix was

$$P_\theta = E\left\{(\hat{\theta}_N - \theta^*)(\hat{\theta}_N - \theta^*)^T\right\} \tag{12.9}$$
$$= 10^{-4} \cdot \begin{bmatrix} 0.7366 & -0.0502 \\ -0.0502 & 0.0096 \end{bmatrix}. \tag{12.10}$$

This corresponds nicely with the estimated covariance matrix for θ, given a fixed denominator model structure. Given the results in Section 10.3.1 on page 159, $\hat{P}_N$ was found as

$$\hat{P}_N = \Omega C_\nu \Omega^T \tag{12.11}$$
$$= 10^{-4} \cdot \begin{bmatrix} 0.7419 & -0.0505 \\ -0.0505 & 0.0096 \end{bmatrix}. \tag{12.12}$$

Furthermore, Ω is deterministic such that the covariance estimate (12.12) applies for any realizations of the noise. Note that $\hat{P}_N$ is estimated from just 75 samples. Of course, we are cheating here since we assume that we know C_ν. However, we can use $\hat{P}_N$ to evaluate the frequency domain uncertainty bounds obtained with a classic approach.

The distributions of the θ-estimate is shown in Figure 12.1. As seen the distributions correspond very nicely to the standard normal distribution as predicted by (9.33).

12.2 Error Bounds with a Classical Approach

Given a realization of the noise, we may now compute frequency domain uncertainty ellipses around the estimated Nyquist. Since $G(q,\theta)$ is linear in θ we may write

$$G(q,\theta) = \Lambda(q)\theta \tag{12.13}$$

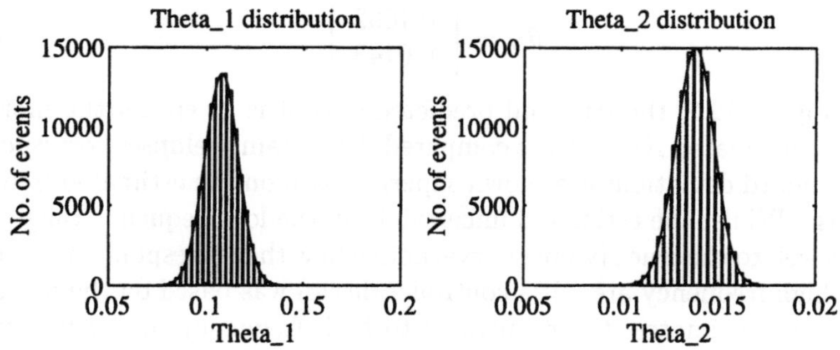

Fig. 12.1. *Distribution of Laguerre model parameter estimates. Also shown are scaled normal distributions with computed mean and standard deviation.*

where $\Lambda(q)$ for the present example is given by

$$\Lambda(q) = \left[\begin{array}{cc} \dfrac{q^{-1}}{1+\xi q^{-1}} & \dfrac{q^{-1}(1-\xi^{-1}q^{-1})}{1+2\xi q^{-1}+\xi^2 q^{-2}} \end{array} \right] . \tag{12.14}$$

Then $\tilde{g}(e^{j\omega T_s})$, defined as before by

$$\tilde{g}(e^{j\omega T_s}) = \left[\begin{array}{c} \Re e \left\{ G\left(e^{j\omega T_s}, \theta^*\right) - G\left(e^{j\omega T_s}, \hat{\theta}_N\right) \right\} \\ \Im m \left\{ G\left(e^{j\omega T_s}, \theta^*\right) - G\left(e^{j\omega T_s}, \hat{\theta}_N\right) \right\} \end{array} \right] , \tag{12.15}$$

will be given by

$$\tilde{g}(e^{j\omega T_s}) = \Lambda(e^{j\omega T_s}) \left(\theta^* - \hat{\theta}_N \right) . \tag{12.16}$$

Under our standard assumptions, $\tilde{g}(e^{j\omega T_s})$ will be normal distributed with covariance $P_{\tilde{g}}(\omega)$:

$$\tilde{g}(e^{j\omega T_s}) \in \mathcal{N}(0, P_{\tilde{g}}(\omega)) \tag{12.17}$$

where $P_{\tilde{g}}(\omega)$ is given by

$$P_{\tilde{g}}(\omega) = \Lambda(e^{j\omega T_s}) P_\theta \Lambda^T(e^{j\omega T_s}) . \tag{12.18}$$

Thus the scalar

$$z(\omega) = \tilde{g}^T(e^{j\omega T_s}) \left[\Lambda(e^{j\omega T_s}) P_\theta \Lambda^T(e^{j\omega T_s}) \right]^{-1} \tilde{g}(e^{j\omega T_s}) \tag{12.19}$$

will have χ^2 distribution with 2 degrees of freedom and may be used to draw confidence ellipses in the complex plane for the frequency response estimate $G(e^{j\omega T_s}, \hat{\theta}_N)$. Notice, however, that $G(q, \theta^*)$ does not represent the true system $G_T(q)$ due to undermodeling. Thus we will expect the estimated error bounds to be inaccurate.

Given a particular realization of the noise $\nu(k)$, the least squares estimate for θ was found as

$$\hat{\theta}_N = \begin{bmatrix} 0.1092 \\ 0.0143 \end{bmatrix} . \tag{12.20}$$

In Figure 12.2, the true and predicted output is given. Furthermore, the true and estimated Nyquist are compared. Uncertainty ellipses corresponding to 2 standard deviations are shown superimposed on the estimated frequency response. Whilst the estimated uncertainty in the low frequency range corresponds well to the true Nyquist curve, notice how the correspondence vanishes in the high frequency area. If a controller design was based on the illustrated uncertainty estimates, the robustness to high frequency model uncertainty could very well be quite small.

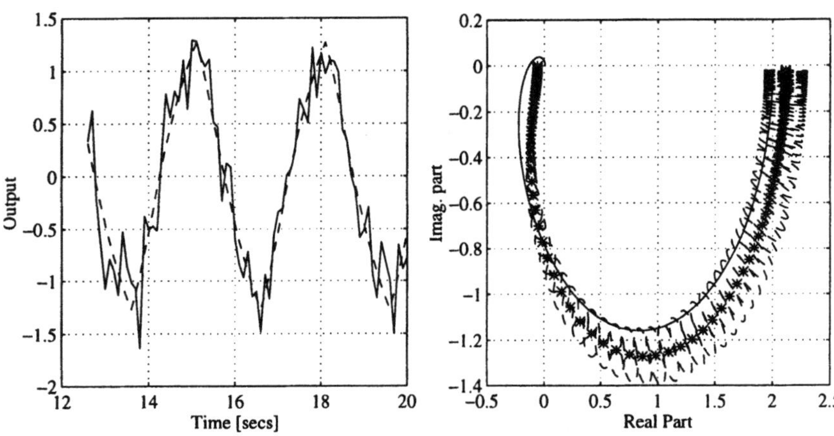

Fig. 12.2. *Results from classical identification on Rohrs Counterexample. On the left are true and estimated output and on the right are true and estimated Nyquist with error bounds.*

In the next section we will use stochastic embedding of the bias in order to improve the classical error bounds.

12.3 Error Bounds with Stochastic Embedding Approach

Estimating the total model error, the probability density functions for the noise and the undermodeling shall be identified. The parameterization of the noise was chosen as in Equation (11.112) on page 182 (white measurement noise). The 3 different parameterizations of the undermodeling presented in Section 11.3.2 on page 183 were investigated. The transformation matrix R was chosen through QR factorization of Φ as outlined in Section 11.2.3 on page 181. The loglikelihood function $\ell(\varpi|U,\zeta)$ was maximized using a Marquardt search algorithm.

12.3.1 Case 1, A Constant Undermodeling Impulse Response

First the undermodeling parameterization in Equation (11.117) on page 184 (a constant undermodeling impulse response variance) will be investigated. The parameter vector ζ is consequently given by:

$$\zeta = \begin{bmatrix} \alpha \\ \sigma_e^2 \end{bmatrix} . \tag{12.21}$$

1000 realizations of the process noise $\nu(k)$ were processed. 702 of the 1000 search procedures converged. The remainder mainly approached $[0, \hat{\sigma_e^2}]$ corresponding to zero uncertainty[1].

The convergent estimates of ζ is displayed in Figure 12.3 together with the corresponding maximums of $\ell(\varpi|U,\zeta)$.

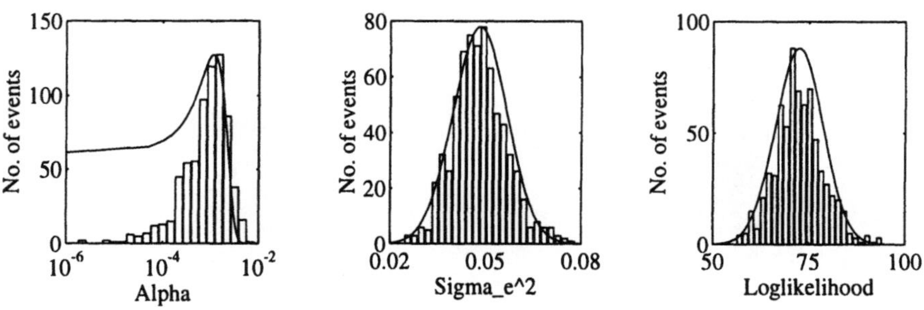

Fig. 12.3. *Case 1: Montecarlo testing of the maximum likelihood estimation of ζ. 1000 realizations of $\nu(k)$ were investigated, 702 converged. Shown are histograms for α, σ_e^2 and $\max\{\ell(\varpi|U,\zeta)\}$ together with the corresponding scaled normal distributions.*

The mean value and standard deviation on ζ was estimated as:

$$\hat{\zeta}_N = \begin{bmatrix} \hat{\alpha} \\ \hat{\sigma_e^2} \end{bmatrix} = \begin{bmatrix} 1.165 \cdot 10^{-3} \\ 4.816 \cdot 10^{-2} \end{bmatrix} \tag{12.22}$$

$$\hat{\sigma}_\zeta = \begin{bmatrix} 0.963 \cdot 10^{-3} \\ 0.851 \cdot 10^{-2} \end{bmatrix} . \tag{12.23}$$

In Figure 12.3, also scaled normal distributions corresponding to the estimated means and standard deviations are shown. We have not made any assumptions with regard to the distribution of ζ. However, if ζ is normal distributed, then we would expect the estimation procedure to be in some sense "well-conditioned". Notice that the α-estimate is not normal distributed

[1] Notice that $\alpha < 0$ corresponds to a negative definite covariance matrix for the undermodeling. Consequently a lower bound of 10^{-10} was enforced on α.

whereas both σ_e^2 and the loglikelihood $\ell(\varpi|U,\zeta)$ corresponds well to a normal distribution. Notice also that an accurate estimate of the noise variance $\sigma_e^2 = 0.05$ is obtained. The loglikelihood mean was found to be:

$$\mu_{\max\{\ell(\varpi|U,\zeta)\}} = 72.43 . \tag{12.24}$$

Now let us consider a particular realization of the noise. We will use the same realization as in Section 12.2. The corresponding estimate of ζ was found as:

$$\hat{\zeta} = \begin{bmatrix} \hat{\alpha} \\ \hat{\sigma_e^2} \end{bmatrix} = \begin{bmatrix} 2.365 \cdot 10^{-3} \\ 5.830 \cdot 10^{-2} \end{bmatrix} . \tag{12.25}$$

Notice that this realization is representative with respect to ζ, see Figure 12.3.

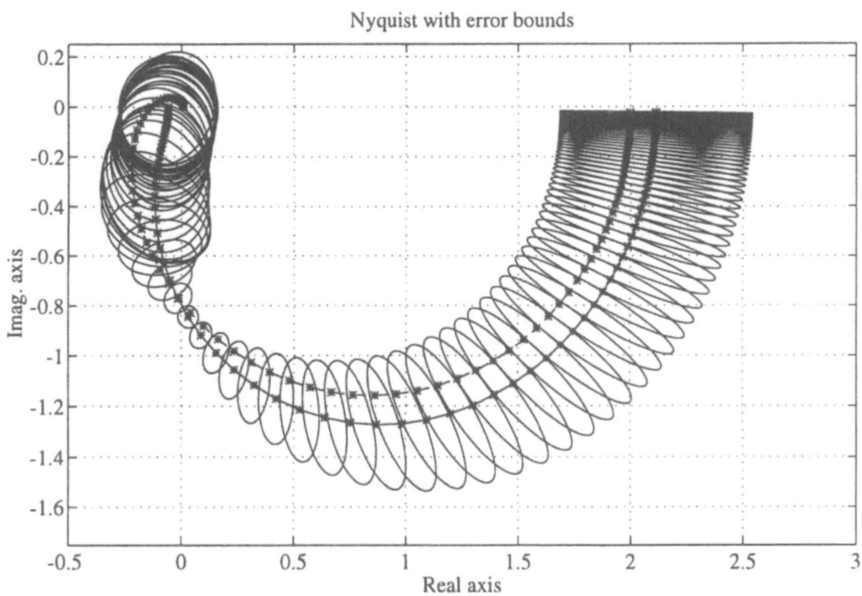

Fig. 12.4. *Case 1: Nominal Nyquist (solid) with 1 standard deviation error bounds. Also shown is the true Nyquist (dashed). Each frequency point is marked (*) to provide insight in the quality of the uncertainty estimates.*

The results of the uncertainty estimation is displayed in Figure 12.4 and 12.5. Notice from Figure 12.4 that a reasonable uncertainty estimate is obtained through this simple parameterization of the noise and undermodeling. The uncertainty estimate is slightly conservative and with little phase information in the high frequency area. Small uncertainty is correctly predicted around the fundamental frequency of the square wave input. In

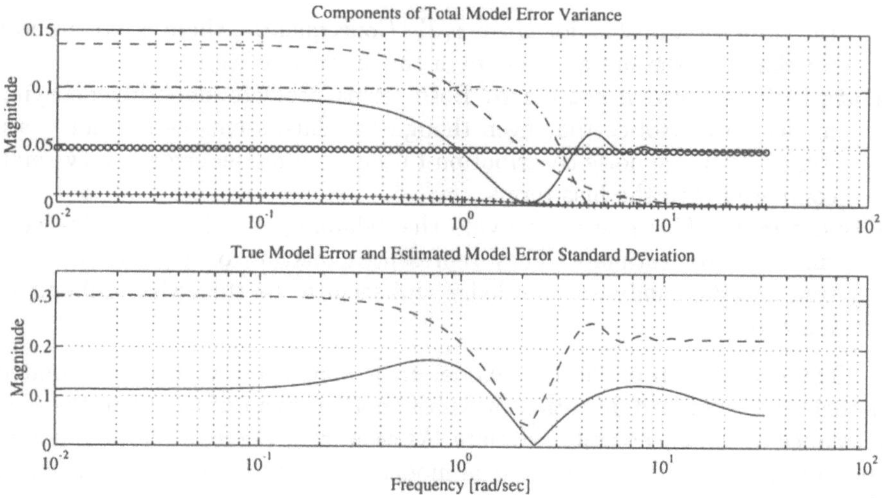

Fig. 12.5. *Case 1: Total model error components and comparison with true model error. Upper: The total model error (solid), a priori estimate (o), data induced term (dashed), cross terms (dash-dotted) and noise term (+). Lower: True model error (solid) and estimated total model error standard deviation (dashed).*

Figure 12.5, the different contributions to the error estimate is shown. Recall from Equation (11.86) on page 177 that the total model error estimate is given by:

$$
\begin{aligned}
E\left\{\left|G\left(e^{j\omega T_s},\hat{\theta}_N\right)-G_{\mathrm{T}}\left(e^{j\omega T_s}\right)\right|^2\right\} \\
&= (\Pi-\Lambda\Omega X)\,C_\eta\,(\Pi-\Lambda\Omega X)^* + \Lambda\Omega C_\nu \Omega^T\Lambda^* \qquad (12.26)\\
&= \Pi C_\eta \Pi^* + \Lambda\Omega X C_\eta X^T\Omega^T\Lambda^* - \Lambda\Omega X C_\eta \Pi^* - \\
&\quad \Pi C_\eta X^T\Omega^T\Lambda^* + \Lambda\Omega C_\nu \Omega^T\Lambda^* \,. \qquad (12.27)
\end{aligned}
$$

The components of (12.27) will be denoted as follows:

- A priori estimate: $\Pi C_\eta \Pi^*$.
- Data induced term: $\Lambda\Omega X C_\eta X^T\Omega^T\Lambda^*$.
- Cross terms: $\Lambda\Omega X C_\eta \Pi^* + \Pi C_\eta X^T\Omega^T\Lambda^*$.
- Noise term: $\Lambda\Omega C_\nu \Omega^T\Lambda^*$.

From Figure 12.5 the following observations may be made:

A priori estimate. The frequency response of the a priori estimate is independent of frequency. This is due to the simple parameterization of the undermodeling, see Section 11.3.2 on page 184. The total model error estimate converges to the a priori estimate as the frequency ω approaches the Nyquist frequency π/T_s. This is reasonable since the information in the data-set vanishes for $\omega \to \pi/T_s$. Consequently, there is no support for changing the initial estimate in this frequency region.

Data induced and noise term. Both terms predicts smoothly decaying model error. Notice that even though the data seems rather noisy, see Figure 12.2, the noise contribution to the total model error is very small compared with the bias terms.

Cross terms. These terms provide the "shaping" of the otherwise very smooth error estimate. They represent a mixture of a priori information and data induced knowledge and seem to be somewhat difficult to interpret.

The comparison of the true model error magnitude with the estimated model error standard deviation provides another mean of assessing the performance of the uncertainty estimate. As seen from Figure 12.5 there is reasonable agreement between the true model error and the estimate although the error estimate is somewhat conservative.

12.3.2 Case 2: An Exponentially Decaying Undermodeling Impulse Response

Now the undermodeling parameterization in Equation (11.121) on page 185 (an exponentially decaying undermodeling impulse response variance) will be investigated. The parameter vector ζ is consequently given by:

$$\zeta = \begin{bmatrix} \alpha \\ \lambda \\ \sigma_e^2 \end{bmatrix} . \tag{12.28}$$

Again, 1000 realizations of the noise $\nu(k)$ were processed. This time, however, only 429 converged. The convergence "hit-rate" for this undermodeling parameterization is consequently considerably lower than for the constant undermodeling impulse response parameterization examined in the previous section. 241 realizations did not converge due to the lower bound on α, see footnote (1) on page 193. The remaining realizations diverged.

The convergent estimates of ζ is displayed in Figure 12.6 together with the corresponding maximums of $\ell(\varpi|U,\zeta)\}$. The mean value and standard deviation on ζ were estimated as:

$$\hat{\zeta}_N = \begin{bmatrix} \hat{\alpha} \\ \hat{\lambda} \\ \hat{\sigma_e^2} \end{bmatrix} = \begin{bmatrix} 4.830 \cdot 10^{-2} \\ 8.581 \cdot 10^{-1} \\ 4.744 \cdot 10^{-2} \end{bmatrix} \tag{12.29}$$

$$\hat{\sigma}_\zeta = \begin{bmatrix} 41.12 \cdot 10^{-2} \\ 3.336 \cdot 10^{-1} \\ 0.824 \cdot 10^{-2} \end{bmatrix} . \tag{12.30}$$

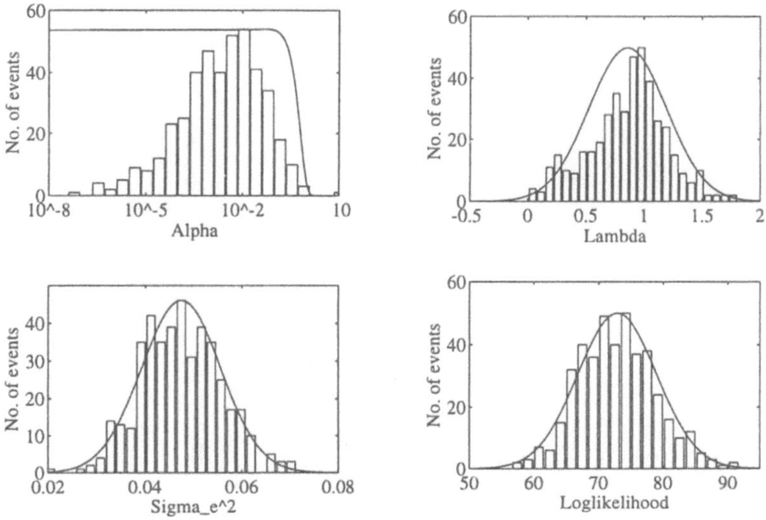

Fig. 12.6. *Case 2: Montecarlo testing of the maximum likelihood estimation of ζ. 1000 realizations of $\nu(k)$ were investigated, 429 converged. Shown are histograms for α, λ, σ_e^2 and $\max\{\ell(\varpi|U,\zeta)\}$ together with the corresponding scaled normal distributions.*

The corresponding normal distributions are also shown in Figure 12.6. Notice that approximately half of the realizations converged to $\lambda > 1$ corresponding to an exponentially rising undermodeling impulse response. This is equivalent to an unstable transfer function $G_\Delta(q)$ for an infinite impulse sequence. However, since only 20 samples of the impulse response is used as approximation for $G_\Delta(q)$ this is generally not the case here.

Notice that the estimate of the noise variance σ_e^2 again corresponds nicely with the true value. The loglikelihood mean was found to be:

$$\mu_{\max\{\ell(\varpi|U,\zeta)\}} = 72.91 \tag{12.31}$$

that is, slightly better than for the undermodeling parameterization in Section 12.3.1.

We now turn to the particular realization of the process noise introduced above. The corresponding estimate of ζ was determined as:

$$\hat{\zeta} = \begin{bmatrix} \hat{\alpha} \\ \hat{\lambda} \\ \hat{\sigma_e^2} \end{bmatrix} = \begin{bmatrix} 2.282 \cdot 10^{-3} \\ 1.0035 \\ 5.830 \cdot 10^{-2} \end{bmatrix} . \tag{12.32}$$

The results of the uncertainty estimation is displayed in Figure 12.7 and 12.8. Notice that the uncertainty estimate presented in Figure 12.7 is almost exactly identical to the estimate in Section 12.3.1, see Figure 12.4. This is not surprising since $\lambda \approx 1$ and the estimates of α and σ_e^2 are very similar, compare Equation (12.32) and (12.25).

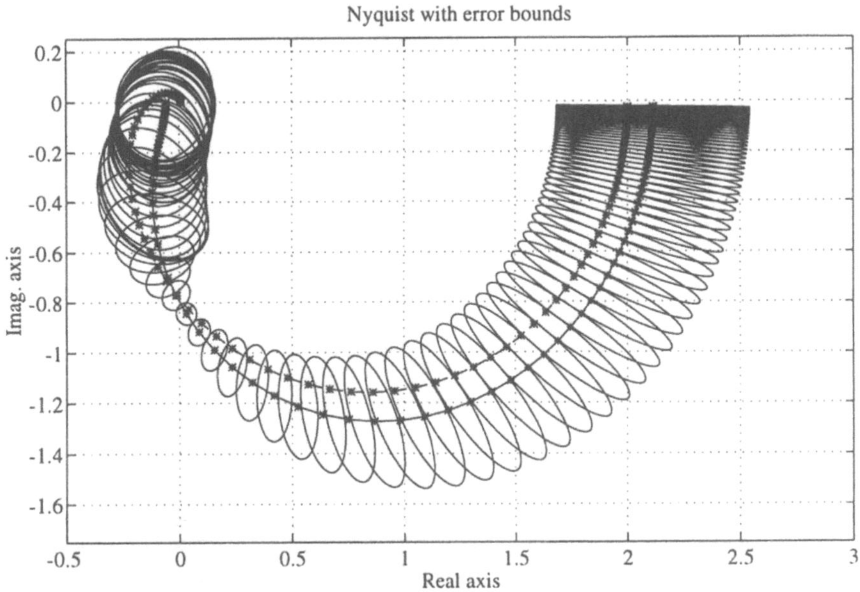

Fig. 12.7. *Case 2: Nominal Nyquist (solid) with 1 standard deviation error bounds. Also shown is the true Nyquist (dashed). Each frequency point is marked (*) to provide insight in the quality of the uncertainty estimates.*

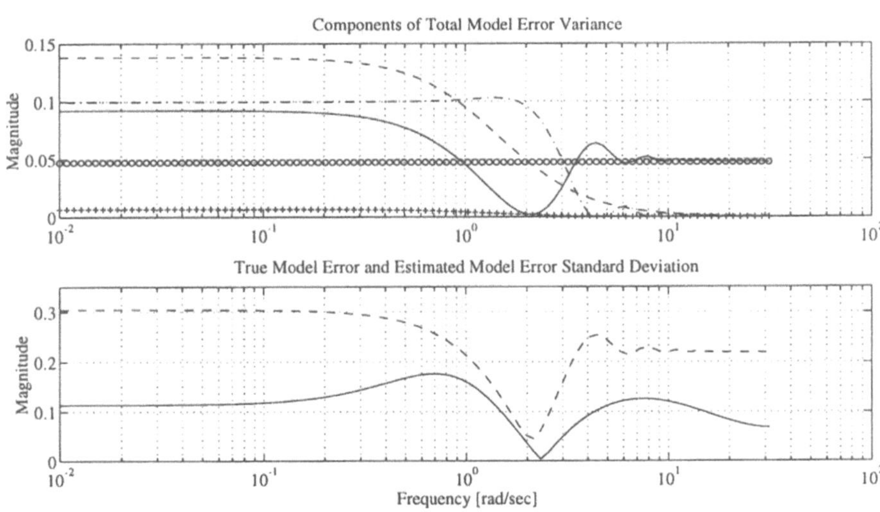

Fig. 12.8. *Case 2: Total model error components and comparison with true model error. Upper: The total model error (solid), a priori estimate (o), data induced term (dashed), cross terms (dash-dotted) and noise term (+). Lower: True model error (solid) and estimated total model error standard deviation (dashed).*

The different components of the total model error, see Figure 12.8, are also almost identical with the results from Section 12.3.1 and hence the conclusions presented there also apply for the current parameterization of the undermodeling. The additional parameter λ introduced in the exponentially decaying parameterization of the undermodeling consequently is of little use in the current example. Furthermore, the convergence rate of exponentially decaying parameterization of the undermodeling was rather poor compared with the simple constant parameterization introduced in Case 1.

12.3.3 Case 3: A First Order Decaying Undermodeling Impulse Response.

Finally, the undermodeling parameterization (11.125) on page 186 (a first order decaying undermodeling impulse response) will be investigated. The parameter vector is still given by (12.28) and again 1000 realizations of the noise were processed. This time 773 of the realizations converged; the highest hit-rate of the three different undermodeling parameterizations. The remaining maximizations collided with the lower bound on α.

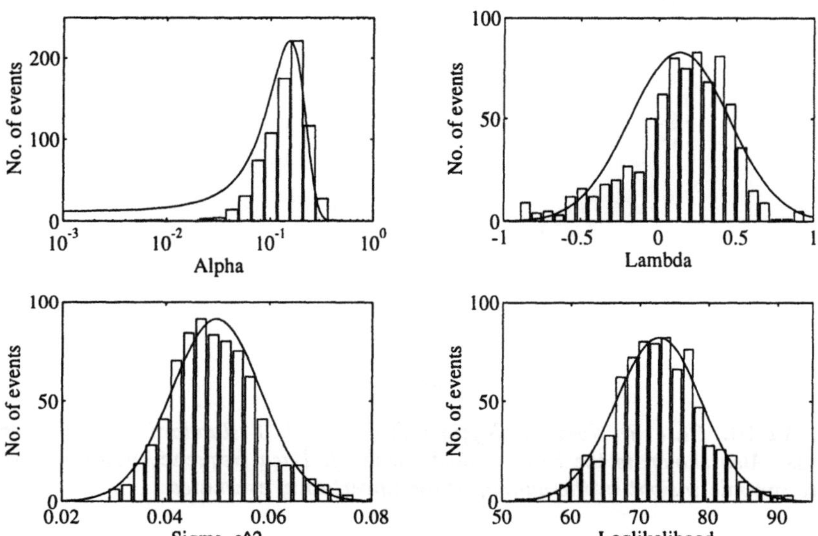

Fig. 12.9. *Case 3: Montecarlo testing of the maximum likelihood estimation of ζ. 1000 realizations of $\nu(k)$ were investigated, 773 converged. Shown are histograms for α, λ, σ_e^2 and $\max\{\ell(\varpi|U,\zeta)\}$ together with the corresponding scaled normal distributions.*

The resulting estimates of ζ is displayed in Figure 12.9 together with the achieved loglikelihood of ϖ and the computed normal distributions. The mean and standard deviation of ζ were found as:

$$\hat{\zeta}_N = \begin{bmatrix} \hat{\alpha} \\ \hat{\lambda} \\ \hat{\sigma}_e^2 \end{bmatrix} = \begin{bmatrix} 1.533 \cdot 10^{-1} \\ 1.357 \cdot 10^{-1} \\ 4.979 \cdot 10^{-2} \end{bmatrix} \tag{12.33}$$

$$\hat{\sigma}_\zeta = \begin{bmatrix} 0.629 \cdot 10^{-1} \\ 3.193 \cdot 10^{-1} \\ 0.861 \cdot 10^{-2} \end{bmatrix}. \tag{12.34}$$

Notice that the spread of λ is rather large, specifically λ takes negative values corresponding to that neighbor points in the undermodeling impulse response have different signs. This is a somewhat odd interpretation, but does not violate our identification scheme since the covariance $C_\eta(\alpha, \lambda)$ is still positive definite.

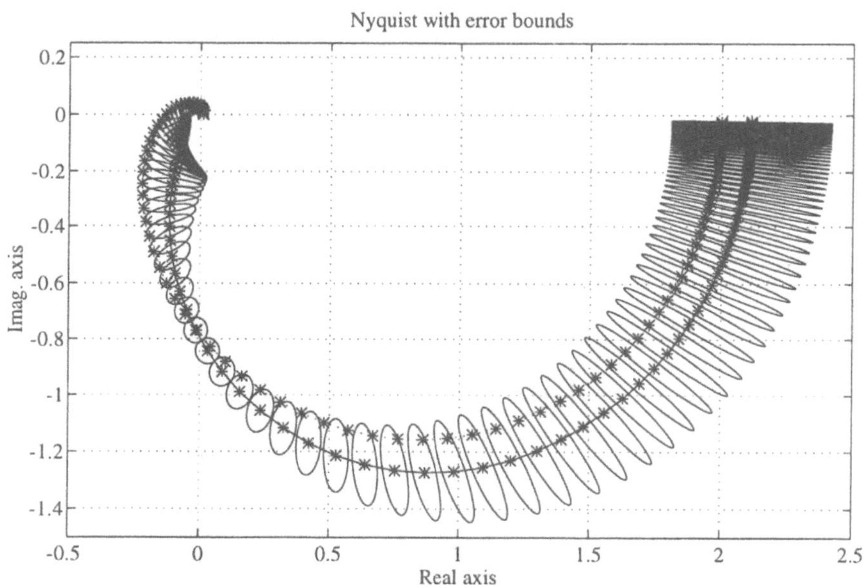

Fig. 12.10. *Case 3: Nominal Nyquist (solid) with 1 standard deviation error bounds. Also shown is the true Nyquist (dashed). Each frequency point is marked (*) to provide insight in the quality of the uncertainty estimates.*

Again a fine estimate of the noise covariance is obtained. The mean value of the obtainable loglikelihood was:

$$\mu_{\max\{\ell(\varpi|U,\zeta)\}} = 72.75 \tag{12.35}$$

that is, close to the means for the two other undermodeling parameterizations. It may consequently be concluded that level of the performance surfaces for the different parameterizations of the undermodeling are quite similar. For a given input-output realization, a considerable difference in the obtainable loglikelihood may be used to distinguish and classify the goodness of different noise and undermodeling parameterizations.

The results of the uncertainty estimate for the particular realization discussed before is displayed in Figure 12.10 and 12.11. The corresponding estimate of ζ was found to be:

$$\hat{\zeta} = \begin{bmatrix} \hat{\alpha} \\ \hat{\lambda} \\ \hat{\sigma_e^2} \end{bmatrix} = \begin{bmatrix} 1.182 \cdot 10^{-1} \\ 2.523 \cdot 10^{-1} \\ 6.769 \cdot 10^{-2} \end{bmatrix} . \tag{12.36}$$

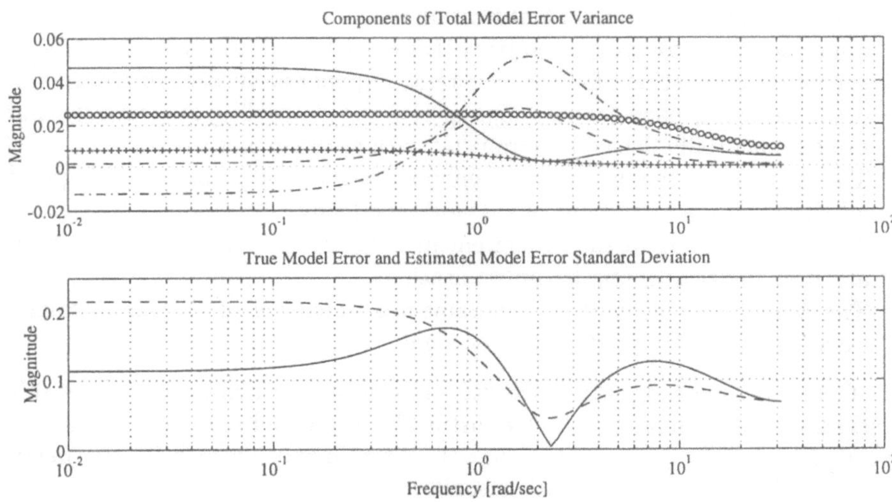

Fig. 12.11. *Case 3: Total model error components and comparison with true model error. Upper: The total model error (solid), a priori estimate (o), data induced term (dashed), cross terms (dash-dotted) and noise term (+). Lower: True model error (solid) and estimated total model error standard deviation (dashed).*

Now the uncertainty estimate is considerably improved compared with the previous estimates. It is less conservative and the phase information of the error estimates have substantially improved. Furthermore, notice from Figure 12.11 that the a priori estimate $\Pi C_\eta \Pi$ is now decaying with frequency. The data induced term $\Lambda \Omega X C_\eta X^T \Omega^T \Lambda^*$ is concentrated around the fundamental frequency of the input square wave, where the information in the data is large. The noise contribution is of approximately similar size as in connection with the previous undermodeling parameterizations. However, the noise part of the total model error is now greater since the uncertainty estimate is less conservative.

12.4 Summary

Estimation of model uncertainty was performed for a second order Laguerre model applied in connection with a third order true system. From the above presented results the following conclusions were drawn:

- The estimated uncertainty regions using a classical approach and a second order Laguerre representation did not capture the true model uncertainty in the high frequency area.
- The estimated uncertainty regions using the stochastic embedding approach gave a fair indication of the model uncertainty, especially for the first order undermodeling impulse response parameterization.
- The maximum likelihood identification of the parameters in the noise and undermodeling covariance matrices converged for 70%, 43% and 77% of the noise realizations for the constant, exponentially decaying and first order decaying parameterizations of the undermodeling respectively. For the exponentially decaying parameterization, this is not satisfactory.
- The variance of the maximum likelihood estimate of ζ was rather high for all three parameterizations of the undermodeling.
- The covariance σ_e^2 of the measurement noise was accurately predicted for all three undermodeling parameterizations.
- The uncertainty around the fundamental frequency of the input square wave was correctly predicted to be small.

Generally speaking, the stochastic embedding method produced reliable uncertainty estimates, but the "robustness"[2] of the method seems somewhat poor. The problem is probably that the performance surface for the maximum likelihood estimate is rather shallow due to the small number of measurements. As noted in Section 11.4.1 on page 188 the maximum likelihood procedure unfortunately becomes intractable for large data series.

[2] By robustness we here mean the probability of finding the global maximum in the maximum likelihood search for ζ.

PART III

A SYNERGISTIC CONTROL SYSTEMS DESIGN PHILOSOPHY

CHAPTER 13

COMBINING SYSTEM IDENTIFICATION AND ROBUST CONTROL

So far we have mainly considered robust control and system identification independently. This reflects the traditional segregation of the two fields within the automatic control community. In this part of the book we will suggest various schemes which combines the results from robust control presented in Part I with the results from system identification presented in Part II. The presentation will fall naturally into three parts. Section 13.1 summarizes the presented results on estimation of frequency domain model error bounds. Section 13.2 discusses how frequency domain uncertainty ellipses may be represented in a robust control framework and finally Section 13.3 addresses the question of posing an identification based robust control design strategy; the main "message" in this book. In the next chapter we will provide an example to illustrate the proposed design philosophy.

13.1 System Identification for Robust Control

When we use system identification for robust control we generally have two aims. The first aim, of course, is to obtain a useful description of the system. By the term useful we emphasize that this description is aimed at subsequent use for control design. Thus it need not be a very accurate detailed model which is valid throughout the entire physical envelope of the system. Rather it should a reasonable simple model which is valid (that is, describes the real system as accurately as "necessary") in the vicinity of the normal operating point of the system. In particular, it is important to obtain a model which is accurate in the frequency region around the desired closed loop bandwidth. Thus, if we have the freedom of designing the input signal used for the identification experiment, then the energy in the input signal should be concentrated around this frequency range. A very popular input signal for system identification is a fundamental square wave. Hence, if such a signal is applied for an identification experiment, the fundamental frequency ω_{sq} should be chosen close to the desired closed loop bandwidth. Experiment design for control purposes have received considerable attention over the last decade. Similar effects may be obtained by pre-filtering of the data-set prior to the identification step. Zang *et al.* [ZBG91] have, for example, proposed an iterative scheme of weighted least squares identification and LQ control design which obtain that the model uncertainty is minimized precisely in the frequency region where it is demanded to optimize the control design.

The second aim using system identification for robust control is to get an estimate of the model quality. In fact, in many practical control designs we would rather have a simple low-order model together with a quantification of the involved model uncertainty than a high-order very accurate model with very little uncertainty. The reason for this is that the order of the controller, using most modern control design methods, equals at least that of the generalized plant. Implementation of high-order controllers can be quite difficult, eg due to numerics, available memory etc. Since we may often overbound the uncertainty of a simple model with a simple first or second order perturbation model, we can usually obtain low-order controllers for simple plant models.

Increasing performance demands will necessitate decreasing uncertainty levels and thus more accurate descriptions of the plant. Consequently, it is in some sense the desired closed loop performance which determine the degree of accurateness with which we must know the plant. The robust control approach provides the designer with the opportunity of weighing out robustness and performance demands.

13.1.1 Bias and Variance Errors

Let us recapitulate what we mean by bias errors and variance errors. Decompose, at each frequency point, the total model error as

$$G_{\mathrm{T}}(e^{j\omega T_s}) - G(e^{j\omega T_s}, \hat{\theta}_N) = G_{\mathrm{T}}(e^{j\omega T_s}) - G(e^{j\omega T_s}, \theta^*) + G(e^{j\omega T_s}, \theta^*) - G(e^{j\omega T_s}, \hat{\theta}_N) \quad (13.1)$$

where θ^* denotes the asymptotic (for $N \to \infty$) estimate of θ as before. The first contribution $G_{\mathrm{T}}(e^{j\omega T_s}) - G(e^{j\omega T_s}, \theta^*)$ is denoted the *bias error*. The bias error is a deterministic quantity. Clearly, it is non-zero, at least at some frequencies, unless there exists a parameter vector θ^* such that $G_{\mathrm{T}}(q) = G(q, \theta^*)$. If so $\theta_0 = \theta^*$ is denoted the true parameter vector. The second contribution $G(e^{j\omega T_s}, \theta^*) - G(e^{j\omega T_s}, \hat{\theta}_N)$ is denoted the *variance error*. The variance error is a random variable with respect to the probability space of the noise distribution. It vanishes when there is no noise or when the number of data tends to infinity.

13.1.2 What Can We Do with Classical Techniques

In Table 13.1 we have summarized some of the results on classical system identification presented in Chapter 9 and 10.

There is a few points we would like to emphasize:

- Local minima in the performance surface $V_N(\theta, Z^N)$ can be a severe problem if the parameter estimate cannot be found analytically, eg for the

Table 13.1. *Some properties of classical system identification techniques.*

	Model structure			
	PEM	ARX	OE	Fixed denominator
Predictor	non-linear in θ	linear in θ	non-linear in θ	linear in θ
Least squares estimate	numeric	analytic	numeric	analytic
Regressors $\phi(k, \theta)$	stochastic	stochastic	deterministic	deterministic
Model gradient $\psi(k, \theta)$	stochastic	stochastic	deterministic	deterministic
$\hat{P}_N$ consistent if	$G_{\mathrm{T}}(q) = G(q, \theta_0)$ $H_{\mathrm{T}}(q) = H(q, \theta_0)$	$G_{\mathrm{T}}(q) = G(q, \theta_0)$ $H_{\mathrm{T}}(q) = H(q, \theta_0)$	$G_{\mathrm{T}}(q) = G(q, \theta_0)$	always
Expression for $\tilde{g}(e^{j\omega T_s})$	1st order Taylor	1st order Taylor	1st order Taylor	exact
Model validation	residuals $\epsilon(k, \hat{\theta}_N)$ = white noise ?	residuals $\epsilon(k, \hat{\theta}_N)$ = white noise ?	$u(k)$ and $\epsilon(k, \hat{\theta}_N)$ correlated ?	$u(k)$ and $\epsilon(k, \hat{\theta}_N)$ correlated ?

general PEM model and for output error (OE) models. This is, of course, especially true when the number of estimated parameters gets large. Various approaches can be made to circumvent this problem. The most obvious way is to repeat the minimization procedure (eg the Marquardt search) with different initial conditions. Alternatively, having determined a (local) minimum, we may step away from it again in such a way that the predictor is still stable and observe whether it will return to the same minimum. Different minima may of course be compared by the corresponding values of $V_N(\theta, Z^N)$.

- Clearly the data estimate, $\hat{P}_N$, of the parameter covariance matrix P_θ must be consistent, i.e converge for $N \to \infty$, if we are to obtain a proper estimate of the frequency domain model uncertainty $\tilde{g}(e^{j\omega T_s})$.
- To obtain a consistent estimate of the parameter covariance matrix P_θ for the general PEM structure, we must assume that both the deterministic and the stochastic part of the true system can be represented within our model set. Thus, we need a complete description of the true system. Of course, due to eg non-linear effects this is often (always) impossible in practice. However, using high-order descriptions we can usually obtain a quite accurate model at least around a specified working point. This, however, implies that many parameters must be estimated in the identification procedure. In order to perform this estimation with adequate accuracy, a large data set will be needed, usually several thousand measurements.
  We can check the model quality by white noise analysis of the residuals $\{\epsilon(k, \hat{\theta}_N)\}$. However, as illustrated by the wind-turbine example in Section 9.6, if we use high-order noise filters, some parts of the output $y(k)$ which are truly due to the deterministic part of the system $G_{\mathrm{T}}(q)$ may be explained by filtered white noise and thus represented in the noise filters. The residual tests were not capable of rejecting the hypothesis that the residuals were not white. This situation usually occur when the excitation in the input in the given frequency range is too small.
- The above applies for ARX models as well.
- With the output error model, a consistent estimate of the covariance matrix P_θ can be obtained if the true deterministic system $G_{\mathrm{T}}(q)$ can be represented within the model set $G(q, \theta)$. Thus we need not an exact description of the noise filter. The estimate for P_θ is then given by

$$\hat{P}_N = N \left(\Psi_N^T \Psi_N\right)^{-1} \Psi_N^T C_\nu \Psi_N \left(\Psi_N^T \Psi_N\right)^{-1} \tag{13.2}$$

where C_ν is the covariance matrix for the noise $\nu(k) = H_{\mathrm{T}}(q)e(k)$ and $\Psi_N = \Psi(\hat{\theta}_N, Z^N)$ is the model gradient matrix, see Section 10.2.1 on page 157. Assuming that $\{\nu(k)\}$ is a stationary stochastic process, we may estimate C_ν from the residuals as described on page 157. This makes the output error approach an appealing method for estimation of model uncertainty. One difficulty is that the parameter estimate $\hat{\theta}_N$ cannot be obtained analytically. Another problem might be that the uncertainty regions gets

too large since we are not trying to use knowledge of the noise filter $H(q)$ to reduce the error bounds. However, if knowledge of $H(q)$ *is* available, eg from an identification run with a general PEM model, we may fix $H(q)$ and perform an output-error identification ***with fixed noise model.*** In this way we may reduce the uncertainty levels without loosing the nice properties of the output-error approach.
We may use cross-correlation analysis between the input $\{u(k)\}$ and the residuals $\{\epsilon(k,\hat{\theta}_N)\}$ to evaluate whether the structure on $G(q,\theta)$ is sufficient to describe the real system.

– With a fixed denominator model structure, a consistent estimate of the parameter covariance P_θ can always be obtained since the estimate on page 159

$$\hat{P}_N = N\left(\Phi^T\Phi\right)^{-1}\Phi^T C_\nu \Phi\left(\Phi^T\Phi\right)^{-1} \tag{13.3}$$

will work *even in the case of undermodeling and with finite data.* Thus fixed denominator models seem an obvious choice for system identification for robust control. However, if $G_{\mathrm{T}}(q) \neq G(q,\theta^*)$ then the parameter covariance will only reflect the variance part of the total model error, see (13.1). Thus it must generally be assumed that the bias part of the error is zero; thus that the true deterministic system can be represented within the model set. From a traditional point of view this is unrealistic, even if the true system is stable and FDLTI, since we have a fixed denominator structure such that the poles of the system is determined prior to the identification procedure. However, if the basis functions $\Lambda_k(q)$, $k = 1,2,\cdots$ forms an orthonormal basis in $\mathcal{H}_2^*(\mathbf{C},\mathbf{C})$, by taking the order n of the model high enough we can make the error arbitrarily small.

13.1.3 The Stochastic Embedding Approach

The stochastic embedding approach is treated in Chapter 11. The key idea is to treat the bias error as if it was a stochastic process. This is done by decomposing the true system as

$$G_{\mathrm{T}}(q) = G(q,\theta_0) + G_\Delta(q) \tag{13.4}$$

and assuming that the undermodeling frequency response $G_\Delta(e^{j\omega T_s})$ is a realization of a stochastic process with zero mean

$$E\left\{G_\Delta(e^{j\omega T_s})\right\} = 0 \tag{13.5}$$

where $E\{\cdot\}$ means expectation over different realizations of the undermodeling. Then, if we use a fixed denominator model structure for $G(q,\theta)$, we may extend the results reviewed above on fixed denominator models to the situation where undermodeling is present. The parameter covariance then becomes

$$P_\theta = N\left(\Phi^T\Phi\right)^{-1}\Phi^T\left(XC_\eta X^T + C_\nu\right)\Phi\left(\Phi^T\Phi\right)^{-1} \tag{13.6}$$

where X is given by (11.26) and C_η is the covariance matrix for the undermodeling impulse response vector

$$C_\eta = E\left\{\eta\eta^T\right\} . \tag{13.7}$$

Compare (13.3) and (13.6). It is seen that the noise contribution to the parameter covariance for the stochastic embedding approach equals the classical covariance result for the fixed denominator structure. However, (13.6) also contains a contribution from the bias. Assuming that the undermodeling impulse response is zero mean Gaussian distributed, we may map the parameter covariance into the frequency domain using a similar approach to the classical one with due attention to extra terms, see Section 11.2.2 on page 180.

As for the classical fixed denominator case we then need an estimate of the covariance matrices for the undermodeling C_η and noise C_ν. Again we will use the residuals to estimate C_ν and C_η. However, since we cannot readily separate the bias and variance contributions, it becomes more complicated for the stochastic embedding approach. Fortunately, it turns out that if we parameterize C_η and C_ν we may estimate the parameters through maximum likelihood methods. The drawback of this approach is that it is only numerically efficient for small data series ($<$ 300 data points). Thus currently, the stochastic embedding approach is only tractable for small data series.

The stochastic embedding approach can be viewed as an extension of the classical results on fixed denominator model structures to the case where $G(q,\theta)$ cannot provide a correct description of the true system.

13.1.4 Proposed Approach

The approach we will propose for identification of frequency domain error bounds is twofold:

- If large data series are available, it is recommended that the classical output error approach is used. Since we have a large number of measurements, we may use quite high-order estimates $G(q,\theta)$ with reasonable little variance. A noise covariance estimate for computing the parameter covariance can be obtained from the residuals. Note that we will not obtain a parameterized model for the noise filter $H_{\mathrm{T}}(q)$ in this case. If the main purpose of the subsequent control design is disturbance attenuation it is sometimes desirable to include an input weight on the noise (this is the weight denoted W_{p1} in Section 4.2). The obvious choice for this weight is then (an approximation to) the estimated noise filter $H(q,\theta)$. In that case the general PEM approach or a combined PEM/OE approach as described above could be alternatives. If the identification experiment can be designed a priori, due attention must be taken to assure that the energy in the input is concentrated in the frequency range where small uncertainty is important in the subsequent control design, i.e typically around the desired closed loop bandwidth. Probably, identification and control design need to be iterated for optimal results.

- If only small data series are available, we recommend that the stochastic embedding approach is used to estimate the model error. Different parameterizations for the undermodeling and the noise covariances should be investigated. The above comments on input design is, of course, equally important in this case.

13.2 Robust Control from System Identification

Regardless of the system identification approach taken the result will be (for scalar systems) a nominal model and a set of frequency domain uncertainty ellipses around the nominal frequency response. The purpose of this section is to discuss how we may represent these uncertainty ellipses in a robust control framework. Generally, we will represent the uncertain system as

$$G_{\mathrm{T}}(z) = G(z, \hat{\theta}_N) + G_{\Delta}(z) \tag{13.8}$$

where $G_{\Delta}(z)$ is some unknown, but norm bounded transfer function for the model uncertainty. We will thus consider the robust stability problem with a norm bounded additive uncertainty. The set-up can be viewed as in Figure 13.1 where the problem then is to determine the most suitable structure on $\Delta(z)$ and $N(z)$ such that an adequate formulation of the control problem is obtained.

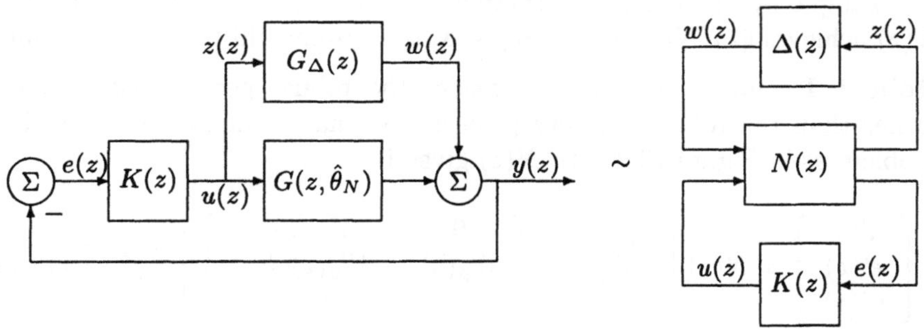

Fig. 13.1. *Approximating estimated frequency domain uncertainty ellipses with a norm bounded perturbation set.*

13.2.1 The $\mathcal{H}_\infty$ Approach

In the $\mathcal{H}_\infty$ solution to the robust stability problem it is assumed that $\Delta(z)$ is an unstructured full complex block bounded in norm by $\bar{\sigma}(\Delta(e^{j\omega T_s})) \leq 1$. This means that we can represent uncertainty circles in the complex plane. In particular let

$$G_\Delta(z) = W_\mathrm{u}(z)\Delta(z) \tag{13.9}$$

with

$$\Delta(z) = \delta^\mathrm{c}\,, \qquad \delta^\mathrm{c} \in \mathbf{C},\ |\delta^\mathrm{c}| \leq 1\,. \tag{13.10}$$

Then the uncertainty description will corresponds to frequency domain uncertainty circles with radius $|W_\mathrm{u}(e^{j\omega T_s})|$ at each frequency ω. Consequently the $\mathcal{H}_\infty$ approach could be outlined as follows. Given the estimated frequency domain uncertainty ellipses, approximate each ellipse with a circumscribed circle. Fit a stable transfer function $W_\mathrm{u}(z)$ to these circles such that

$$\left|W_\mathrm{u}\left(e^{j\omega T_s}\right)\right| \approx R(\omega) \tag{13.11}$$

where $R(\omega)$ denotes the radius of the circle evaluated at frequency ω. Construct the augmented system

$$N(z) = \begin{bmatrix} 0 & W_\mathrm{u}(z) \\ -1 & -G(z,\hat{\theta}_N) \end{bmatrix} \tag{13.12}$$

and solve the optimal robust stability problem

$$K(z) = \arg \min_{K(z)\in\mathcal{K}_\mathrm{S}} \|F_\ell(N(z),K(z))\|_{\mathcal{H}_\infty} \tag{13.13}$$

for example by bilinear transformation to continuous-time. If the controller $K(z)$ achieves $\|F_\ell(N(z),K(z))\|_{\mathcal{H}_\infty} < 1$, the system will be robustly stable.

Robust Performance. If we consider the robust performance problem rather than the robust stability problem, we may formulate a 2×2 block problem as in Figure 13.2 with $N(z)$ given by

$$\begin{bmatrix} u'(z) \\ e'(z) \\ u'(z) \end{bmatrix} = N(z)\begin{bmatrix} r(z) \\ u(z) \end{bmatrix} = \begin{bmatrix} 0 & W_\mathrm{u}(z) \\ W_\mathrm{p}(z) & -W_\mathrm{p}(z)G(z) \\ 1 & -G(z) \end{bmatrix}\begin{bmatrix} r(z) \\ u(z) \end{bmatrix}. \tag{13.14}$$

It is then easy to show that

$$\begin{bmatrix} e'(z) \\ u'(z) \end{bmatrix} = F_\ell(N(z),K(z))r(z) = \begin{bmatrix} W_\mathrm{p}(z)S(z) \\ W_\mathrm{u}(z)M(z) \end{bmatrix} r(z) \tag{13.15}$$

where $W_{\mathrm{p}}(z)$ is a sensitivity specification[1] and $S(z)$ and $M(z)$ are the sensitivity and control sensitivity respectively. Thus we can formulate an $\mathcal{H}_\infty$ problem as

$$K(z) = \arg \min_{K(z)\in\mathcal{K}_\mathrm{S}} \|F_\ell(N(z), K(z))\|_{\mathcal{H}_\infty} . \tag{13.16}$$

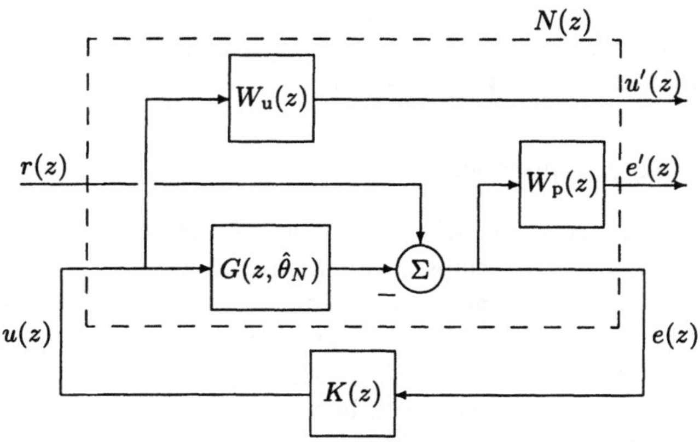

Fig. 13.2. 2 × 2 *block problem used for robust performance with $\mathcal{H}_\infty$ approach.*

For scalar systems, it is easy to show that a *necessary and sufficient* condition for robust performance given an additive uncertainty and a sensitivity specification is

$$\sup_\omega \left(\left|W_\mathrm{p}\left(e^{j\omega T_s}\right) S\left(e^{j\omega T_s}\right)\right| + \left|W_\mathrm{u}\left(e^{j\omega T_s}\right) M\left(e^{j\omega T_s}\right)\right|\right) < 1 . \tag{13.17}$$

Let us investigate how this corresponds to the $\mathcal{H}_\infty$ solution. Omitting the dependency on ω we have

$$\bar\sigma\left(\begin{bmatrix} W_\mathrm{p}S \\ W_\mathrm{u}M \end{bmatrix}\right) = \sqrt{\lambda\left(\begin{bmatrix} W_\mathrm{p}S \\ W_\mathrm{u}M \end{bmatrix}^* \begin{bmatrix} W_\mathrm{p}S \\ W_\mathrm{u}M \end{bmatrix}\right)} = \sqrt{|W_\mathrm{p}S|^2 + |W_\mathrm{u}M|^2}$$
$$= \begin{cases} \frac{1}{\sqrt{2}}\left(|W_\mathrm{p}S| + |W_\mathrm{u}M|\right) & \text{, if } |W_\mathrm{p}S| = |W_\mathrm{u}M| \\ \\ |W_\mathrm{p}S| + |W_\mathrm{u}M| & \text{, if } |W_\mathrm{p}S| = 0 \vee |W_\mathrm{u}M| = 0 \end{cases} . \tag{13.18}$$

Thus, if we tighten our robust performance condition slightly to

$$\sup_\omega \bar\sigma\left(\begin{bmatrix} W_\mathrm{p}S \\ W_\mathrm{u}M \end{bmatrix}\right) < \frac{1}{\sqrt{2}} \tag{13.19}$$

[1] Note that we do not distinguish between the input and output sensitivity since they are identical for scalar systems.

we obtain in the above extreme cases

$$\sup_{\omega} |W_{\mathrm{p}}S| + |W_{\mathrm{u}}M| < 1\,, \qquad \text{if } |W_{\mathrm{p}}S| = |W_{\mathrm{u}}M| \tag{13.20}$$

$$\sup_{\omega} |W_{\mathrm{p}}S| + |W_{\mathrm{u}}M| < \frac{1}{\sqrt{2}}\,, \qquad \text{if } |W_{\mathrm{p}}S| = 0 \text{ or if } |W_{\mathrm{u}}M| = 0 \tag{13.21}$$

and robust performance will thus be guaranteed. The $\mathcal{H}_\infty$ solution will be non-conservative if the supremum is reached for $|W_{\mathrm{p}}S| = |W_{\mathrm{u}}M|$ and in general it will be conservative only up to a factor $\sqrt{2}$. Compare with the multivariable results in Example 4.3 on page 49.

13.2.2 The Complex μ Approach

In the $\mathcal{H}_\infty$ approach above two types of conservatism were introduced. Firstly, the frequency domain uncertainty ellipses were approximated by the circumscribed circle. If the ratio of the principal axis of a given ellipse is close to one, the ellipse will be "round" and may be quite accurately approximated by a circle. However, if the principal axis ratio is much larger that one, the circle approximation will be potentially very conservative. Secondly, the "usual" $\mathcal{H}_\infty$ conservatism in connection with robust performance problems cannot be avoided. However, for scalar systems the conservatism will be less than a factor $\sqrt{2}$ and for most designs it will in fact be quite small since the supremum for $|W_{\mathrm{p}}S| + |W_{\mathrm{u}}M|$ is usually obtained in the frequency area where $|W_{\mathrm{p}}S| \approx |W_{\mathrm{u}}M|$.

Let us now investigate what can be gained by a complex μ approach. We thus allow the perturbation set $\Delta(z)$ to be structured. However, only complex perturbations are allowed. The sad fact is then that this *does not* enhance our description of the uncertainty ellipses. Using a complex perturbation set we may still only describe circles in the complex plane. Thus for the robust stability problem the complex μ approach reduce to a standard $\mathcal{H}_\infty$ problem and nothing is gained.

Robust Performance. The robust performance problem may be addressed non-conservatively since it may be formulated as a complex μ problem with the augmented perturbation block

$$\tilde{\Delta}(z) = \mathrm{diag}\{\delta^{\mathrm{c}}, \delta^{\mathrm{c}}_{\mathrm{p}}\}, \qquad \delta^{\mathrm{c}}, \delta^{\mathrm{c}}_{\mathrm{p}} \in \mathbf{C}, |\delta^{\mathrm{c}}| \leq 1, |\delta^{\mathrm{c}}_{\mathrm{p}}| \leq 1\,. \tag{13.22}$$

The $N\Delta K$ formulation can then be given as in Figure 13.3 with $N(z)$ given by

$$N(z) = \begin{bmatrix} 0 & 0 & W_{\mathrm{u}}(z) \\ -1 & 1 & -G(z,\hat{\theta}_N) \\ -W_{\mathrm{p}}(z) & W_{\mathrm{p}}(z) & -W_{\mathrm{p}}(z)G(z,\hat{\theta}_N) \end{bmatrix}. \tag{13.23}$$

The control problem is thus one of finding a controller which satisfies

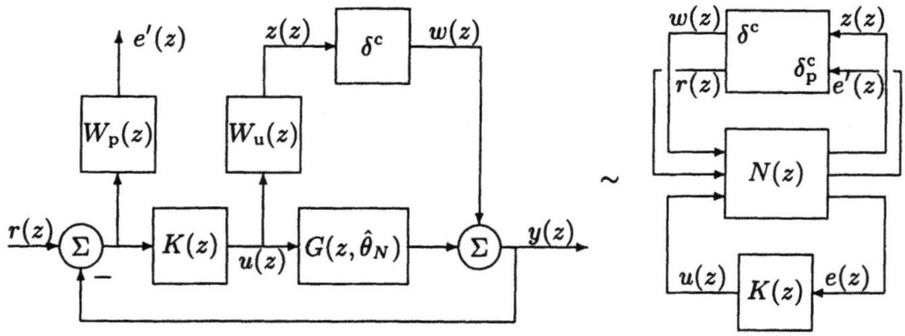

Fig. 13.3. *$N\Delta K$ framework used for robust performance with complex μ approach.*

$$K(z) = \arg \min_{K(z) \in \mathcal{K}_S} \left\| \mu_{\tilde{\Delta}}(F_\ell(N(z), K(z))) \right\|_\infty . \tag{13.24}$$

The optimal complex μ controller (13.24) can then be found using *D-K* iteration as outlined in Procedure 5.1 on page 75. If the controller obtains $\|\mu_{\tilde{\Delta}}(F_\ell(N(z), K(z)))\|_\infty < 1$ the closed loop system has robust performance. Even though the robust performance problem may be specified without conservatism using a complex μ formulation, the enhancement in comparison with the $\mathcal{H}_\infty$ approach will be less than a factor $\sqrt{2}$. We find that the μ formulation is more natural due to the $N\Delta K$ framework. However, actual computation of the optimal complex μ controller is much more involved than the standard $\mathcal{H}_\infty$ solution. Furthermore the order of the controller is higher for the μ approach. The achievements using a complex μ approach over the standard $\mathcal{H}_\infty$ solution thus seems to be very marginal.

13.2.3 The Mixed μ Approach

As shown above, a structured, but purely complex perturbation set does not allow for an accurate description of the frequency domain uncertainty ellipses obtained from the system identification procedure. We will now show that using a mixed perturbation set with one real and one complex perturbation, we may, in fact, construct an reasonable accurate approximation to the uncertainty ellipses. The conservatism will be bounded by a factor $4/\pi$.

The idea is as follows. In connection with the complex μ approach we noted that using a complex scalar perturbation we describe uncertainty circles in the frequency domain. If we combine this description with a real scalar perturbation, we may thus describe a region in the complex plane constructed by successive circles placed along a straight line of some given length and some given angle. We may then use this as an approximation for an ellipse, see Figure 13.4.

We will thus approximate the additive model uncertainty $G_\Delta(z)$ with the perturbation set

$$L(z) = W_{u,c}(z)\delta^c + W_{u,r}(z)\delta^r \tag{13.25}$$

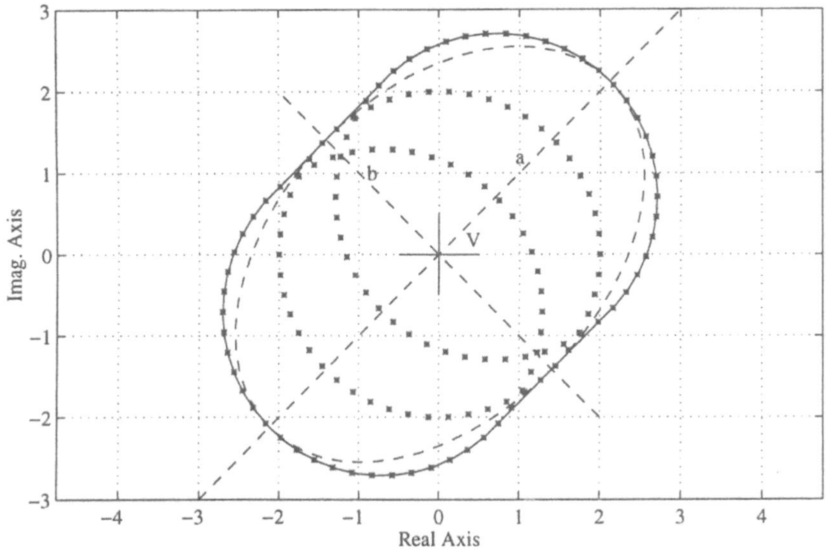

Fig. 13.4. *Fitting an uncertainty ellipse (dashed) with a mixed perturbation set (solid). A, a and b denote the angle of rotation and major and minor principal axis of the ellipse respectively.*

where δ^c and δ^r are a complex respectively real perturbation:

$$\delta^c \in \mathbf{C}, \qquad |\delta^c| \leq 1 \tag{13.26}$$

$$\delta^r \in \mathbf{R}, \qquad -1 \leq \delta^r \leq 1 \tag{13.27}$$

and $W_{u,c}(z)$ and $W_{u,r}(z)$ are stable weighting functions. If A, a and b denote the angle of rotation and major and minor principal axis of the ellipse respectively, the weighting functions $W_{u,c}(z)$ and $W_{u,r}(z)$ should be fitted to the ellipses in such a way that

$$\left|W_{u,c}\left(e^{j\omega T_s}\right)\right| \approx b \tag{13.28}$$

$$\left|W_{u,r}\left(e^{j\omega T_s}\right)\right| \approx a - b \tag{13.29}$$

$$\arg\left(W_{u,r}\left(e^{j\omega T_s}\right)\right) \approx A\,. \tag{13.30}$$

In Appendix L it is shown how the relevant information about A, a and b can easily be extracted from the form matrix $P_{\tilde{g}}(\omega)$ for the uncertainty ellipse. Notice that the weighting function $W_{u,r}(z)$ must be fitted both in amplitude and phase. This is not a trivial task. However, in the MATLAB *μ-Analysis and Synthesis Toolbox* the script file `fitsys.m` performs such a fit. The routine works for continuous-time transfer functions only, but, of course, the result may be transformed into discrete-time time using, for example, a pre-warped Tustin approximation. The other weight $W_{u,c}(z)$ needs only to be fitted in magnitude[2].

[2] `fitmag.m` from the μ toolbox may be used to perform the fit.

The quality of the approximation assuming perfect realizations of the weighting functions can easily be checked by investigation of the ratio between the surface of the approximation and the surface of the ellipse:

$$\Gamma_{\mathrm{A}} = \frac{A_{\mathrm{L}}}{A_{\mathrm{ell}}} \tag{13.31}$$

$$= \frac{4|W_{\mathrm{u,c}}||W_{\mathrm{u,r}}| + \pi|W_{\mathrm{u,c}}|^2}{\pi|W_{\mathrm{u,c}}|(|W_{\mathrm{u,r}}||W_{\mathrm{u,c}}|)} \tag{13.32}$$

$$= \frac{4|W_{\mathrm{u,r}}| + \pi|W_{\mathrm{u,c}}|}{\pi(|W_{\mathrm{u,r}}| + |W_{\mathrm{u,c}}|)} . \tag{13.33}$$

For the two extremes $|W_{\mathrm{u,r}}| = 0$ and $|W_{\mathrm{u,c}}| = 0$ we obtain

$$|W_{\mathrm{u,r}}| = 0 \Rightarrow A_{\mathrm{L}} = A_{\mathrm{ell}} \tag{13.34}$$

$$|W_{\mathrm{u,c}}| = 0 \Rightarrow A_{\mathrm{L}} = \frac{4}{\pi} A_{\mathrm{ell}} . \tag{13.35}$$

Thus for perfect realizations of the weightings

$$A_{\mathrm{ell}} \leq A_{\mathrm{L}} \leq 1.27 A_{\mathrm{ell}} . \tag{13.36}$$

Robust Performance. The robust performance problem may then be considered with the augmented perturbation block

$$\tilde{\Delta}(z) = \mathrm{diag}\{\delta^{\mathrm{c}}, \delta^{\mathrm{r}}, \delta^{\mathrm{c}}_{\mathrm{p}}\} , \qquad \delta^{\mathrm{c}}, \delta^{\mathrm{c}}_{\mathrm{p}} \in \mathbf{C}, \delta^{\mathrm{r}} \in \mathbf{R} \tag{13.37}$$

with $|\delta^{\mathrm{c}}| \leq 1$, $|\delta^{\mathrm{c}}_{\mathrm{p}}| \leq 1$ and $-1 \leq \delta^{\mathrm{r}} \leq 1$. The $N\Delta K$ formulation can be given as in Figure 13.5 with $N(z)$ given by

$$N(z) = \begin{bmatrix} 0 & 0 & 0 & W_{\mathrm{u,c}}(z) \\ 0 & 0 & 0 & W_{\mathrm{u,r}}(z) \\ -1 & -1 & 1 & -G(z, \hat{\theta}_N) \\ -W_{\mathrm{p}}(z) & -W_{\mathrm{p}}(z) & W_{\mathrm{p}}(z) & -W_{\mathrm{p}}(z) G(z, \hat{\theta}_N) \end{bmatrix} . \tag{13.38}$$

The optimal controller is thus given by

$$K(z) = \arg \min_{K(z) \in \mathcal{K}_{\mathrm{S}}} \left\| \mu_{\tilde{\Delta}} \left(F_{\ell}(N(z), K(z)) \right) \right\|_{\infty} . \tag{13.39}$$

No known solution to (13.39) exists but an approximation to the robust performance problem can be considered using μ-K iteration as outlined in Procedure 5.3. If the final controller achieves $\|\mu_{\tilde{\Delta}}(F_{\ell}(N(z), K(z)))\|_{\infty} < 1$ the closed loop system will have robust performance.

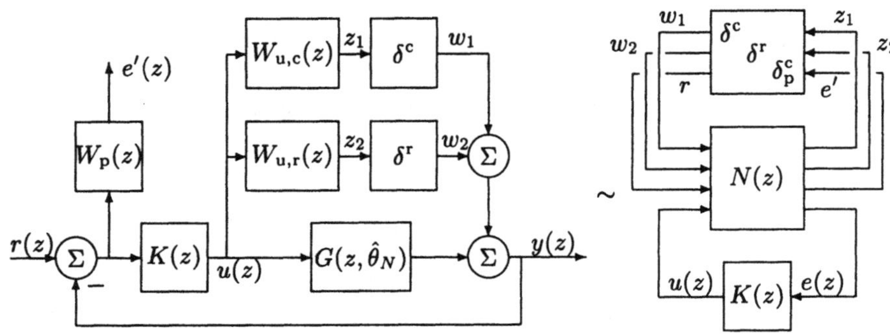

Fig. 13.5. *$N\Delta K$ formulation for robust performance problem with a mixed μ approach.*

13.3 A Synergistic Approach to Identification Based Control

As shown above we can address the control problem with very little conservatism using a mixed μ approach. The price we must pay is a significant increase in the complexity of the solution since we have to solve a mixed μ control problem. The optimal mixed μ controller will usually be of very high order and even though the number of controller states normally can be significantly reduced, the reduced controller will often have more states than the augmented plant. This may increase the costs of the actual implementation since, for example, more memory will be needed in the process computer. Consequently, the increase in controller performance must be weighed out with the additional complexity of the control design and implementation. A sensible design methodology for system identification based robust control can thus be outlined as in Procedure 13.1.

Procedure 13.1 (A Synergistic Approach to Robust Control).

1. *Using either of the two proposed system identification methods outlined in Section 13.1.4 on page 210 estimate a nominal model $G(q,\hat{\theta}_N)$ and frequency domain uncertainty ellipses at a set of chosen frequency points ω. Thus, if large data series are available, use a classical approach. Use an output error model structure if a noise model is not important for the subsequent control design. If disturbance attenuation is of primary importance in the control design, use a more general model structure to obtain a noise model $H(q,\hat{\theta}_N)$ as well. If only small data sets are available, use the stochastic embedding paradigm to extend the classical results on fixed denominator models to the case with both bias and variance errors.*
2. *Design a $\mathcal{H}_\infty$ optimal controller for the system using the approach outlined in Section 13.2.1 on page 212. Thus, approximate the uncertainty ellipses determined in Step 1 with circumscribed circles. Then pose the 2×2 block problem shown in Figure 13.2 on page 213 where the stable uncertainty weight $W_{\mathrm{u}}(z)$ has been fitted in magnitude to $R(\omega)$, the radius*

of the circle evaluated at frequency ω. Use a standard sensitivity specification for the performance weight $W_p(z)$. Of course more specialized performance requirements can be incorporated into the design. However, the $\mathcal{H}_\infty$ solution may then become more conservative.

3. *Evaluate the $\mathcal{H}_\infty$ control design. Thus check for robust performance by (13.19). Also check, for example, the open loop Nyquist with uncertainty ellipses, the transient response and any other performance demands in question. If these tests are passed satisfactorily, stop. If not, continue to Step 4.*
4. *Design a mixed μ optimal controller for the system using the approach outlined in Section 13.2.3 on page 215. Thus, approximate the uncertainty ellipses with a mixed real and complex perturbation set and pose the control problem as in Figure 13.5. The uncertainty weight $W_{u,r}(z)$ must be fitted in magnitude to the difference between the major and minor principal axis of the ellipses and in phase to the ellipses angle as indicated in Figure 13.4. The uncertainty weight $W_{u,c}(z)$ must be fitted in magnitude to the minor principal axis of the ellipses. Use a standard sensitivity specification for the performance weight $W_p(z)$. If more specialized performance requirements should be incorporated into the design this can be done without conservatism as long as they can be specified within the $N\Delta K$ framework. Solve the control problem using μ-K iteration as outlined in Procedure 5.3 on page 82.*
5. *Evaluate the mixed μ design. Thus, check for robust performance using $\mu_{\tilde{\Delta}}(F_\ell(N(z), K(z)))$. Additional checks may be performed as in Step 3. Compare with the $\mathcal{H}_\infty$ solution to assess any increase in performance obtained with the mixed μ approach. If the performance increase is significant and of importance in connection with the actual operation of the plant, select the mixed μ controller. If not, use the $\mathcal{H}_\infty$ controller determined in Step 2 or redo the design with modified uncertainty and/or performance specifications. For example, a new set of plant measurements for the identification procedure may be needed to reduce the uncertainty level. Also the performance requirements perhaps must be reduced in order to fulfill the robust performance criterion.*
6. *Finally, when necessary, use model reduction techniques to iteratively reduce the number of states in the controller until just before a significant increase in $\|\mu_{\tilde{\Delta}}(F_\ell(N(z), K(z)))\|_\infty$ will occur.*

The above procedure is a central result of this book. It outlines a complete identification based robust control design approach for scalar systems. The extension to multivariable systems are straightforward with the exception of the following points. The stochastic embedding approach is only well developed for scalar systems. Furthermore the mapping of parametric uncertainty into the frequency domain for multivariable systems is not straightforward.

Of course, the design procedure need not be terminated in Step 3 even though the $\mathcal{H}_\infty$ design is satisfactory. The mixed μ controller can be computed to investigate the possible increase in controller performance. Generally, if the uncertainty ellipses contain little phase information (are "round") we will expect the $\mathcal{H}_\infty$ design to have similar control performance to the mixed μ controller. On the other hand, if the ellipses are narrow we will generally expect the mixed μ design to be significantly superior to the $\mathcal{H}_\infty$ design.

13.4 Summary

The combination of the results presented in Part I and Part II was addressed. At first, system identification for robust control was considered. A brief discussion of the classical techniques was given. Specifically we reviewed under which circumstances the data estimate $\hat{P}_N$ of the parameter covariance matrix will be consistent. For the general PEM structure, the requirement is that the models for *both* the deterministic and stochastic part of the true system will admit an exact description. The same is true for the popular ARX model. However, for output error models, the parameter covariance estimate will be consistent even if the noise model $(H(q, \theta) = 1)$ is inadequate. For fixed denominator models the parameter covariance estimate will be consistent even in the case of undermodeling, but then the corresponding frequency domain uncertainty regions will represent the variance error only. The conclusion was that if large data sets are available, the classical output error approach may be applied if a disturbance model is unimportant in connection with the control design. We will then probably require the model order to be quite large in order to obtain residuals which are uncorrelated with the input signal. This could be impractical in connection with standard $\mathcal{H}_\infty$ or μ design since the order of the controller will increase accordingly. Fixed order $\mathcal{H}_\infty$ algorithms may be a solution to this problem. However, this has not been considered in this book.

If only short data series are available, the classical approach does not seem adequate. Then we propose that the stochastic embedding approach is used to estimate a fixed denominator model and corresponding frequency domain uncertainty ellipses. Different parameterizations of the noise and the undermodeling should be investigated.

Next, robust control from system identification was considered. For both the classical and the stochastic embedding approach the result is a nominal model and frequency domain uncertainty ellipses. Thus we need to synthesize a control design methodology which non-conservatively can capture these uncertainty templates. Here, three different design approaches were investigated: An $\mathcal{H}_\infty$ approach, a complex μ approach and a mixed real and complex μ approach. It was illustrated how both $\mathcal{H}_\infty$ and complex μ will allow only frequency domain uncertainty circles. With proper formulation of a mixed sensitivity problem, the conservatism in the $\mathcal{H}_\infty$ approach compared with the complex μ approach can be bounded by $\sqrt{2}$. Since the design of the complex μ controller is considerably more involved and will result in a controller with more states, the $\mathcal{H}_\infty$ design seems to be the most appropriate of the two. However, using a mixed μ approach we may obtain a quite accurate description of the uncertainty ellipses through a mixed perturbation set with one real and one complex scalar. The conservativeness in this description was bounded by $4/\pi = 1.27$. This approach furthermore has the advantage that more complicated performance measures may be included without conservatism.

Finally, a synergistic system identification based robust control design algorithm was outlined. It was recommended that an initial $\mathcal{H}_\infty$ design was applied and checked. If this design was not satisfactory, the mixed μ approach could be applied instead.

CHAPTER 14
CONTROL OF A WATER PUMP

The purpose of this chapter is to provide an applications example of the proposed system identification based robust design approach, see Procedure 13.1. The design procedure has been applied in control of a small domestic water supply system consisting of a centrifugal pump driven by a 840 Watt frequency modulated micro-computer controlled asynchronous type induction motor. The pump was a typical domestic water supply pump with a maximum capacity of approximately 3 m^3/hr at 2.5 bar. A 3.5 liter (0.92 gallon) rubber membrane buffer tank was placed in the outlet pipe from the pump. The size of the buffer tank is much smaller compared with more traditional domestic installations, where the volume of the buffer tank is usually around 60 liters (15.9 gallons).

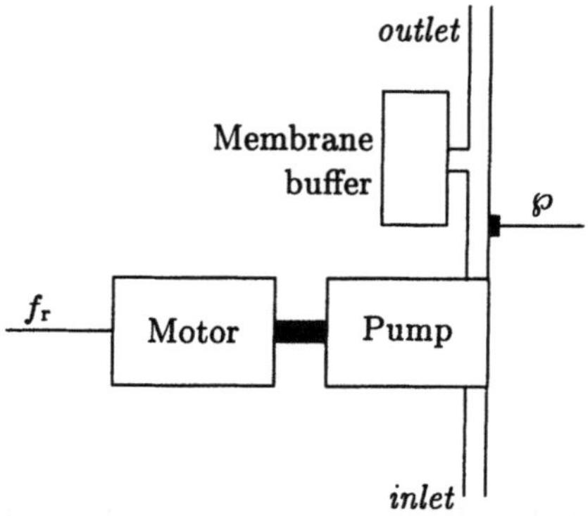

Fig. 14.1. *The water supply system. The induction motor speed is controlled via the frequency f_r. Outlet water pressure $\wp$ is recorded as output.*

The small buffer tank makes it possible to design a very compact water supply unit which takes up much less space than a traditional system. On the other hand, the small buffer tank increase the performance demands to the control system significantly since the buffer tank have only very limited operating range. The traditional on/off control will then have unsatisfactory performance. Therefore, the electric motor has been equipped with frequency modulation in order to control the velocity of the pump. The input to the system was the frequency reference f_r [Hz] for the motor and the recorded output was the outlet water pressure $\wp$ [bar]. The system is sketched in Figure 14.1.

The data acquisition and control of the system were performed using an A/D,D/A data acquisition card and a 486 PC interface to the micro-computer.

14.1 Identification Procedure

The system was sampled with sampling frequency 50 Hz and the input used for identification was a 0.4 Hz fundamental square wave. 500 samples were collected, 300 of which were used to get rid of initial condition effects. The last 200 samples were used for identification. In Figure 14.2, the data sequence is shown. The working point of the system is given in Table 14.1.

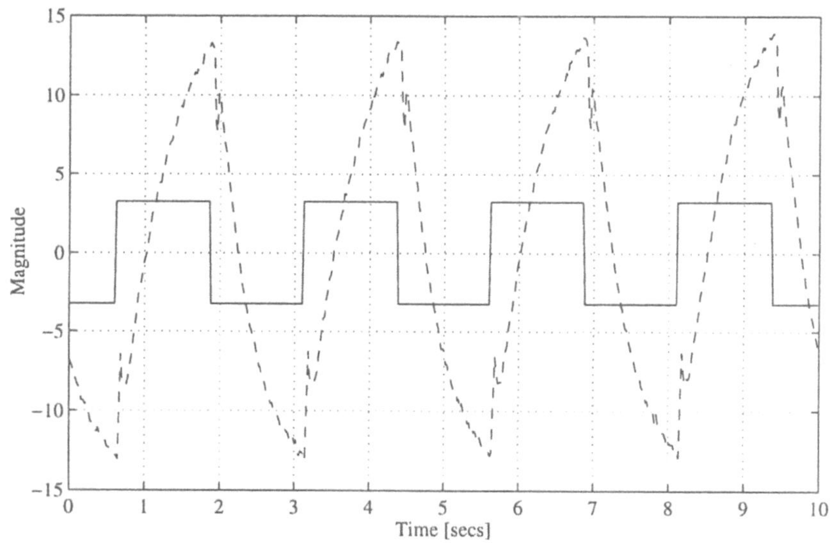

Fig. 14.2. *Input/output sequence used for identification of pump system. Shown are working point deviations for the motor frequency reference f_r [Hz] (solid) and for the outlet water pressure $\wp$ [0.01 bar].*

The system dynamics was dominated by a low frequency first order component with a time constant of approximately 1 sec. The low frequency component originates from the membrane buffer tank. Furthermore smaller high frequency components can also be identified from Figure 14.2. These originates mainly from the inertia in the induction motor and centrifugal pump. A second order Laguerre model

$$G(q,\theta) = \frac{\theta_1 q^{-1}}{1+\xi q^{-1}} + \frac{\theta_2 q^{-1}\left(1-\xi^{-1}q^{-1}\right)}{\left(1+\xi q^{-1}\right)^2} \tag{14.1}$$

Table 14.1. *Working point values for the data sequence used for estimation.*

	Pressure $\wp$ [bar]	Motor frequency f_r [Hz]	Flow Q [m^3/hr]
Working point	2.33	90.7	0.90

was fitted to the data using a standard least squares estimate. The Laguerre pole $-\xi$ was chosen as 0.96, corresponding to a time constant $\tau \approx 0.5$ secs which is slightly faster than the dominating time constant of the true system. The least squares estimate of the parameter vector θ was found as

$$\hat{\theta}_N = \begin{bmatrix} 0.6845 \\ 0.1213 \end{bmatrix} \tag{14.2}$$

corresponding to a zero at $z = 0.9433$. In Figure 14.3, the true and simulated output is shown together with the estimated Nyquist. Here, also measured discrete frequency point estimates are shown. These frequency point estimates were obtained from the true system by applying pure sine inputs at different frequencies and measuring the gain and phase shift in the output.

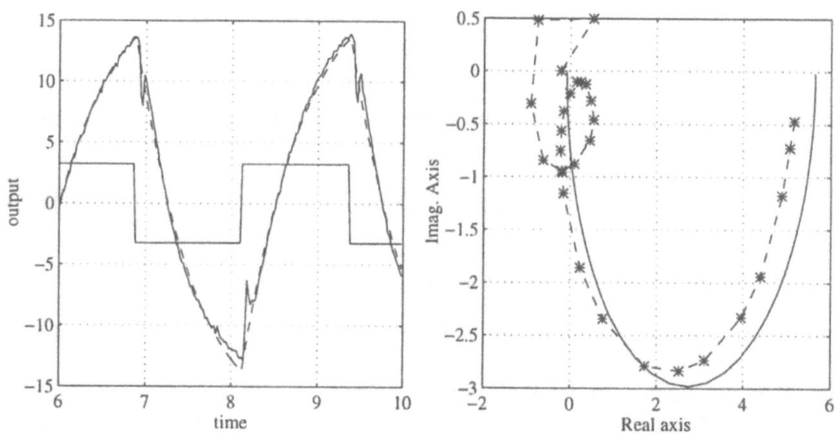

Fig. 14.3. *True (solid) and simulated (dashed) output together with estimated Nyquist (solid) and measured discrete frequency points (∗).*

Notice how a fairly accurate estimate of the low frequency behavior of the pump system is obtained with the Laguerre model. However, in the high frequency range considerable deviations between the nominal Nyquist and the measured frequency points can be identified.

14.1.1 Estimation of Model Uncertainty

The stochastic embedding approach were used for estimating the model uncertainty. The order of the FIR model describing the undermodeling was chosen as $L = 50$. The following parameterizations for the noise and undermodeling covariance matrices were chosen

$$C_{\nu}^{i,j}(\gamma) = \begin{cases} \sigma_e^2 \left(1 + \dfrac{(c-a)^2}{1-a^2}\right), & i = j \\ \dfrac{\sigma_e^2(-a)^M \left(1 + c^2 - a * c - c/a\right)}{1-a^2}, & |j-i| = M \end{cases} \tag{14.3}$$

$$C_{\eta}(\beta) = \operatorname*{diag}_{1 \leq k \leq L} \alpha\lambda^k \tag{14.4}$$

corresponding to a first order ARMA noise model and an exponentially decaying undermodeling impulse response. Thus, the combined parameter vector ζ for the covariances is given by

$$\zeta = \begin{bmatrix} \beta \\ \gamma \end{bmatrix} = \begin{bmatrix} \alpha \\ \lambda \\ a \\ c \\ \sigma_e^2 \end{bmatrix}. \tag{14.5}$$

An estimate of ζ was found using the maximum likelihood approach outlined in Section 11.2.1 on page 179:

$$\hat{\zeta} = \arg\max_{\zeta} \mathcal{L}(\varpi | U, \zeta) \tag{14.6}$$

where U is the input vector and ϖ is the transformed residuals

$$\varpi = R^T \epsilon . \tag{14.7}$$

The transformation matrix R ensures that the distribution on ϖ in non-singular. R was chosen through QR-factorization as described in Section 11.2.2 on page 180. The result of the maximum likelihood estimation is displayed in Figure 14.4 where the parameter tracks are shown together with the loglikelihood function.

The estimate of ζ was found as

$$\hat{\zeta} = \begin{bmatrix} \hat{\alpha} \\ \hat{\lambda} \\ \hat{a} \\ \hat{c} \\ \hat{\sigma_e^2} \end{bmatrix} = \begin{bmatrix} 0.3145 \\ 0.6264 \\ -0.9687 \\ -0.5095 \\ 0.0794 \end{bmatrix}. \tag{14.8}$$

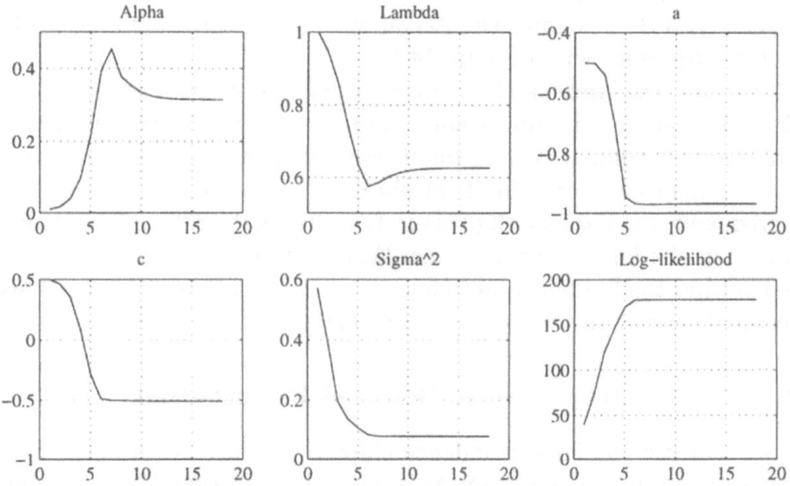

Fig. 14.4. ***Results from the maximum likelihood estimation of ζ. Shown are parameter tracks for each parameter and the expiration of the loglikelihood function.***

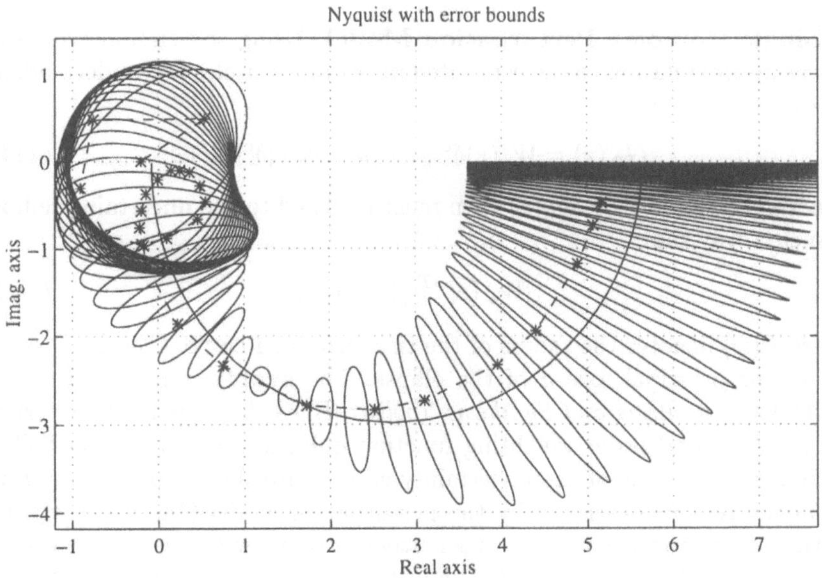

Fig. 14.5. ***Nominal Nyquist with 90% uncertainty ellipses. Measured frequency response estimates also shown (*).***

The covariance functions C_ν and C_η were then replaced by their corresponding estimates $C_\nu(\gamma)$ and $C_\eta(\beta)$ in the expression for $P_{\tilde{g}}(\omega)$, see Equation (11.71). A 90% confidence interval was chosen corresponding to the uncertainty ellipses shown in Figure 14.5.

Generally, the uncertainty ellipses provide a useful description of the model uncertainty. The uncertainty estimates seem rather conservative at low frequencies compared with the discrete frequency point estimates. This will, as we shall show later, limit the obtainable closed loop performance. The high frequency uncertainty ellipses nicely captures the rather large model uncertainty in the high frequency area. Furthermore, little uncertainty is predicted around the fundamental frequency of the square wave input. This also corresponds well with the measured frequency points.

14.1.2 Constructing A Norm Bounded Perturbation

Now we will use the 90% confidence ellipses to construct a norm bounded additive uncertainty description as discussed in Section 13.2 on page 211. As suggested in Procedure 13.1 on page 218 we will consider two design approaches, namely an $\mathcal{H}_\infty$ design and a mixed μ design. We thus need two perturbation models; one based on a single complex perturbation corresponding to circumscribed circles and one based on a mixed perturbation set corresponding to the "oval" uncertainty region shown in Figure 13.4 on page 216.

A Single Complex Perturbation Model. Using the circumscribed circles as approximation to the estimated uncertainty circles the undermodeling $G_\Delta(z)$ is described as

$$G_\Delta(z) = W_u(z)\delta^c\,, \qquad \delta^c \in \mathbf{C},\ |\delta^c| \le 1. \tag{14.9}$$

The weighting function $W_u(z)$ then must be fitted to the uncertainty ellipses in such a way that

$$\left|W_u\left(e^{j\omega T_s}\right)\right| \approx R(\omega) \tag{14.10}$$

where $R(\omega)$ denotes the radius of the circumscribed circle. Clearly, R equals the major principal axis a of the ellipse. The `sysfit.m` routine from the MATLAB *μ-Analysis and Synthesis Toolbox* [BDG+93] was used to perform the fit. A second order weighting function quite accurately matched $R(\omega)$. In Figure 14.6, some of the uncertainty ellipses are shown together with the frequency regions described by the perturbation model (14.9). Clearly, this perturbation model is rather conservative for narrow ellipses. However, the orientation of the ellipses are also important in assessing the conservatism. In the low frequency area, for example, it seems that the circumscribed circles are very conservative. However, since the sign of the gain is known, the low frequency uncertainty regions will not destabilize the closed loop plant. Furthermore, assume that we use a controller without pure integral action (finite real steady-state frequency response) and that our performance specification bounds the sensitivity function $S(z)$. A sensitivity specification corresponds to requiring that the open loop Nyquist $G(e^{j\omega T_s})K(e^{j\omega T_s})$ for each frequency ω remains outside a circle centered at the Nyquist point $(-1,0)$ and with radius $|S_p^{-1}(e^{j\omega T_s})|$ where $S_p(z)$ denotes the upper bound on the sensitivity function. In this case, the low frequency uncertainty circles will *not* be conservative from a performance point of view since the distance from the open loop uncertainty circles[1] and the open loop uncertainty ellipses to the Nyquist point will be approximately the same close to steady-state.

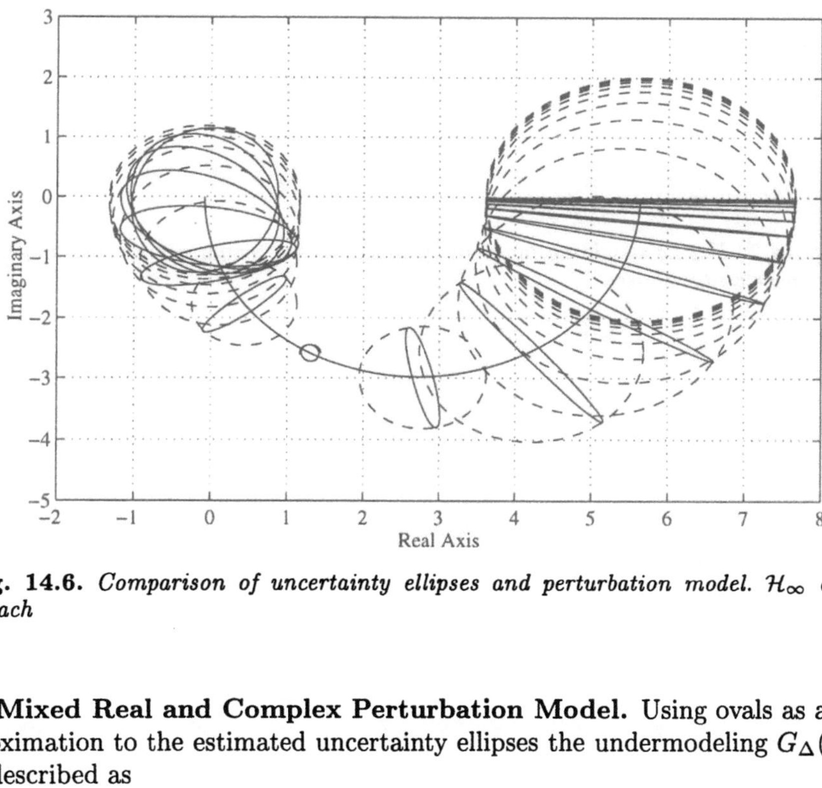

Fig. 14.6. *Comparison of uncertainty ellipses and perturbation model. $\mathcal{H}_\infty$ approach*

A Mixed Real and Complex Perturbation Model. Using ovals as approximation to the estimated uncertainty ellipses the undermodeling $G_\Delta(z)$ is described as

$$G_\Delta(z) = W_{u,c}(z)\delta^c + W_{u,r}(z)\delta^r\ , \qquad \delta^c \in \mathbf{C},\ \delta^r \in \mathbf{R},\ |\delta^{c,r}| \leq 1. \quad (14.11)$$

The weighting functions $W_{u,c}(z)$ and $W_{u,r}(z)$ must be fitted to the uncertainty ellipses in such a way that

[1] By open loop uncertainty we mean the plant uncertainty multiplied by the controller.

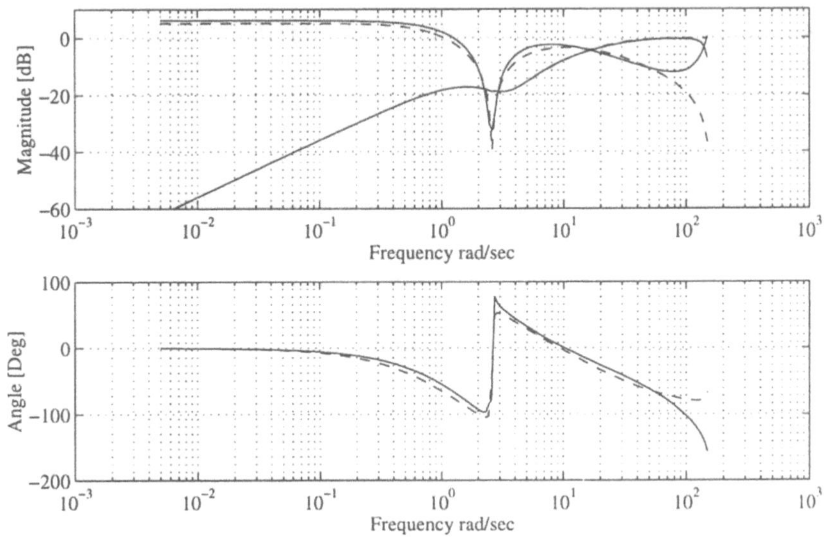

Fig. 14.7. *Fitting weighting functions (dashed) to the principal axis and angles of the uncertainty ellipses*

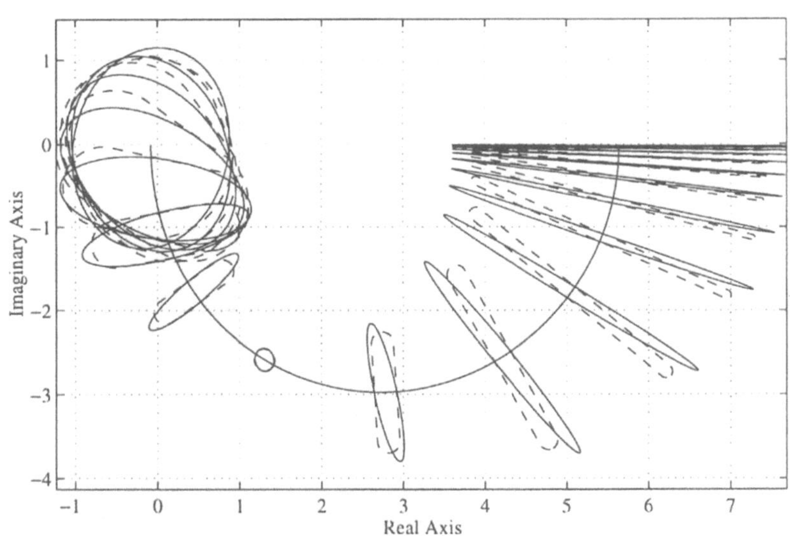

Fig. 14.8. *Comparison of uncertainty ellipses (solid) and frequency domain regions described by the weights (dashed).*

$$\left|W_{\mathrm{u,c}}\left(e^{j\omega T_s}\right)\right| \approx b \tag{14.12}$$

$$\left|W_{\mathrm{u,r}}\left(e^{j\omega T_s}\right)\right| \approx a - b \tag{14.13}$$

$$\arg\left(W_{\mathrm{u,r}}\left(e^{j\omega T_s}\right)\right) \approx A \tag{14.14}$$

where a and b denote major and minor principal axis respectively and A the angle of the ellipse. A third order weighting function was used for $W_{\mathrm{u,r}}(z)$ and a fifth order weight for $W_{\mathrm{u,c}}(z)$. The results of the fitting procedure is shown in Figure 14.7 and 14.8. In Figure 14.7 a comparison is shown of the principal axis and angles extracted from the ellipses with the frequency response of the weighting functions according to Equation (14.12)–(14.14). Furthermore, in Figure 14.8 a comparison is made between the estimated ellipses and the frequency domain uncertainty regions corresponding to the weighting functions. It can be seen that with the given orders of the weighting functions a reasonable accurate fit is obtained.

The identification part of the design has then been completed. We obtained a second order Laguerre model of the pump system and 90% confidence ellipses were estimated using the stochastic embedding approach. These uncertainty ellipses were then approximated with a purely complex perturbation set for $\mathcal{H}_\infty$ design and a mixed real and complex perturbation set for mixed μ design.

14.2 Robust Control Design

We then turn to the controller design. First we will perform an $\mathcal{H}_\infty$ design as outlined in Section 13.2.1 on page 212 and secondly a mixed μ control design will be made and compared with the $\mathcal{H}_\infty$ design.

14.2.1 Performance Specification

We will consider the robust performance problem. Thus, a performance specification must be constructed. There exists generally no explicit formalisms for obtaining such specifications. We have used time domain demands on the pump pressure response towards sudden changes in water flow Q, to formulate a maximum sensitivity bound. A standard step of $\Delta Q = 2/3\ \mathrm{m}^3/\mathrm{h}$ was used as the performance measure. The time domain demands on the outlet pressure $\wp(t)$ given a standard step on $\Delta Q(t)$ were formulated as:

- maximum transient error: 0.4 bar,
- max 0.1 bar settling time: 2 sec,
- max stationary error: 0.1 bar.

These demands originates in design goals from a major Danish pump producer. It was observed that step disturbances on the flow ΔQ acted approximately through a first order system to $d_\wp$, the disturbance on the outlet pressure. For a standard step $\Delta Q(s) = K_s s^{-1}$ with $K_s = 2/3$ we then have[2]:

$$d_{\wp,\text{step}}(s) = \frac{K}{s + \tau^{-1}} \cdot \frac{K_s}{s} \tag{14.15}$$

where K and τ are the gain and time constant of the first order filter respectively. $\tilde{K} = KK_s$ and τ were measured on the pump set-up as $\tilde{K} = 1.3$ bar/(m^3/h) and $\tau = 0.75$ sec.

Given a sensitivity specification $S(s) = \wp(s)/d_\wp(s)$, the corresponding time domain response $\wp_{\text{step}}(t)$ may then be computed. We will then use the heuristic assumption that our time demands will be fulfilled for a given compensated system having a sensitivity which falls below the specification for all frequencies. This is probably not guaranteed to be true for all systems, but it seems to work well in practice.

Here a first order discrete-time sensitivity specification was chosen:

$$S_p(z) = \frac{1.413z - 1.413}{z - 0.951} \ . \tag{14.16}$$

As seen in Figure 14.9 the corresponding time domain response $\wp_{\text{step}}(t)$ fulfill the stated demands.

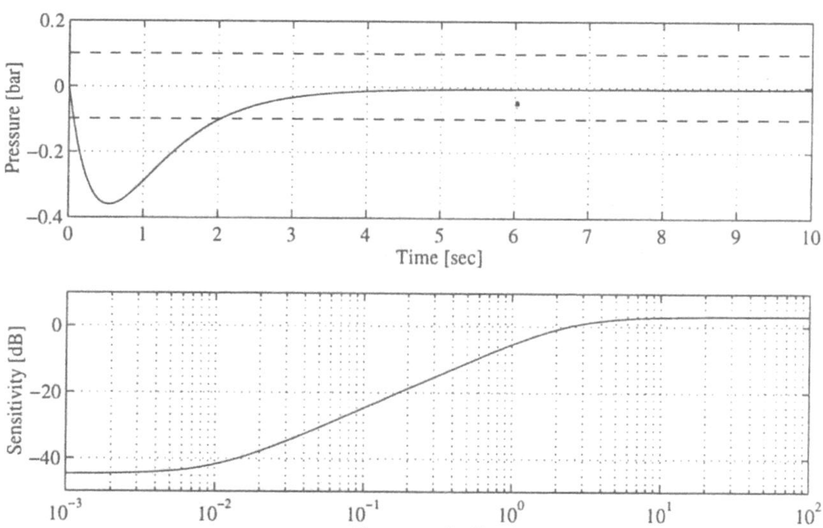

Fig. 14.9. *Time domain response $\wp_{\text{step}}(t)$ (upper) corresponding to the chosen performance specification $S_\wp(z)$ (lower).*

The standard $\mathcal{H}_\infty$ performance specification, see eg Theorem 5.3 on page 69, puts a unity bound on the transfer function from the normalized

---

[2] In order to make the approach more transparent we will carry out the derivation for a continuous-time system and simply transform the results to discrete-time.

disturbances d' to the normalized errors e'. Letting $d' = d_{\wp}$ and $e' = W_{\mathrm{p}} e_{\wp}$ we have the performance specification:

$$\frac{e_{\wp}(z)}{d_{\wp}(z)} \leq S_{\mathrm{p}}(z) \Rightarrow W_{\mathrm{p}}(z) = S_{\mathrm{p}}^{-1}(z) \, . \tag{14.17}$$

14.2.2 $\mathcal{H}_\infty$ Design

Let us then consider the $\mathcal{H}_\infty$ control design. The control problem is then defined as the 2×2 block problem shown in Figure 13.2 on page 213. The optimal $\mathcal{H}_\infty$ controller is given by

$$K(z) = \arg \min_{K(z) \in \mathcal{K}_S} \sup_{\omega} \bar{\sigma} \left(F_l(N(e^{j\omega T_s}), K(e^{j\omega T_s})) \right) \, . \tag{14.18}$$

$N(z)$ is given by Equation (13.14) on page 212 with $G(z)$ being the Laguerre model (14.1) on page 224. `dhfsyn.m` from the MATLAB μ toolbox was used to computed the optimal $\mathcal{H}_\infty$ controller which achieved

$$\sup_{\omega} \bar{\sigma} \left(F_l(N(e^{j\omega T_s}), K(e^{j\omega T_s})) \right) = 0.775 \, . \tag{14.19}$$

Remember that in order to guarantee robust performance the robust performance condition had to be strengthen to $1/\sqrt{2} = 0.707$. Thus the optimal $\mathcal{H}_\infty$ did just miss guaranteed robust performance. In Figure 14.10 and 14.11 we have checked for robust stability and nominal performance.

In Figure 14.10 the nominal Nyquist with open loop uncertainty ellipses are shown. In Appendix M it is shown how the form matrix for an open loop ellipse is obtained. Furthermore, the discrete frequency points obtained by multiplying the measured plant frequency response with the controller are displayed. Since none of the uncertainty ellipses includes the Nyquist point $(-1, 0)$ we conclude that the system is robust stable.

In Figure 14.11 the nominal sensitivity function $S(z)$ is shown together with 90% confidence bounds. The error bounds on the sensitivity were calculated as the inverse of the minimum respectively maximum distance from the open loop uncertainty ellipses to the Nyquist point $(-1, 0)$. Also the upper bound on the sensitivity and a measured sensitivity estimate is shown. The latter was obtained by multiplying the measured discrete frequency point estimates by the controller, adding one and inverting. Note that the error bounds exceed the performance specification for frequencies below 0.5 rad/sec and around 10 rad/sec thus indicating a violation of the robust performance criteria. We may also check for robust performance by Equation (13.17) on page 213. In Figure 14.12 we have plotted $|W_{\mathrm{p}}(e^{j\omega T_s}) S(e^{j\omega T_s})| + |W_{\mathrm{u}}(e^{j\omega T_s}) M(e^{j\omega T_s})|$ versus frequency ω. Note the correspondence between Figure 14.11 and 14.12. In the high frequency area the robust performance check in Figure 14.12 does not match the corresponding

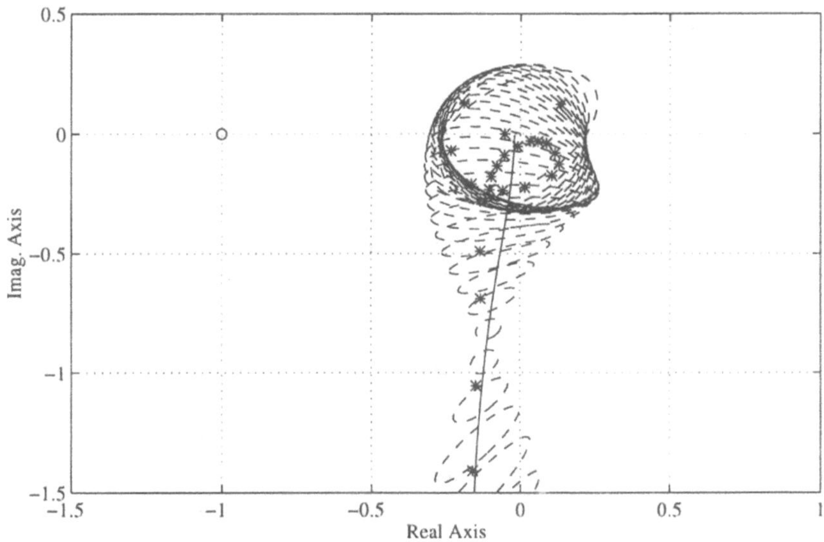

Fig. 14.10. *$\mathcal{H}_\infty$ approach: Nominal Nyquist with error bounds. Shown are also the measured frequency response estimates multiplied by the controller.*

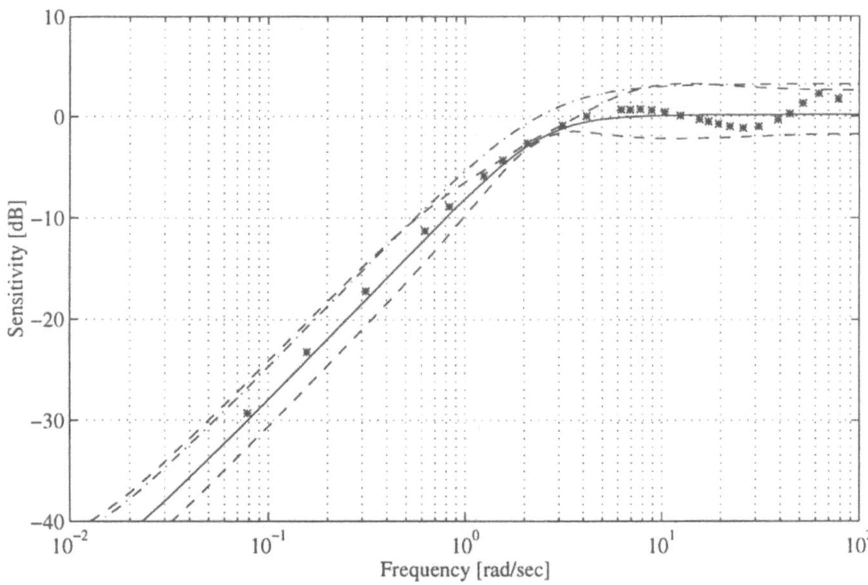

Fig. 14.11. *$\mathcal{H}_\infty$ approach: Nominal sensitivity with error bounds. Shown are also the upper bound for the sensitivity function $S(e^{j\omega T_s})$ (dash-dotted) and the measured sensitivity estimates.*

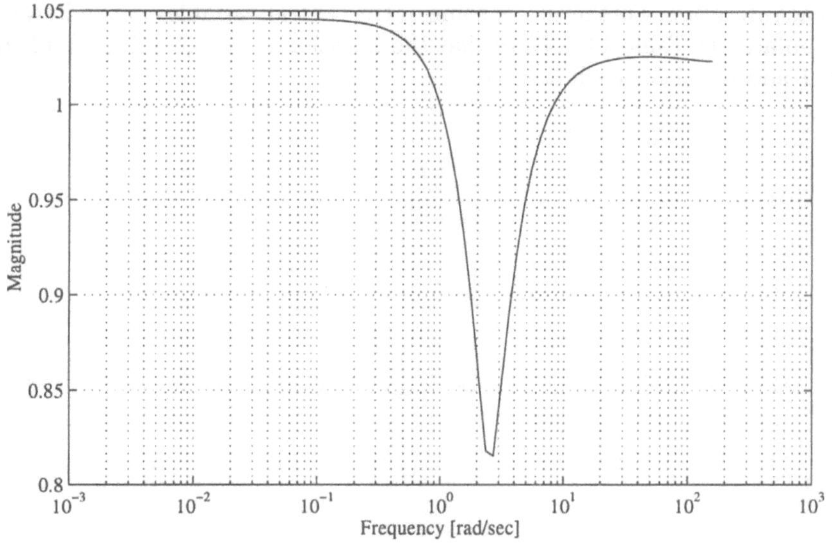

Fig. 14.12. *Robust performance check for the $\mathcal{H}_\infty$ controller. Shown are* $|W_p(e^{j\omega T_s})S(e^{j\omega T_s})| + |W_u(e^{j\omega T_s})M(e^{j\omega T_s})|$ *versus frequency* ω.

bounds on the sensitivity function in Figure 14.11. This is due to the conservativeness in the perturbation approximation and, possible, to imperfect realizations of the weighting functions.

Even though the $\mathcal{H}_\infty$ design does not obtain robust performance, generally the design is quite satisfactory.

14.2.3 Mixed μ Design

Then let us turn to the mixed μ control design. The $N\Delta K$ formulation is thus given as in Figure 13.5 on page 218 with $N(z)$ given by Equation (13.38) on page 217 and $G(z, \hat{\theta}_N)$ given by the Laguerre model (14.1) on page 224. The optimal mixed μ control problem is thus

$$K(z) = \arg \min_{K(z)\in\mathcal{K}_S} \sup_{\omega} \mu_{\tilde{\Delta}}(F_l(N(e^{j\omega T_s}), K(e^{j\omega T_s}))) \,. \tag{14.20}$$

As noted several times, the above problem cannot be solved directly since only upper and lower bounds for μ can be computed. Our approach here will be μ-K iteration as outlined in Procedure 5.3. Of course, the iteration will be performed in discrete-time[3]. The results of the μ-K iteration is shown in Figure 14.13 where the upper bound for $\mu_{\tilde{\Delta}}(F_l(N(e^{j\omega T_s}), K(e^{j\omega T_s})))$ is shown

[3] The discrete-time μ-K iteration is identical to the continuous-time iteration with s and $j\omega$ replaced by z and $e^{j\omega T_s}$. Since some of the MATLAB μ toolbox routines can be applied only for continuous-time systems them , if they are used, some bilinear transformations from continuous-time to discrete-time must be employed.

for each iteration. Furthermore, some of the central variables in the iteration is tabulated in Table 14.2. Note that the iteration converge quickly and that $\bar{\alpha} = 1$ and $\kappa = 1$ for each iteration.

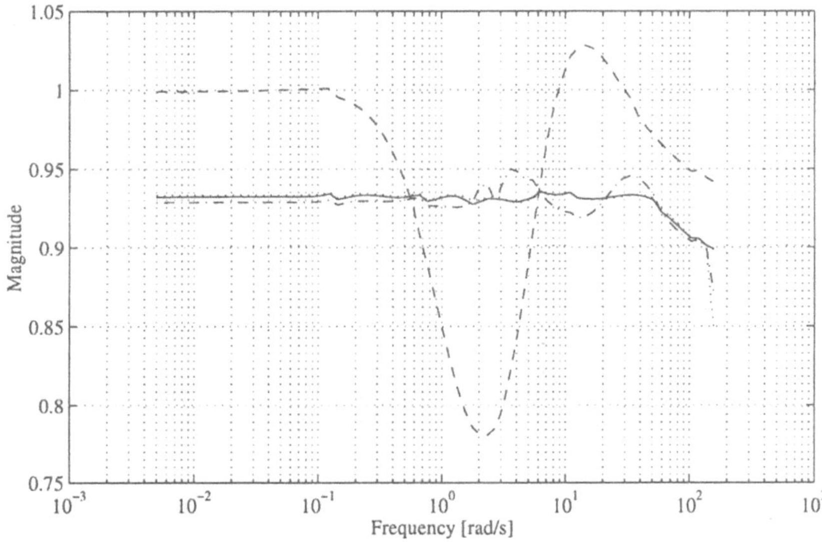

Fig. 14.13. *Result of μ-K iteration on the pump system. Shown are $\mu_{\tilde{\Delta}}(F_l(N(e^{j\omega T_s}), K(e^{j\omega T_s})))$ for 1st (dashed), 2nd (dash-dotted), 3rd (dotted) and 4th (solid) iteration.*

Table 14.2. *Results from μ-K iteration on the pump system.*

	μ-K iteration				
Iteration No.	1	2	3	4	Red. order
$\|\mu_{\tilde{\Delta}}(F_l(N(z), K(z)))\|_\infty$	1.028	0.950	0.935	0.935	0.936
$\sup_\omega \beta_i(\omega)$	0.114	0.0248	0.0133	0.0130	–
$\|F_l(P_i(z), K_i(z))\|_\infty$	1.302	1.031	0.966	0.935	–
$\inf_\omega \bar{\alpha}_i(\omega)$	1	1	1	1	–
κ	1	1	1	1	–

Since $\mu_{\tilde{\Delta}}(F_l(N(e^{j\omega T_s}), K(e^{j\omega T_s})))$ peaks at 0.935 for the final full order controller robust performance was achieved. The final controller was of very high order (45), but it was possible to reduce the number of states down to 6 with very little increase in $\mu_{\tilde{\Delta}}(F_l(N(e^{j\omega T_s}), K(e^{j\omega T_s})))$, see Table 14.2.

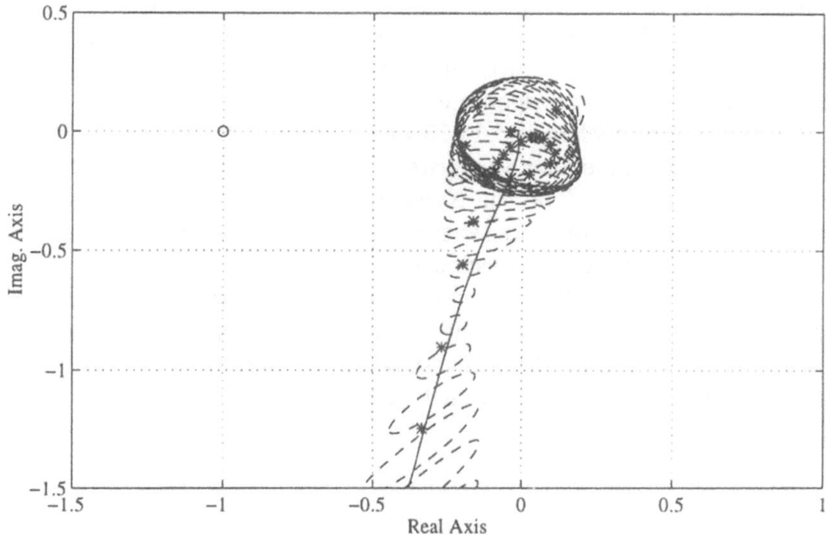

Fig. 14.14. *μ approach: Nominal Nyquist with error bounds. Shown are also the measured frequency response estimates multiplied by the controller.*

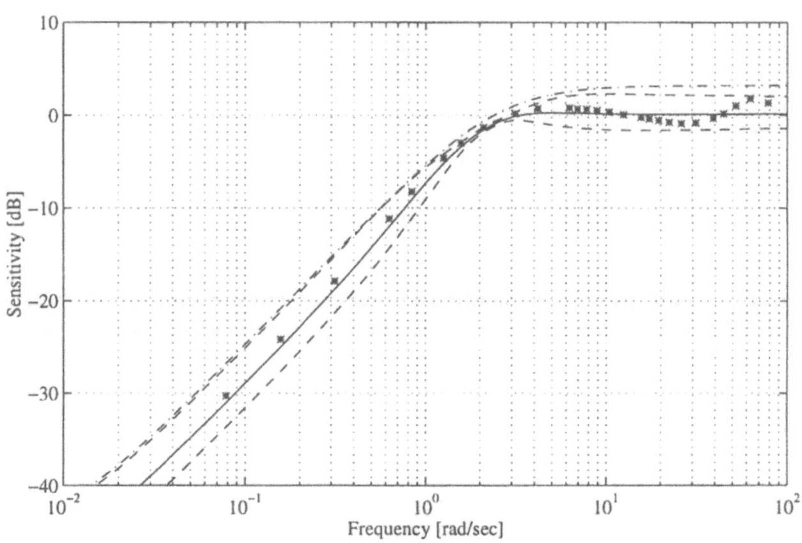

Fig. 14.15. *μ approach: Nominal sensitivity with error bounds. Shown are also the upper bound for the sensitivity function $S(e^{j\omega T_s})$ (dash-dotted) and the measured sensitivity estimates.*

In Figure 14.14 and 14.15, the nominal Nyquist with open loop uncertainty ellipses and the nominal sensitivity function with error bounds are shown respectively. The reduced order controller has been used in computing the results given in both figures. In both plots we have furthermore included the measured frequency response estimates for comparison.

Compare Figure 14.14 and 14.15 with Figure 14.10 and 14.11. From Figure 14.14 we conclude that the system is robustly stable since none of the uncertainty ellipses include the Nyquist point. In Figure 14.15, notice how smooth the mixed μ controller shapes the perturbed sensitivity function according to the performance specification.

Finally, in Figure 14.16 the results of implementing the sixth order controller on the pump system are shown. Here the system response to standard flow steps $\Delta Q = 2/3\ \mathrm{m}^3/\mathrm{hr}$ is investigated. As seen we quite easily comply with the demands given in Section 14.2.1 on page 231.

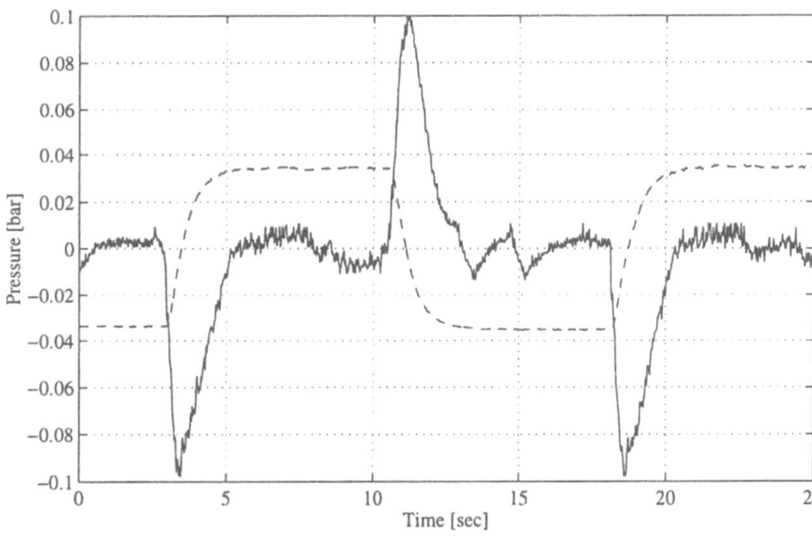

Fig. 14.16. *Results from implementing the reduced order mixed μ controller on the pump system. Shown are the pressure response (solid) to standard flow steps (dashed)*

14.3 Summary

A robust control design for a domestic water supply unit was considered. The design methodology outlined in Procedure 13.1 on page 218 was applied to the system. A second order Laguerre model was identified and frequency domain uncertainty ellipses were estimated using stochastic embedding of the bias. Discrete frequency response measurement obtained by applying pure sinusoids at the input verified the obtained uncertainty estimates. Two control designs were then performed; an $\mathcal{H}_\infty$ design and a mixed μ design. Both designs produced satisfactory results even though the $\mathcal{H}_\infty$ controller did not quite achieve robust performance. The mixed μ controller very smoothly shaped the closed loop sensitivity function such that the perturbed sensitivity fell below the performance specification. The mixed μ controller was of very high order (45). However, it could be reduced down to sixth order with very little increase in μ. The sixth order controller was implemented on the system with satisfactory results.

PART IV
CONCLUSIONS

CHAPTER 15
CONCLUSIONS

Now we will summarize the main results from the book. The main purpose of this work has been to develop a coherent design procedure for identification based robust control design.

15.1 Part I, Robust Control – Theory and Design

In Part I we presented the robust control design framework necessary for our approach. In Chapter 3 some of the important norms and spaces used in robust control design were discussed. It was shown how the familiar vector 1,2 and ∞ norms induce corresponding matrix norms. Furthermore, the norm concept was extended to operator norms or norms on sets of functions both in the time- and frequency domain. Of particular interest is the matrix 2-norm since we can use it to extend the classical Bode plot for scalar systems to multivariable frequency responses. Let eg $G(s)$ be a multivariable transfer function matrix. Then

$$\|G(j\omega)\|_2 = \sup_{u\in\mathbf{C}^n, u\neq 0} \frac{\|G(j\omega)u\|_2}{\|u\|_2} = \bar{\sigma}(G(j\omega)) \tag{15.1}$$

where $\|u\|_2$ is the usual Euclidean length of the vector u and $\bar{\sigma}(\cdot)$ denotes the maximum singular value. Thus the matrix 2-norm $\|G(j\omega)\|_2$ measures the maximum possible "gain" of $G(j\omega)$ in terms of the 2-norm of the input vector before and after multiplication by $G(j\omega)$. Performing a singular value decomposition, see Lemma 3.1, of $G(j\omega)$ also the minimum singular value:

$$\underline{\sigma}(G(j\omega)) = \inf_{u\in\mathbf{C}^n, u\neq 0} \frac{\|G(j\omega)u\|_2}{\|u\|_2} \tag{15.2}$$

can be determined. Thus the gain of $G(j\omega)$ is bounded by its maximum and minimum singular values as the input vector varies over all possible directions. By plotting the singular values $\bar{\sigma}(G(j\omega))$ and $\underline{\sigma}(G(j\omega))$ for each frequency ω we obtain the multivariable generalization of the classical magnitude Bode plot. The peak of the singular value Bode plot equals the transfer function $\mathcal{H}_\infty$-norm:

$$\|G(s)\|_{\mathcal{H}_\infty} = \sup_\omega \bar{\sigma}(G(j\omega)) \,. \tag{15.3}$$

$\|G(s)\|_{\mathcal{H}_\infty}$ thus measures the maximum possible gain of $G(j\omega)$ for all frequencies and all possible input directions. Another important interpretation of the transfer function $\mathcal{H}_\infty$-norm is that it also measures the maximum amplification of the input in terms of the *time-domain* operator 2-norm:

$$\|G(s)\|_{\mathcal{H}_\infty} = \sup_{u(t)\in\mathbf{R},u\neq 0} \frac{\|y(t)\|_2}{\|u(t)\|_2} \,. \tag{15.4}$$

The space of all stable transfer functions for which (15.3) is finite is the Hardy space $\mathcal{H}_\infty(\mathbf{C}, \mathbf{C}^{n\times m})$.

In Chapter 4, the "classical" $\mathcal{H}_\infty$ control theory is reviewed. Robust control design is essentially a question of weighing out robustness and performance in an optimal way. In $\mathcal{H}_\infty$ control design the uncertainty is assumed to be in the form of a single unstructured complex perturbation, eg like

$$G_{\mathrm{T}}(s) = G(s) + \tilde{\Delta}(s) \tag{15.5}$$

where $G_{\mathrm{T}}(s)$ and $G(s)$ denote the true system and nominal model respectively and $\tilde{\Delta}(s) \in \mathbf{C}^{m\times n}$ is some unknown perturbation which is bounded by its maximum singular value:

$$\bar{\sigma}(\tilde{\Delta}(j\omega)) \leq \ell(\omega) \,. \tag{15.6}$$

Usually diagonal weighting matrices are introduced to normalize $\Delta(s)$ to norm one

$$\tilde{\Delta}(s) = W_{\mathrm{u2}}(s)\Delta(s)W_{\mathrm{u1}}(s) \tag{15.7}$$

where $\bar{\sigma}(\Delta(j\omega)) \leq 1$. There are two main reasons for picking such a perturbation model. First of all, it captures well the effects of high-frequency unmodeled dynamics, non-linearities, time-delays, etc. and secondly, it leads to tractable expressions for robust stability. In particular, using the Small Gain Theorem on page 42 it is straightforward to show, see Theorem 4.4, that for the additive perturbation model (15.5), a necessary and sufficient condition for robust stability of the closed loop system is:

$$\bar{\sigma}(W_{\mathrm{u1}}(j\omega)M(j\omega)W_{\mathrm{u2}}(j\omega)) < 1 \,, \qquad \forall\omega \geq 0 \tag{15.8}$$

where $M(s) = K(s)(I + G(s)K(s))^{-1}$ is usually denoted the control sensitivity. It is a common mistake to believe that (15.8) is a conservative stability condition. This is *not* true. The condition (15.8) is tight such that there exists a perturbation $\Delta^*(s)$ with $\bar{\sigma}(\Delta^*(j\omega)) \leq 1$ for which the closed loop system becomes unstable unless (15.8) is satisfied. The conservatism in $\mathcal{H}_\infty$ does not lie in the condition (15.8), but in the assumed structure of the perturbation. We will return to this shortly. Let us first, however, look at the performance measures in $\mathcal{H}_\infty$ robust control. Generally, in any control design there are two main performance goals, namely command following (tracking problem) and disturbance attenuation (regulation problem). However, the tracking problem can to a large extent be considered by two-degree-of-freedom control concepts using a pre-filter to improve the transient response of the system. Thus, the performance demand for the feedback controller is mainly one of attenuating disturbances. A typical performance specification uses the output sensitivity function $S_o(s) = (I + G(s)K(s))^{-1}$. Since $S_o(s)$ is the transfer function matrix from disturbances on the output $y(s)$ to the control error $e(s)$ a performance demand can be written

$$\bar{\sigma}(W_{p2}(j\omega)S_o(j\omega)W_{p1}(j\omega)) \leq 1 \qquad \forall\omega \geq 0 \tag{15.9}$$

where $W_{p1}(s)$ and $W_{p2}(s)$ are weighting functions, see Figure 4.3. The input weight $W_{p1}(s)$ is used to transform the normalized input vector $\delta'(s)$ to the physical inputs $\delta(s) = W_{p1}(s)d'(s)$. The normalized inputs are assumed to belongs to the bounded set

$$\mathcal{D}' = \left\{ \delta'(s) \,\middle|\, \|\delta'\|_2 = \frac{1}{2\pi}\sqrt{\int_{-\infty}^{\infty} \delta(j\omega)^* \delta(j\omega) d\omega} \leq 1 \right\} . \tag{15.10}$$

The output weight $W_{p2}(s)$ is used to trade off the relative importance of the individual components of the control error $e(s)$ and to specify the attenuation level across frequency. An important interpretation of the performance specification is that the time-domain 2-norm of the weighted control error $e'(t)$ will be bounded by $\|W_{p2}(s)S_o(s)W_{p1}(s))\|_{\mathcal{H}_\infty}$:

$$\sup_{d'(t)\in\mathcal{D}'_t} \|e'(t)\|_2 = \sup_{d'(t)\in\mathcal{D}'_t} \sqrt{\int_0^\infty e'(t)^T e'(t) dt} \tag{15.11}$$

$$= \|W_{p2}(s)S_o(s)W_{p1}(s))\|_{\mathcal{H}_\infty} . \tag{15.12}$$

If a controller is designed such that (15.9) is fulfilled, the closed loop system is said to have nominal performance. In Chapter 4, it is then shown how a 2×2 block problem may be posed both for the robust stability problem and for the nominal performance problem. For both control problems the optimal design problem will be of the form

$$K(s) = \arg \min_{K(s)\in\mathcal{K}_S} \|F_\ell(N(s), K(s))\|_{\mathcal{H}_\infty} \tag{15.13}$$

where $N(s)$ is the augmented plant. (15.13) is the standard $\mathcal{H}_\infty$ optimal control problem. An iterative state space solution with $K(s)$ having the same number of states as $N(s)$ were presented by Doyle, Glover, Khargonekhar and Francis in 1988.

We may also consider robust stability and nominal performance simultaneously within the framework of the 2×2 block problem. However, often our main objective is robust performance, that is, our performance demands should be satisfied for all possible plants $G_T(s)$. Unfortunately, as shown in Section 4.4 on page 46 nominal performance and robust stability does *not* necessarily imply robust performance. In fact, if the plant is ill-conditioned, then robust performance can be arbitrarily poor even though we have robust stability and nominal performance. On the other hand, if the plant is well-conditioned, robust performance *can not* be arbitrarily poor if we have robust stability and nominal performance. Generally, if an unstructured uncertainty description is tight, we believe that robust performance can be considered non-conservatively in the $\mathcal{H}_\infty$ framework if a little care is taken when formulating the control problem, i.e when choosing the performance and uncertainty weights. We believe that the main problem with the $\mathcal{H}_\infty$ approach is that we can only handle full complex perturbation blocks. Often much more detailed uncertainty descriptions will be available and they may consequently only be handled with conservatism in the $\mathcal{H}_\infty$ framework.

Fortunately theory exists which can handle structured uncertainty in a non-conservative manner. Using the structured singular value μ we may address robust stability problem with mixed real and complex perturbation structures where the perturbation has a block diagonal structure. We thus can consider structured uncertainty which enters the nominal model in a linear fractional manner. The permitted perturbations in the μ framework are much more detailed than those permitted for $\mathcal{H}_\infty$ problems. Using μ we can then derive necessary and sufficient conditions for robust stability. Furthermore, the robust performance problem can be considered non-conservatively by augmenting the perturbation structure with a full complex perturbation block.

Very general control problems can be naturally formulated within the $N\Delta K$ framework introduced in Section 5.1.2 on page 68 and simple conditions for nominal performance, robust stability and robust performance can be given with μ.

Unfortunately, μ cannot be directly computed unless in some very restrictive special cases. This has naturally hampered the practical use of of μ theory. However, reasonable tight upper and lower bounds for μ can be effectively computed, and today commercially available hardware exists which support this. Thus, from a control engineering point of view, the mathematical problems concerning μ are more or less insignificant. Of much greater importance is whether we can solve the optimal μ control problem:

$$K(s) = \arg \min_{K(s)\in\mathcal{K}_S} \|\mu_{\tilde{\Delta}}(F_\ell(N(s), K(s)))\|_\infty \tag{15.14}$$

Since we cannot normally compute μ it is clear that the problem (15.14) is not tractable. However, we may formulate the control problem in terms of the upper bound on μ instead. The upper bound problem is, however, also an yet unsolved problem. However, for purely complex perturbation sets an approximation to μ synthesis can be made through a series of minimizations, first over the controller $K(s)$ and then over the scalings $D(\omega)$ involved in the upper bound. This procedure, known as *D-K iteration*, see Section 5.2.1 on page 74, seems to work quite well in practice even though the iteration cannot be guaranteed to converge. Thus, for systems with purely complex perturbations, a well-documented and software-supported design approach exists for optimal μ design.

Unfortunately, the same is not true for mixed real and complex perturbation sets. During the past 5 years, the mixed μ synthesis problem has received considerable interest in the automatic control community. Peter Young [You93] was probably the first who presented a solution to this problem. However, the procedure outlined by Young, denoted *D,G-K iteration*, see Section 5.2.2 on page 76, is much more involved than *D-K* iteration. It involves a lot of spectral factorizations theory and unlike *D-K* iteration, the scaling matrices must be fitted both in phase and magnitude. Since the *G*-scalings are purely imaginary, this severely hampers the practical use of *D,G-K* iteration.

One of the major results of this book is a new approach for mixed real and complex μ synthesis. By sacrificing some of the convergence properties in *D,G-K* iteration we have obtained an iterative procedure denoted *μ-K iteration* where we only need to fit scaling matrices in magnitude.

In Chapter 6, a case study is performed in order to compare the results of the *D,G-K* iteration with our new approach. For this particular system it turns out that μ-*K* iteration performs 20% better in terms of achieved μ levels.

In Chapter 7, a complex μ design for an ASTOVL aircraft was performed. This particular system was very ill-conditioned so that a special performance weight had to be introduced. The final complex μ design gave significant improvement over the existing classical design.

15.2 Part II, System Identification and Estimation of Model Error Bounds

In Part II, the problem of estimating a nominal model and frequency domain error bounds was considered. In robust control theory, the uncertainty bounds are usually simply assumed given a priori. However, the determination of these bounds are by no means a trivial problem. It is therefore appealing to think that they may be estimated using system identification methods. However, as shown in Chapter 9, classical identification techniques will only produce consistent estimates of the model error provided the true system can be described by the chosen model structure. For the general model structure (9.1), in order to obtain consistent estimated of the parameter covariance, we must require that both the deterministic and stochastic part of the true system can be described within our model set. In Chapter 10, different special cases are reviewed. In particular, we notice that with an output error model structure we will obtain consistent estimates of the parameter covariance if the deterministic part of the true system can be represented within our model set regardless whether this is true for the stochastic part as well. Furthermore, using a fixed denominator model we may obtain consistent estimates of the parameter covariance even in the case of undermodeling, that is, even though the true system cannot be represented within our model set. Since the least-squares parameter estimate for fixed denominator structures furthermore are analytical, there are strong motivations for using such model structures. During the last 5 years intensive research in fixed denominator model structures have been performed eg by Bo Wahlberg. Unfortunately, even though our parameter covariance estimates are consistent if the true system cannot be represented within the model set, then the frequency domain model error bounds will be misleading. Thus, to obtain useful estimates of the frequency domain uncertainty with the classical approach, we still need to assume that the true system can be represented within the model set.

Unfortunately, this is often an inadequate assumption in connection with robust control since we will like to have simple plant descriptions in order to avoid high order controllers. Consequently, the last decade have witnessed a growing awareness that classical identification methods cannot provide us with the frequency domain uncertainty bounds that we need for robust control design. As a result, intensive research has been done on developing new techniques for estimation of model error bounds.

However, estimating of model error bounds from finite noisy data is a very difficult problem if the true system cannot be described within the model set. In that case, the model residuals will contain two parts; a deterministic part due to the undermodeling and a stochastic part due to the noise. However, the approach taken by most researchers has been to assume either that the noise is deterministic (unknown, but bounded in magnitude) or that the undermodeling is stochastic (non-stationary and correlated with the input).

It is then well-known that some a priori knowledge of the noise and the undermodeling must be available in order to compute the model error bounds. However, as shown in Chapter 11, the *stochastic embedding approach* provides the opportunity of estimating the quantitative part of the necessary a priori knowledge. We feel that this makes the stochastic embedding approach superior to the other new approaches for estimation of model uncertainty. In Chapter 11, a thorough introduction to the stochastic embedding approach is provided. The main idea is to assume that the undermodeling is a realization of a zero mean stochastic process. We may then derive results for fixed denominator models which can be viewed as an extension to the classical results.

Our main contribution in this part of the book is investigation of new parameterizations for the covariance matrices for the noise and the undermodeling. In Chapter 12, a case study is performed where different parameterizations for the undermodeling are investigated. It is shown that we may obtain reasonable accurate estimates of the model error using the stochastic embedding approach.

15.3 Part III, A Synergistic Control Systems Design Methodology

Finally, in Part III, we have combined the results from Part I and Part II into a coherent design approach for identification based robust control design for scalar systems. If a large number of measurements are available, we suggest that a classical output error approach is used for system identification. If only few measurements are available the stochastic embedding approach is suggested. In any case, the result will be a nominal model and frequency domain uncertainty ellipses.

It is then shown how a mixed perturbation set may be used to approximate the estimated uncertainty ellipses. The corresponding control problem will thus be a mixed μ problem. We suggest that μ-K iteration is applied in order to solve the control problem. A full step-by-step design procedure is outlined in Section 13.3 on page 218. This design procedure is the main new contribution in the book. The design approach has been applied in control of a compact domestic water supply unit, see Chapter 14. The final mixed μ controller was implemented on the pump system with satisfactory results.

15.4 Future Research

The design methodology presented in this book can be applied only to scalar systems. However, modern design methods like $\mathcal{H}_\infty$ and μ based synthesis algorithms have potential in particular for multivariable systems where the classical design methods become more "unpredictable". The results presented on robust control design in Chapter 4 and Chapter 5 thus apply for multivariable systems in general. On the other hand, it seems that the emphasis in system identification has been placed mostly on scalar systems. It is possible to extend the classical PEM approach to multivariable systems without much difficulty. However, the asymptotic properties of the estimate seem not to have attracted much attention. One exception is the work by Zhu and coworkers, see e.g. [ZBE91]. These methods, however, are asymptotic not only in the number of data N, but also in the model order n. Thus the error bounds will only be valid for very high order models and for large data series. The stochastic embedding approach produce promising results for scalar systems. However, the maximum likelihood estimation of the covariance matrices for the noise and the undermodeling is only developed for scalar systems and only tractable for small data series.

In our future research on the subject we will try to extend the results presented on system identification in Chapter 9, 10 and 11 to multivariable systems.

Another area for future research is the proposed new algorithm for mixed real and complex μ synthesis, *μ-K iteration.* We established two necessary conditions for convergence of μ-K iteration. Firstly, the iteration must be monotonically non-increasing in the $\mathcal{H}_\infty$-norm of the augmented closed loop system $F_\ell(N_{D\Gamma_i}(s), K(s))$. Thus

$$\|F_\ell(N_{D\Gamma_i}(s), K_i(s))\|_{\mathcal{H}_\infty} \leq \|F_\ell(N_{D\Gamma_{i-1}}(s), K_{i-1}(s))\|_{\mathcal{H}_\infty} , \quad \forall i . \tag{15.15}$$

Secondly, the iteration must be monotonically non-increasing in the ∞-norm of $\beta_i(\omega)$. Thus

$$\|\beta_i(\omega)\|_\infty \leq \|\beta_{i-1}(\omega)\|_\infty , \quad \forall i . \tag{15.16}$$

$\beta_i(\omega)$ is a measure of how well the augmented closed loop system approximates the mixed μ upper bound. We were able to prove the first of the two conditions above. We believe that we cannot always achieve the second condition. However, this is certainly an area which requires more research.

BIBLIOGRAPHY

[AA94] N. Amann and F. Allgöwer. μ-suboptimal design of a robustly performing controller for a chemical reactor. *Int. J. Control*, 59(3):665–687, 1994.

[ATCP94] P. Andersen, S. Tøffner-Clausen, and T.S. Pedersen. Estimation of frequency domain model uncertainties with application to robust controller design. In *Proc. SYSID'94*, volume 3, pages 603–608, Copenhagen, Denmark, July 1994.

[ÅW84] K.J. Åström and B. Wittenmark. *Conputer Controlled Systems: Theory and Design*. Prentice-Hall Information and System Sciences Series. Prentice-Hall Inc., Englewood Cliffs, NJ, 1984.

[ÅW89] K.J. Åström and B. Wittenmark. *Adaptive Control*. Number 09720 in Addison-Wesley Series in Electrical and Computer Engineering: Control Engineering. Addison-Wesley, 1989.

[Bai92] J. Baillieul, editor. *IEEE Trans. Aut. Contr., Special Issue on System Identification for Robust Control*, volume 37, No. 7. IEEE Control Systems Society, July 1992.

[BATC94] M. Blanke, P. Andersen, and S. Tøffner-Clausen. Modelling and uncertainty. Lecture Note for EURACO Young Researchers Week, Dublin. R94-4065, Dept. of Control Engineering, Aalborg University, Frederik Bajers Vej 7, DK-9220 Aalborg Ø, Denmark, Aug. 1994.

[BD92] T.C.P.M. Backx and A.A.H. Damen. Identification for the control of mimo industrial processes. *IEEE Trans. Aut. Contr.*, 37(7):980–986, July 1992.

[BDG+93] G.J. Balas, J.C. Doyle, K. Glover, A. Packard, and R. Smith. *μ-Analysis and Synthesis Toolbox*. The MathWorks Inc., Natick, Mass., USA, 2nd edition, July 1993.

[Ber94] B. Bernhardsson. The $\mathcal{H}_\infty$ approach. In *EURACO Network: Robust and Adaptive Control Tutorial Workshop*, University of Dublin, Trinity College, 1994. Lecture 2.2.

[BG94a] S. G. Breslin and M. J. Grimble. Multivariable control of an ASTOVL aircraft. Part 1 : Analysis and design. Technical Report ICC/90/94, Industrial Control Centre, University of Strathclyde, 1994.

[BG94b] S. G. Breslin and M. J. Grimble. Multivariable control of an ASTOVL aircraft. Part 2 : MATLAB Toolbox. Technical Report ICC/91/94, Industrial Control Centre, University of Strathclyde, 1994.

[BGFB94] S. Boyd, L. El Ghaoui, E. Feron, and V. Balakrishnan. *Linear Matrix Inequalities in System and Control Theory*. Number 15 in SIAM Studies in Applied Mathematics. Society for Industrial and Applied Mathematics, Philadelphia, Pennsylvania, 1994.

[BGW90] R.R. Bitmead, M. Gevers, and V. Wertz. *Adaptive Optimal Control, The Thinking Man's GPC*. Prentice-Hall International Series in Systems and Control Engineering. Prentice Hall of Australia Pty Ldt, 1990.

[BYM92] D.S. Bayard, Y. Yam, and E. Mettler. A criterion for joint optimization of identification and robust control. *IEEE Trans. Aut. Contr.*, 37(7):986–991, July 1992.

[CS92] R.Y. Chiang and M.G. Safonov. *Robust Control Toolbox*. The MathWorks Inc., Natick, Mass., USA, Aug. 1992.

[Dai90] R. Lane Dailey. Lecture notes for the workshop on $\mathcal{H}_\infty$ and μ methods for robust control. American Control Conf., San Diego, California, May 1990.

[dB93] A.C. den Brinker. Adaptive orthonormal filters. In *Proc. IFAC World Congress*, volume 5, pages 287–292, Sydney, Australia, July 1993.

[DC85] J.C. Doyle and C.-C. Chu. Matrix interpolation and $\mathcal{H}_\infty$ performance bounds. In *Proc. American Control Conf.*, pages 129–134, Boston, MA., 1985.

[Dem88] J. Demmel. On structured singular values. In *Proc. American Control Conf.*, pages 2138–2143, Austin, Texas, Dec. 1988.

[DFT92] J.C. Doyle, B.A. Francis, and A.R. Tannenbaum. *Feedback Control Theory*. Maxwell Macmillan, Singapore, 1992.

[DGKF89] J.C. Doyle, K. Glover, P.P. Khargonekar, and B.A. Francis. State space solutions to standard $\mathcal{H}_2$ and $\mathcal{H}_\infty$ control problems. *IEEE Trans. Aut. Contr.*, AC-34(8):831–847, 1989.

[dMGG91] B. de Moor, M. Gevers, and G.C. Goodwin. Overbiased, underbiased and unbiased estimation of transfer functions. In *Proc. 9th IFAC/IFORS Symposium on System Identification and Parameter Estimation*, pages 946–951, Budapest, Hungary, July 1991.

[Doy81a] J.C. Doyle. Limitations on achievable performance of multivariable feedback systems. AGARD Lectures Series No. 117 on Multivariable Analysis and Design Techniques, Sept. 1981.

[Doy81b] J.C. Doyle. Multivariable design techniques based on singular value generalizations of classical control. AGARD Lectures Series on Multivariable Analysis and Design Techniques, Sept. 1981.

[Doy82] J.C. Doyle. Analysis of feedback systems with structured uncertainties. In *IEE Proceedings*, volume 129, Part D, No. 6, pages 242–250, November 1982.

[Doy85] J.C. Doyle. Structured uncertainty in control system design. In *Proc. 24th Conf. on Decision and Control*, pages 260–265, 1985.

[DP87] J.C. Doyle and A. Packard. Uncertain multivariable systems from a state space perspective. In *Proc. American Control Conf.*, pages 2147–2152, Minneapolis, MN, 1987.

[DPZ91] J.C. Doyle, A. Packard, and K. Zhou. Review of LFTs, LMIs and μ. In *Proc. 30th Conf. on Decision and Control*, pages 1227–1232, Brighton, England, December 1991. IEEE.

[DS79] J.C. Doyle and G. Stein. Robustness with observers. *IEEE Trans. Aut. Contr.*, AC-24(4):607–611, August 1979.

[DS81] J.C. Doyle and G. Stein. Multivariable feedback design: Concepts for a classical/modern synthesis. *IEEE Trans. Aut. Contr.*, AC-26(1):4–16, February 1981.

[DV75] C.A. Desoer and M. Vidyasagar. *Feedback Systems: Input-Output Properties*. Academic Press, New York, 1975.

[FD93] Y. Fu and G.A. Dumont. On determination of laquerre filter pole through step or impulse response data. In *Proc. IFAC World Congress*, volume 5, pages 303–307, Sydney, Australia, July 1993.

[Fra87] B.A. Francis. *A Course in $\mathcal{H}_\infty$ Control Theory*, volume 88 of *Lecture Notes in Control and Information Sciences*. Springer Verlag, Berlin, 1987.

[Fre89a] J.S. Freudenberg. Analysis and design for ill-conditioned plants. part 1. lower bounds on the structured singular value. *Int. J. Control*, 49(3):851–871, 1989.

[Fre89b] J.S. Freudenberg. Analysis and design for ill-conditioned plants. part 2. directionally uniform weightings and an example. *Int. J. Control*, 49(3):873–903, 1989.

[FTD91] M.K.H. Fan, A.L. Tits, and J.C. Doyle. Robustness in the presence of mixed parametric uncertainty and unmodeled dynamics. *IEEE Trans. Aut. Contr.*, 36(1):25–38, Jan. 1991.

[Gev91] M. Gevers. Connecting identification and robust control: A new challenge. In *Proc. IFAC Symposium on Identification and System Parameter Estimation*, Budapest, 1991. IFAC.

[GGM91] G.C. Goodwin, M. Gevers, and D.Q. Mayne. Bias and variance distribution in transfer function estimation. In *Proc. 9th IFAC Symposium on Identification and System Parameter Estimation*, pages 952–957, Budapest, 1991. IFAC.

[GGN90] G.C. Goodwin, M. Gevers, and B. Ninness. Optimal model order selection and estimation of model uncertainty for identification with finite data. Technical report, Louvain University, Belgium, 1990.

[GGN91] G.C. Goodwin, M. Gevers, and B. Niness. Optimal model order selection and estimation of model uncertainty for identification with finite data. In *Proc. 30th Conf. on Decision and Control*, pages 285–290, Brighton, England, December 1991. IEEE.

[GGN92] G.C. Goodwin, M. Gevers, and B. Ninness. Quantifying the error in estimated transfer functions with application to model order selection. *IEEE Trans. Aut. Contr.*, 37(7):913–928, July 1992.

[GJ88] M.J. Grimble and M.A. Johnson. *Optimal Control and Stochastic Estimation, Theory and Applications*, volume 1 and 2. John Wiley & Sons, 1988.

[GK92] G. Gu and P.P. Khargonekar. Linear and nonlinear algorithms for identification in $\mathcal{H}_\infty$ with error bounds. *IEEE Trans. Aut. Contr.*, 37(7):953–963, July 1992.

[GM94] L. Giarrè and M. Milanese. $\mathcal{H}_\infty$ identification with mixed parametric and nonparametric models. In *Proc. SYSID '94*, volume 3, pages 255–259, Copenhagen, Denmark, July 1994. IFAC.

[GNS90] G.C. Goodwin, B. Ninness, and M.E. Salgado. Quantification of uncertainty in estimation. In *Proc. American Control Conf.*, pages 2400–2405, San Diego, USA, May 1990.

[Gri86] M.J. Grimble. Optimal $\mathcal{H}_\infty$ robustness and the relationship to LQG design problems. *Int. J. Control*, 43(2):351–372, 1986.

[Gri88] Micheal J. Grimble. Optimal $\mathcal{H}_\infty$ multivariable robust controllers and the relationship to LQG design problems. *Int. J. Control*, 48(1):33–58, 1988.

[Gri94] M.J. Grimble. *Robust Industrial Control*. Systems and Control Engineering. Prentice Hall, 1994.

[GS89a] G.C. Goodwin and M.E. Salgado. Quantification of uncertainty in estimation using an embedding principle. In *Proc. American Control Conf.*, pages 1416–1421, Pittsburgh, PA, 1989.

[GS89b] G.C. Goodwin and M.E. Salgado. A stochastic embedding approach for quantifying uncertainty in the estimation of restricted complexity models. *Int. J. Adaptive Control and Signal Processing*, 3(4):333–356, 1989.

[GSM89] G.C. Goodwin, M.E. Salgado, and D.Q. Mayne. A baysian approach to estimation with restricted complexity models. Technical report, Newcastle University, Dept. of Electrical & Computer Engineering, NSW 2308 Australia, 1989.

[HdHB93] P.S.C. Heuberger, P.M.J. Van den Hof, and O.H. Bosgra. Modelling linear dynamical systems through generalized orthonormal basis functions. In *Proc. IFAC World Congress*, volume 5, pages 283–286, Sydney, Australia, July 1993.

[Hja90] Håkon Hjalmarson. On estimation of model quality in system identification. Licentiate Thesis LIU-TEK-LIC-1990:51, Linköping University, Department of Electrical Engineering, S-581 83 Linköping, Sweden, October 1990.

[Hja93] Håkon Hjalmarson. *Aspects on Incomplete Modeling in System Identification.* PhD thesis, Linköping University, Department of Electrical Engineering, S-581 83 Linköping, Sweden, 1993.

[HJN91] A.J. Helmicki, C.A. Jakobson, and C.N. Nett. Control oriented system identification: A worst-case/deterministic approach in $\mathcal{H}_\infty$. *IEEE Trans. Aut. Contr.*, 36(10):1161–1176, October 1991.

[HL90a] H. Hjalmarson and L. Ljung. How to estimate model uncertainty in the case of undermodelleing. In *Proc. Americal Control Conf.*, pages 323–324, San Diego, 1990.

[HL90b] H. Hjalmarson and L. Ljung. How to estimate model uncertainty in the case of undermodelleing. LITH-ISY-I 1067, Linköping University, Department of Electrical Engineering, Marts 1990.

[HL92] Håkon Hjalmarson and L. Ljung. Estimating model variance in case of undermodelling. *IEEE Trans. Aut. Contr.*, 37(7):1004–1008, july 1992.

[HL94] H. Hjalmarson and L. Ljung. A unifying view of disturbances in identification. In *Proc. SYSID '94*, volume 2, pages 73–78, Copenhagen, Denmark, July 1994. IFAC.

[Hol94] A.M. Holohan. A tutorial on mu-analysis. In *EURACO Network: Robust and Adaptive Control Tutorial Workshop*, University of Dublin, Trinity College, 1994. Lecture 2.5.

[Kal60] R.E. Kalman. Contributions to the theory of optimal control. *Bol. Soc. Mat. Mex.*, 5:102–119, 1960.

[Kal64] R.E. Kalman. When is a linear control system optimal? *Journal of Basic Engineering (Trans. ASME D).*, 86:51–60, 1964.

[KD88] K.Glover and J.C. Doyle. State-space formulae for all stabilizing controllers that satisfy an $\mathcal{H}_\infty$-norm bound and relations to risk sensitivity. *Systems & Control Letters*, 11(3):167–172, 1988.

[KHN62] R.E. Kalman, Y.C. Ho, and K.S. Narendra. Controllability of linear dynamic systems. *Contributions to Differential Equations*, 1, 1962.

[KK92] J.M. Krause and P.P. Khargonekar. A comparison of classical stochastic estimation and deterministic robust estimation. *IEEE Trans. Aut. Contr.*, 37(7):994–1000, Juli 1992.

[KLB92] R.L. Kosut, M.K. Lau, and S.P. Boyd. Set-membership identification of systems with parametric and nonparametric uncertainty. *IEEE Trans. Aut. Contr.*, 37(7):929–941, July 1992.

[Knu93] T. Knudsen. Systemidentifikation. Technical Report AUC-PROCES-U-93-4008, Institute of Electronic Systems, Aalborg University, Frederik Bajers Vej 7, DK-9220 Aalborg Ø, Denmark, Jan. 1993.

[Kos93] R.L. Kosut. Determining model uncertainty of identified models for robust control design. In *Proc. IFAC World Congress*, volume 7, pages 373–376, Sydney, Australia, July 1993.

[KR82] M. Knudsen and H. Rasmussen. Systemidentifikation. Technical Report R82-4, Institute of Electronic Systems, Aalborg University, Frederik Bajers Vej 7, DK-9220 Aalborg Ø, Denmark, 1982. Revised version, Jan. 1987.

[Kwa85] H. Kwarkernaak. Minimax frequency domain performance and robustness optimazation of linear feedback systems. *IEEE Trans. Aut. Contr.*, AC-30(10):994–1004, 1985.

[LBKF91] M.K. Lau, S.P. Boyd, R.L. Kosut, and G.F. Franklin. Robust control design for ellipsoidal plants set. In *Proc. 30th Conf. on Decision and Control*, pages 291–296, Brighton, England, December 1991. IEEE.

[Lei89] W.E. Leithead. Control systems for wind turbines. *Wind Engineering*, 13(6):293–301, 1989.

[Let81] N.A. Lethomaki. *Practical Robustness Measures in Multivariable Control System Analysis*. PhD thesis, Dept. of Electrical Eng. and Computer Science, Massachusetts Institute of Technology, Cambridge, MA, 1981.

[Lit93] J. Little. *Control Systems Toolbox*. The MathWorks Inc., Natick, Mass., USA, 2nd edition, 1993.

[Lju85] L. Ljung. Asymtotic variance expressions for identified black-box transfer function models. *IEEE Trans. Aut. Contr.*, AC-30(9):834–844, Sept. 1985.

[Lju87] L. Ljung. *System Identification - Theory for the User*. Prentice Hall Inc., Englewood Cliffs NJ, 1987.

[Lju89] L. Ljung. System identification in a noise free enviroment. In *Proc. Adaptive Systems in Control and Signal Processing*, pages 441–420, Glasgow, UK, 1989. IFAC.

[Lju91] L. Ljung. A discussion of model accuracy in system identification. LiTH-ISY-I 1307, Linköping University, Department of Electrical Engineering, December 1991.

[LPG93] J.-L. Lin, I. Postlethwaite, and D.-W. Gu. $\mu - K$ iteration: A new algorithm for μ-synthesis. *Automatica*, 29(1):219–224, 1993.

[LW93] P. Lindskog and B. Wahlberg. Applications of kautz models in system identification. In *Proc. IFAC World Congress*, volume 5, pages 309–312, Sydney, Australia, July 1993.

[LWH91] L. Ljung, B. Wahlberg, and H. Hjalmarsson. Model quality: The roles of prior knowledge and data information. In *Proc. of the 30th Conf. on Decision and Control*, pages 273–278, Brighton, England, December 1991. IEEE.

[M9̈1] P.M. Mäkilä. Laguerre methods and $\mathcal{H}^\infty$ identification of continuous-time systems. *Int. J. Control*, 53(3):698–707, 1991.

[Mac89] J.M. Maciejowski. *Multivariable Feedback Design*. Addison-Wesley Series in Electronic Systems Engineering. Addison-Wesley, 1989.

[MPG94] P.M. Mäkilä, J.R. Partington, and T.K. Gustafsson. Robust identification. In *Proc. SYSID '94*, volume 1, pages 45–63, Copenhagen, Denmark, July 1994. IFAC.

[MZ89] M. Morari and E. Zafiriou. *Robust Process Control*. Prentice-Hall Inc., 1989.

[NG94] B. Ninness and G.C. Goodwin. Estimation of model quality. In *Proc. SYSID '94*, volume 1, pages 25–44, Copenhagen, Denmark, July 1994. IFAC.

[Nih93] M.T. Nihtilä. Continous-time order-recursive least-squares identification. In *Proc. IFAC World Congress*, volume 5, pages 299–302, Sydney, Australia, July 1993.

[Nin93] B.M. Niness. *Stochastic and Deterministic Modelling*. PhD thesis, University of Newcastle, New South Wales, Australia, Aug. 1993.

[Nin94] B. Ninness. Orthonormal bases for geometric interpretation of the frequency response estimation problem. In *Proc. SYSID '94*, volume 3, pages 591–596, Copenhagen, Denmark, July 1994. IFAC.

[NS91] M.P. Newlin and R.S. Smith. Model validation and a generalization of μ. In *Proc. 30th Conf. on Decision and Control*, pages 1257–1258, Brighton, England, December 1991. IEEE.

[PB87] P.J. Parker and R.R. Bitmead. Adaptive frequency response identification. In *Proc. 26th Conf. on Decision and Control*, pages 348–353, Los Angeles, CA, Dec. 1987.

[PD93] A. Packard and J.C. Doyle. The complex structured singular value. *Automatica*, 29(1):71–109, 1993.

[PKKU94] A. Patra, U. Keuchel, U. Kiffmeier, and H. Unbehauen. Identification for robust control of an unstable plant. In *Proc. SYSID '94*, volume 3, pages 597–602, Copenhagen, Denmark, July 1994. IFAC.

[PM94] J.R. Partington and P.M. Mäkilä. Analysis of linear methods for robust identification in ℓ_1. In *Proc. SYSID '94*, volume 2, pages 79–84, Copenhagen, Denmark, July 1994. IFAC.

[RI94] M.A. Rotea and T. Iwasaki. An alternative to the D-K iteration ? In *Proc. American Control Conf.*, pages 53–55, Baltimore, Maryland, June 1994.

[RVAS85] C.E. Rohrs, L. Valavani, M. Athans, and G. Stein. Robustness of continuous-time adaptive control algorithms in the presence of unmodeled dynamics. *IEEE Trans. Aut. Contr.*, AC-30(9):881–889, Sept. 1985.

[SA87] G. Stein and M. Athans. The LQG/LTR procedure for multivariable feedback control design. *IEEE Trans. Aut. Contr.*, AC-32(2):105–114, February 1987.

[Sch92] R.J.P. Schrama. Accurate identification for control: The necessity of an iterative scheme. *IEEE Trans. Aut. Contr.*, 37(7):991–994, July 1992.

[SD89] R.S. Smith and J.C. Doyle. Model invalidation: A connection between robust control and identification. In *Proc. American Control Conf.*, pages 1435–1440, Pittsburg, PA, 1989.

[SD90] R.S. Smith and J.C. Doyle. Towards a methodology for robust parameter identification. In *Proc. American Control Conf.*, pages 2394–2399, San Diego, USA, 1990.

[SD91] G. Stein and J.C. Doyle. Beyond singular values and loop shapes. *Journal of Guidance, Control and Dynamics*, 14:5–16, Jan. 1991.

[SD92] R.S. Smith and J.C. Doyle. Model validation: A connection between robust control and identification. *IEEE Trans. Aut. Contr.*, 37(7):942–952, July 1992.

[SD93] R.S. Smith and J.C. Doyle. Closed loop relay estimation of uncertain bounds for robust control models. In *Proc. IFAC World Congress*, volume 9, pages 57–60, Sydney, Australia, July 1993.

[SDMS87] R.S. Smith, J.C. Doyle, M. Morari, and A. Skjellum. A case study using μ: Laboratory process control problem. In *Proc. IFAC World Congress*, pages 403–415, Munich, Germany, 1987.

[SLC89] M.G. Safonov, D.J.N. Limebeer, and R.Y. Chiang. Simplifying the $\mathcal{H}_{infty}$ theory via loop-shifting, matrix-pencil and descriptor concepts. *Int. J. Control*, 50(6):2467–2488, 1989.

[SMD88] S. Skogestad, M. Morari, and J.C. Doyle. Robust control of ill-conditioned plants: High-purity destillation. *IEEE Trans. Aut. Contr.*, 33(12):1092–1105, Dec. 1988.

[Smi90] R.S. Smith. *Model Validation for Uncertain Systems*. PhD thesis, California Institute of Technology, 1990.

[Smi93] R.S. Smith. Model validation for robust control: An experimental process control application. In *Proc. IFAC World Congress*, volume 9, pages 61–64, Sydney, Australia, July 1993.

[sN94] K.J. ström and J. Nilsson. Analysis of a scheme for iterated identification and control. In *Proc. SYSID '94*, volume 2, pages 171–176, Copenhagen, Denmark, July 1994. IFAC.

[SS89] T. Söderström and P. Stoica. *System Identification*. Prentice Hall International Series in System and Control Engineering. Prentice Hall, New York, 1989.

[Sto92] A.A. Stoorvogel. *The $\mathcal{H}_\infty$ Control Problem: A State Space Approach*. Prentice Hall Inc., Englewood Cliffs NJ, 1992.

[TC91] S. Tøffner-Clausen. Linear quadratic optimal design methods, theory & design. Master's thesis, Aalborg University, Institute of Electronic Systems, 1991.

[TC95a] S. Tøffner-Clausen. Identification for control: Quantification of uncertainty. In *Proc. Youth Automation Conference*, pages 155–159, Beijing, China, 1995. IFAC.

[TC95b] S. Tøffner-Clausen. *System Identification and Robust Control - A Synergistic Approach*. PhD thesis, Aalborg University, Institute of Electronic Systems, Department of Control Engineering, Fredrik Bajers Vej 7, DK-9220 Aalborg Ø, Denmark, Oktober 1995.

[TCA93] S. Tøffner-Clausen and P. Andersen. Quantifying frequency domain model uncertainty in estimated transfer functions using a stochastic embedding approach. Research Report W1D-06-001, Reliability and Robustness in Industrial Process Control, Feb. 1993.

[TCA94] S. Tøffner-Clausen and P. Andersen. Identification for control: Estimation of frequency domain model uncertainty. Research Report R94-4054, Aalborg University, Dept. of Control Eng., Frederik Bajers Vej 7, DK-9220 Aalborg , Denmark, Aug. 1994.

[TCA95] S. Tøffner-Clausen and P. Andersen. μ-synthesis – a non-conservative methodology for design of controllers with robustness towards dynamic and parametric uncertainty. In *Proc. EURACO Workshop on Recent Results in Robust and Adaptive Control*, pages 269–303, Florence, Italy, September 1995.

[TCABG95] S. Tøffner-Clausen, P. Andersen, S.G. Breslin, and M.J. Grimble. The application of μ-analysis and synthesis to the control of an ASTOVL aircraft. In *Proc. EURACO Workshop on Recent Results in Robust and Adaptive Control*, pages 304–322, Florence, Italy, September 1995.

[TCASN94a] S. Tøffner-Clausen, P. Andersen, J. Stoustrup, and H.H. Niemann. Estimated frequency domain model uncertainties used in robust controller design — a μ-approach. In *Proc. 3rd IEEE Conf. on Control Applications*, volume 3, pages 1585–1590, Glasgow, Scotland, Aug. 1994.

[TCASN94b] S. Tøffner-Clausen, P. Andersen, J. Stoustrup, and H.H. Niemann. Estimated frequency domain model uncertainties used in robust controller design — a μ-approach. Poster Session, EURACO Young Researchers Week, Dublin, Aug. 1994.

[TCASN95] S. Tøffner-Clausen, P. Andersen, J. Stoustrup, and H.H. Niemann. A new approach to μ-synthesis for mixed perturbation sets. In *Proc. 3rd European Control Conf.*, pages 147–152, Rome, Italy, 1995.

[TCB95] S. Tøffner-Clausen and S.G. Breslin. Classical versus modern control design methods for safety critical control engineering practice. Technical Report ACT/CS08/95, Industrial Control Center, University of Strathclyde, 50 George Street, Glasgow G1 1QE, 1995.

[vdHS94] P.M.J. van den Hof and R.J.P. Schrama. Identification and control – closed loop issues. In *Proc. SYSID '94*, volume 2, pages 1–13, Copenhagen, Denmark, July 1994. IFAC.

[vdKvOvdB94] A.C. van der Klauw, J.E.F. van Osch, and P.P.J. van den Bosch. Closed-loop identification methods for lq control design. In *Proc. SYSID '94*, volume 3, pages 609–613, Copenhagen, Denmark, July 1994. IFAC.

[VH95] D.K. De Vries and P.M.J. Van Den Hof. Quantification of uncertainty in transfer function estimation: a mixed probabilistic – worst-case approach. *Automatica*, 31(4):543–557, 1995.

[Wah87] B. Wahlberg. *On the Identification and Approximation of Linear Systems*. PhD thesis, Linköping University, Department of Electrical Engineering, Sweden, 1987.

[Wah91a] B. Wahlberg. Identification of resonant systems using kautz filters. In *Proc. 30th Conf. on Decision and Control*, pages 2005–2010, Brighton, England, Dec. 1991.

[Wah91b] B. Wahlberg. System identification using laquerre models. *IEEE Trans. Aut. Contr.*, 36(5):551–562, May 1991.

[Wah94] B. Wahlberg. Laguerre and kautz models. In *Proc. SYSID '94*, volume 3, pages 1–12, Copenhagen, Denmark, July 1994. IFAC.

[WBM89] C. Webb, H. Budman, and M. Morari. Identifying frequency domain uncertainty bounds for robust controller design - theory with application to a fixed-bed reactor. In *Proc. American Control Conf.*, pages 1528–1533, Pittsburg, PA, 1989.

[WL90a] B. Wahlberg and L. Ljung. Hard frequency-domain model error bounds from least-squares like identification techniques. LITH-ISY-I 1144, Linköping University, Department of Electrical Engineering, Dec. 1990.

[WL90b] B. Wahlberg and L. Ljung. On estimation of transfer function error bounds. LITH-ISY-I 1063, Linköping University, Department of Electrical Engineering, Feb. 1990.

[WL91] B. Wahlberg and L. Ljung. On estimation of transfer function error bounds. LITH-ISY-I 1186, Linköping University, Department of Electrical Engineering, Feb. 1991.

[WL92] B. Wahlberg and L. Ljung. Hard frequency-domain model error bounds from least-squares like identification techniques. *IEEE Trans. Aut. Contr.*, 37(7):900–912, July 1992.

[YÅ94] P.M. Young and K.J. Åström. μ meets bode. In *Proc. American Control Conf.*, pages 1223–1227, Baltimore, Maryland, June 1994.

[YND91] P.M. Young, M.P. Newlin, and J.C. Doyle. μ analysis with real parametric uncertainty. In *Proc. 30th IEEE Conf. on Decision and Control*, pages 1251–1256, Brighton, England, Dec. 1991.

[YND92] P.M. Young, M.P. Newlin, and J.C. Doyle. Practical computation of the mixed μ problem. In *Proc. American Control Conf.*, volume 3, pages 2190–2194, Chicago, Illinois, June 1992.

[You93] P.M. Young. *Robustness with Parametric and Dynamic Uncertainty*. PhD thesis, California Institute of Technology, Pasadena, California, May 1993.

[You94] P.M. Young. Controller design with mixed uncertainties. In *Proc. American Control Conf.*, pages 2333–2337, Baltimore, Maryland, June 1994.

[ZBE91] Y.C. Zhu, A.C.P.M. Backx, and P. Eykhoff. Multivariable process identification for robust control. EUT Report 91-E-249, Eindhoven University of Technology, Faculty of Electrical Engineering, Eindhoven, The Netherlands, Jan. 1991.

[ZBG91] Z. Zang, R.L. Bitmead, and M. Gevers. $\mathcal{H}_2$ iterative model refinement and control robustness enhancement. In *Proc. 30th Conf. on Decision and Control*, pages 279–284, Brighton, England, December 1991. IEEE.

[ZDG96] K. Zhou, J.C. Doyle, and K. Glover. *Robust and Optimal Control*. Prentice Hall, Inc., Upper Sadle River, New Jersey 07458, 1996.

PART V
APPENDICES

APPENDIX A
THE GENERALIZED NYQUIST CRITERION

The generalized Nyquist stability criterion plays a key role in robust control theory. Like the classical Nyquist criterion for scalar systems, the generalized Nyquist criterion enables us to judge the stability of a closed loop system by inspection of the open loop transfer matrix, typically $G(s)K(s)$.

Assume that the closed loop system is given as in Figure A.1 where $G(s)$ and $K(s)$ denote the plant and controller transfer function matrix respectively.

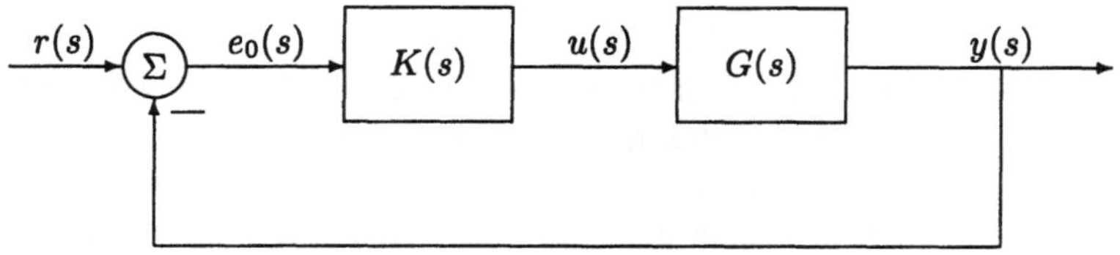

Fig. A.1. *Closed loop system.*

Let the open loop $G(s)K(s)$ have the state-space realization

$$\dot{x}(t) = A_o x(t) + B_o e(t) \tag{A.1}$$
$$y(t) = C_o x(t) + D_o e(t)\ . \tag{A.2}$$

When the loop is closed, $e(t) = r(t) - y(t)$ and

$$y(t) = C_o x(t) + D_o(r(t) - y(t)) \tag{A.3}$$
$$\Leftrightarrow \quad (I + D_o)y(t) = C_o x(t) + D_o r(t) \tag{A.4}$$
$$\Leftrightarrow \quad y(t) = (I + D_o)^{-1} C_o x(t) + (I + D_o)^{-1} D_o r(t) \tag{A.5}$$
$$\Leftrightarrow \quad y(t) = C_c x(t) + D_c r(t) \tag{A.6}$$

where $C_c = (I + D_o)^{-1} C_o$ and $D_c = (I + D_o)^{-1} D_o$. Furthermore,

$$\dot{x}(t) = A_o x(t) + B_o(r(t) - y(t)) \tag{A.7}$$
$$\Leftrightarrow \quad \dot{x}(t) = A_o x(t) + B_o \left(r - (I + D_o)^{-1} C_o x(t) - (I + D_o)^{-1} D_o r(t)\right) \tag{A.8}$$
$$\Leftrightarrow \quad \dot{x}(t) = \left(A_o - B_o(I + D_o)^{-1} C_o\right) x(t) + B_o \left(I - (I + D_o)^{-1} D_o\right) r(t) \tag{A.9}$$
$$\Leftrightarrow \quad \dot{x}(t) = A_c x(t) + B_c r(t) \tag{A.10}$$

where

$$A_c = A_o - B_o(I + D_o)^{-1}C_o \tag{A.11}$$
$$B_c = B_o\left(I - (I + D_o)^{-1}D_o\right) \tag{A.12}$$
$$= B_o\left((I + D_o)^{-1}(I + D_o - D_o)\right) = B_o(I + D_o)^{-1} . \tag{A.13}$$

Let us define the ***open loop characteristic polynomial*** $\phi_{OL}(s)$:

$$\phi_{OL}(s) = \det(sI - A_o) \tag{A.14}$$

and the ***closed loop characteristic polynomial*** $\phi_{CL}(s)$:

$$\phi_{CL}(s) = \det(sI - A_c) . \tag{A.15}$$

Since any transfer matrix $F(s)$ with state-space realization A, B, C, D can be expressed as

$$F(s) = C(sI - A)^{-1}B + D = \frac{1}{\det(sI - A)} D\mathrm{adj}(sI - A)B + D \tag{A.16}$$

where adj denotes the adjoint, then the zeros of $\det(sI - A)$ (the eigenvalues of A) equal the poles of $F(s)$[1] Consequently, the stability of the closed loop system in Figure A.1 is determined by the zeros of the characteristic polynomial $\phi_{CL}(s)$. In order to derive a Nyquist-like criterion we then need to express $\phi_{CL}(s)$ in terms of the open loop $G(s)K(s)$.

For this purpose we define the *return difference* $F(s)$:

$$F(s) = I + G(s)K(s) . \tag{A.17}$$

Furthermore, we will need the following lemma which will be given without proof.

Lemma A.1 (Schur's Formula for Partitioned Determinants). *Let a quadratic matrix P be partitioned like*

$$P = \begin{bmatrix} P_{11} & P_{12} \\ P_{21} & P_{22} \end{bmatrix} . \tag{A.18}$$

Then the determinant for P can be expressed as:

$$\det P = \det P_{11} \det(P_{22} - P_{21}P_{11}^{-1}P_{12}) , \qquad \textit{if } \det P_{11} \neq 0 \tag{A.19}$$

or

$$\det P = \det P_{22} \det(P_{11} - P_{12}P_{22}^{-1}P_{21}) , \qquad \textit{if } \det P_{22} \neq 0. \tag{A.20}$$

[1] In fact, we must require the state-space realization to be *minimal.*

We may express the determinant for the return difference $F(s)$ as:

$$\det F(s) = \det(I + C_o(sI - A_o)^{-1}B_o + D_o) \,. \tag{A.21}$$

Since $F(s)$ is quadratic we may apply Lemma A.1 (Equation (A.19)) with:

$$P_{11} = sI - A_o \qquad P_{12} = -C_o \tag{A.22}$$

$$P_{21} = B_o \qquad P_{22} = I + D_o \,. \tag{A.23}$$

$\det F(s)$ can thus be written

$$\det F(s)\det(sI - A_o) = \det\left(\begin{bmatrix} sI - A_o & B_o \\ -C_o & I + D_o \end{bmatrix}\right) \tag{A.24}$$

$$\Leftrightarrow \quad \det F(s) = \frac{1}{\det(sI - A_o)}\det\left(\begin{bmatrix} sI - A_o & B_o \\ -C_o & I + D_o \end{bmatrix}\right) . \tag{A.25}$$

By applying Schur's formula again we may furthermore show that

$$\det\left(\begin{bmatrix} I_r & -B_o(I + D_o)^{-1} \\ 0 & I_n \end{bmatrix}\right) = \det I_r \det\left(I_n + 0I_rB_o(I + D_o)^{-1}\right) \tag{A.26}$$

$$= \det I_r \det I_n = 1 \tag{A.27}$$

Combining Equation (A.27) with (A.25) we get

$$\det F(s) = \frac{1}{\det(sI - A_o)}\det\left(\begin{bmatrix} I_r & -B_o(I + D_o)^{-1} \\ 0 & I_n \end{bmatrix}\right) \cdot \det\left(\begin{bmatrix} sI - A_o & B_o \\ -C_o & I + D_o \end{bmatrix}\right) \tag{A.28}$$

$$\Rightarrow \quad \det F(s) = \frac{1}{\det(sI - A_o)} \cdot \det\left(\begin{bmatrix} sI - A_o + B_o(I + D_o)^{-1}C_o & 0 \\ -C_o & I + D_o \end{bmatrix}\right) \tag{A.29}$$

$$\Rightarrow \quad \det F(s) = \frac{1}{\det(sI - A_o)}\det(sI - A_c) \cdot \det(I + D_o + c_o(sI - A_c)^{-1}0) \tag{A.30}$$

$$\Leftrightarrow \quad \det F(s) = \frac{1}{\det(sI - A_o)}\det(sI - A_c)\det(I + D_o) \,. \tag{A.31}$$

Since $\lim_{s\to\infty} F(s) = I + D_o$ we finally achieve

$$\det F(s) = \frac{\phi_{\mathrm{CL}}}{\phi_{\mathrm{OL}}}\det F(\infty) \tag{A.32}$$

$$\Leftrightarrow \quad \phi_{\mathrm{CL}} = \frac{\det F(s)}{\det F(\infty)}\phi_{\mathrm{OL}} \,. \tag{A.33}$$

We have now expressed the closed loop characteristic polynomial ϕ_{CL} by the return difference $F(s)$ and the open loop characteristic polynomial ϕ_{OL}. Note that $\det F(\infty)$ is just a constant scaling. For physical systems, the open loop is usually strictly proper and then $\det F(\infty) = 1$. We may now relate the zeros of $\det F(s)$ to the poles in the open loop and closed loop system respectively. From Equation (A.32) it is seen that

- Closed loop poles, i.e zeros of ϕ_{CL}, appear as zeros in $\det F(s)$.
- Open loop poles, i.e zeros of ϕ_{OL}, appear as poles in $\det F(s)$.

Let us then assume that $\det F(s)$ has n_{p} poles and n_{z} zeros in the open right half plane (RHP). Precisely as for scalar systems we then have from the *principle of the argument* that

$$\Delta \arg \det F(s) = -2\pi(n_{\mathrm{z}} - n_{\mathrm{p}}) \tag{A.34}$$

where $\Delta \arg$ denotes the change in the argument (phase) of $\det F(s)$ when s traverses the Nyquist $\mathcal{D}$ contour[2] once. Thus, $\Delta \arg /(2\pi)$ gives the number of counterclockwise encirclements around the origo.

If the closed loop system is to be stable we must have that $n_{\mathrm{z}} = 0$ since n_{z} equals the number of closed loop RHP poles. Thus

$$n_{\mathrm{z}} = 0 \Rightarrow \Delta \arg \det F(s) = 2\pi n_{\mathrm{p}} \,. \tag{A.35}$$

Equation (A.35) then provides the generalized Nyquist stability criterion:

Theorem A.1 (Generalized Nyquist Stability Criterion). *If the open loop transfer function matrix $G(s)K(s)$ has p poles in the right-half s-plane, then the closed loop system is stable if and only if the map of* $\det(I + G(s)K(s))$, *as s traverses the Nyquist $\mathcal{D}$ contour once, encircles the origin p times anti-clockwise assuming no right-half s-plane zero-pole cancelations have occurred forming the product $G(s)K(s)$.*

[2] Remember that the Nyquist $\mathcal{D}$ contour travels up the imaginary axis from origo to infinity, then along a semicircular arc in the right half-plane until it meets the negative imaginary axis, and finally up towards the origin. If any poles are encountered on the imaginary axis the contour is indented so as to exclude these poles.

APPENDIX B

SCALING AND LOOP SHIFTING FOR H_∞

In this appendix, the necessary scalings and loop shifting for transforming a general D state space matrix into the form required by Theorem 4.6 on page 54 will be given. The main idea is that these transformations do not change the $\mathcal{H}_\infty$-norm of the closed loop system. Let the general plant $N(s)$ be given by:

$$N(s) = \left[\begin{array}{c|cc} A & B_1 & B_2 \\ \hline C_1 & D_{11} & D_{12} \\ C_2 & D_{21} & D_{22} \end{array}\right] \tag{B.1}$$

where $A \in \mathbf{R}^{n\times n}$, $B_1 \in \mathbf{R}^{n\times d}$, $B_2 \in \mathbf{R}^{n\times m}$, $C_1 \in \mathbf{R}^{e\times n}$, $C_2 \in \mathbf{R}^{r\times n}$, $D_{11} \in \mathbf{R}^{e\times d}$, $D_{12} \in \mathbf{R}^{e\times m}$, $D_{21} \in \mathbf{R}^{r\times d}$ and finally $D_{22} \in \mathbf{R}^{r\times m}$. Consequently n is the order of the generalized plant and the dimensions of the inputs d' and u and the outputs e' and y are given by:

$$d = \dim\ (d') \tag{B.2}$$

$$m = \dim\ (u) \tag{B.3}$$

$$e = \dim\ (e') \tag{B.4}$$

$$r = \dim\ (y)\ . \tag{B.5}$$

It will be required that D_{12} is "tall" with full column rank and that D_{21} is "fat" with full row rank. This means that:

$$\text{rank}\ (D_{12}) = \dim\ (u) = m \tag{B.6}$$

$$\text{rank}\ (D_{21}) = \dim\ (y) = r\ . \tag{B.7}$$

Now complete the following steps:

Step 1. Use the singular value decomposition (SVD) to factor D_{12} and D_{21}:

$$D_{12} = U_1 \begin{bmatrix} 0_{(e-m)\times m} \\ \Sigma_1 \end{bmatrix} V_1^T \tag{B.8}$$

$$D_{21} = U_2 \begin{bmatrix} 0_{r\times(d-r)} & \Sigma_2 \end{bmatrix} V_2^T \tag{B.9}$$

where $U_1 \in \mathbf{R}^{e\times e}$, $V_1 \in \mathbf{R}^{m\times m}$, $U_2 \in \mathbf{R}^{r\times r}$ and $V_2 \in \mathbf{R}^{d\times d}$ are unitary matrices and $\Sigma_1 \in \mathbf{R}^{m\times m}$ and $\Sigma_2 \in \mathbf{R}^{r\times r}$ are diagonal matrices of singular values. Notice that the above SVD is not in the standard format provided by eg MATLAB, so the columns of U_1 and the rows of V_2 may have to be rearranged to suit this format.

Step 2. Scale D_{11} and partition it into a block 2×2 matrix:

$$\tilde{D}_{11} = U_1^T D_{11} V_2 = \begin{bmatrix} \tilde{D}_{11_{11}} & \tilde{D}_{11_{12}} \\ \tilde{D}_{11_{21}} & \tilde{D}_{11_{22}} \end{bmatrix} \tag{B.10}$$

where $\tilde{D}_{11} \in \mathbf{R}^{e\times d}$, $\tilde{D}_{11_{11}} \in \mathbf{R}^{(e-r)\times(d-m)}$, $\tilde{D}_{11_{12}} \in \mathbf{R}^{(e-r)\times m}$, $\tilde{D}_{11_{21}} \in \mathbf{R}^{r\times(d-m)}$ and $\tilde{D}_{11_{22}} \in \mathbf{R}^{r\times m}$.

Step 3. Let

$$K_\infty = -\left(\tilde{D}_{11_{22}} + \tilde{D}_{11_{21}}\left(\gamma^2 I - \tilde{D}_{11_{11}}^T \tilde{D}_{11_{11}}\right)^{-1} \tilde{D}_{11_{11}}^T \tilde{D}_{11_{12}}\right) \tag{B.11}$$

such that $K_\infty \in \mathbf{R}^{r\times m}$.

Step 4. Calculate

$$\bar{\tilde{D}}_{11} = \tilde{D}_{11} + \begin{bmatrix} 0 & 0 \\ 0 & K_\infty \end{bmatrix} = \begin{bmatrix} \tilde{D}_{11_{11}} & \tilde{D}_{11_{12}} \\ \tilde{D}_{11_{21}} & \tilde{D}_{11_{22}} + K_\infty \end{bmatrix} \tag{B.12}$$

and the transformation matrix

$$\Theta = \begin{bmatrix} \Theta_{11} & \Theta_{12} \\ \Theta_{21} & \Theta_{22} \end{bmatrix} \tag{B.13}$$

$$= \begin{bmatrix} -\bar{\tilde{D}}_{11} & \left(I - \gamma^{-2}\bar{\tilde{D}}_{11}\bar{\tilde{D}}_{11}^T\right)^{1/2} \\ \left(I - \gamma^{-2}\bar{\tilde{D}}_{11}^T\bar{\tilde{D}}_{11}\right)^{1/2} & \gamma^{-2}\bar{\tilde{D}}_{11}^T \end{bmatrix}. \tag{B.14}$$

Step 5. Let

$$\tilde{D}_{12} = \Theta_{12}^{-1} \begin{bmatrix} 0 \\ I \end{bmatrix} \tag{B.15}$$

$$\tilde{D}_{21} = \begin{bmatrix} 0 & I \end{bmatrix} \Theta_{21}^{-1} \tag{B.16}$$

$$\tilde{D}_{22} = \begin{bmatrix} 0 & I \end{bmatrix} \Theta_{22}\Theta_{12}^{-2} \begin{bmatrix} 0 \\ I \end{bmatrix} \tag{B.17}$$

denote the last m columns of Θ_{12}^{-1}, the last r rows of Θ_{21}^{-1} and the last $r \times m$ rows and columns of $\Theta_{22}\Theta_{12}^{-2}$ respectively.

singular value decomposition (SVD) to factor $\tilde{D}_{12}$ and $\tilde{D}_{21}$:

$$\tilde{D}_{12} = U_3 \begin{bmatrix} 0_{(e-m)\times m} \\ \Sigma_3 \end{bmatrix} V_3^T \tag{B.18}$$

$$\tilde{D}_{21} = U_4 \begin{bmatrix} 0_{r\times(d-r)} & \Sigma_4 \end{bmatrix} V_4^T \tag{B.19}$$

where $U_3 \in \mathbf{R}^{e\times e}$, $V_3 \in \mathbf{R}^{m\times m}$, $U_4 \in \mathbf{R}^{r\times r}$ and $V_4 \in \mathbf{R}^{d\times d}$ are unitary matrices and $\Sigma_3 \in \mathbf{R}^{m\times m}$ and $\Sigma_4 \in \mathbf{R}^{r\times r}$ are diagonal matrices of singular values.

Step 7. Calculate the combined transformation above (T_1) and below (T_2) $N(s)$, see Figure B.1.

$$T_1 = \begin{bmatrix} U_3^T\Theta_{11}V_4 & U_3^T\Theta_{12}U_1^T \\ V_2\Theta_{21}V_4 & V_2\Theta_{22}U_1^T \end{bmatrix} \tag{B.20}$$

$$T_2 = \begin{bmatrix} T_{2_{11}} & T_{2_{12}} \\ T_{2_{21}} & T_{2_{22}} \end{bmatrix} \tag{B.21}$$

where

$$T_{2_{11}} = V_1\Sigma_1^{-1}K_\infty\Sigma_2^{-1}U_2^T\,(I-L_1)^{-1} \tag{B.22}$$

$$T_{2_{12}} = V_1\Sigma_1^{-1}\,(I-L_2)^{-1}\,V_3\Sigma_3^{-1} \tag{B.23}$$

$$T_{2_{21}} = \Sigma_4^{-1}U_4^T\Sigma_2^{-1}U_2^T\,(I-L_1)^{-1} \tag{B.24}$$

$$T_{2_{22}} = -\Sigma_4^{-1}U_4^T\left[\Sigma_2^{-1}U_2^TD_{22}V_1\Sigma_1^{-1}\,(I-L_2)^{-1} - \tilde{D}_{22}\right]V_3\Sigma_3^{-1} \tag{B.25}$$

$$L_1 = -D_{22}V_1\Sigma_1^{-1}K_\infty\Sigma_2^{-1}U_2^T \tag{B.26}$$

$$L_2 = -K_\infty\Sigma_2^{-1}U_2^TD_{22}V_1\Sigma_1^{-1}\,. \tag{B.27}$$

Step 8. Form the transformed system:

$$\tilde{N}(s) = F_\ell(T_1, F_\ell(N(s), T_2))\,. \tag{B.28}$$

The D state-space matrix of $\tilde{N}(s)$ can now be shown to satisfy:

$$D_{11} = 0 \tag{B.29}$$

$$D_{12} = \begin{bmatrix} 0 \\ I \end{bmatrix} \tag{B.30}$$

$$D_{21} = \begin{bmatrix} 0 & I \end{bmatrix} \tag{B.31}$$

$$D_{22} = 0\,. \tag{B.32}$$

Step 9. Compute the suboptimal $\mathcal{H}_\infty$ controller $\tilde{K}(s)$ for the transformed system $\tilde{N}(s)$ and back-transform it to the original system $N(s)$. This is done via the linear fractional transformation:

$$K(s) = F_\ell(T_2, \tilde{K}(s)) \tag{B.33}$$

Note that the transformation matrix T_1 is not needed to back-transform the controller.

The scalings and loop shifting operations described above is illustrated in block-diagram form in Figure B.1. Notice that since finding a optimal $\mathcal{H}_\infty$ controller requires iteration on γ, for each iteration the scaling and loop shifting involving γ must be redone.

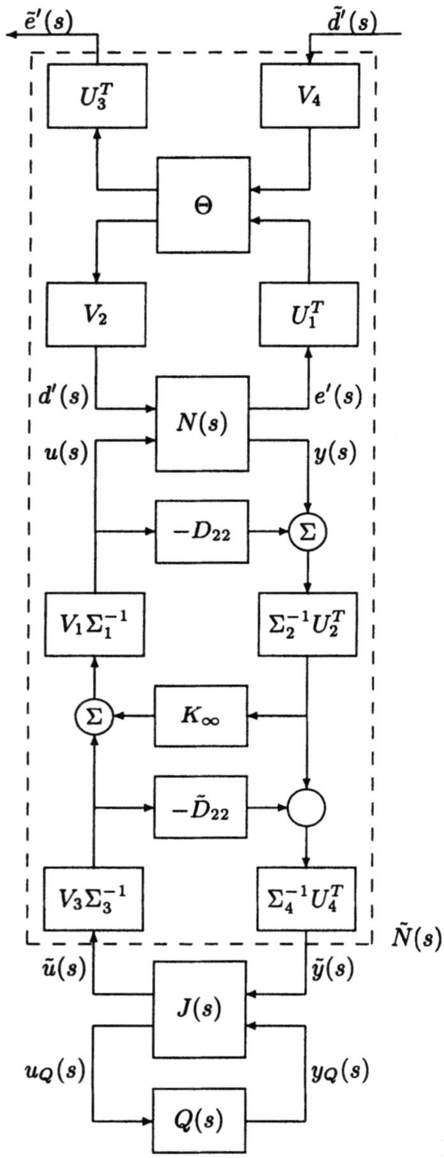

Fig. B.1. *Scaling and loop shifting for $\mathcal{H}_\infty$ suboptimal control.*

APPENDIX C
CONVERGENCE OF μ-K ITERATION

Here proof will be given for Lemma 5.1 on page 85. It will be shown that the minimizations:

$$K_i(s) = \arg \min_{K(s)\in\mathcal{K}_S} \sup_{\omega} \left\{\bar{\sigma}\left(F_\ell\left(N_{\mathrm{D}\Gamma_{i-1}}(j\omega), K(j\omega)\right)\right)\right\} \tag{C.1}$$

$$D_i^*(\omega) = \arg \min_{D\in\mathbf{D}} \bar{\sigma}\left(DF_\ell(N(j\omega), K_i(j\omega))D^{-1}\right), \qquad \forall\omega \tag{C.2}$$

from μ-K iteration are monotonically non-increasing in $\|F_\ell(N_{\mathrm{D}\Gamma_i}, K_i)\|_{\mathcal{H}_\infty}$. Given a controller $K_i(s)$ and scaling matrices $\Gamma_i(s)$ and $D_i(s)$ it is easy to show that:

$$\begin{aligned}
&\|F_\ell(N_{\mathrm{D}\Gamma_i}(s), K_i(s))\|_{\mathcal{H}_\infty} \\
&\quad = \sup_{\omega} \bar{\sigma}\left(F_\ell\left(\Gamma_i(j\omega)D_i(j\omega)N(j\omega)D_i^{-1}(j\omega), K_i(j\omega)\right)\right) && \text{(C.3)}\\
&\quad = \sup_{\omega} |\gamma_i(j\omega)|\bar{\sigma}\left(F_\ell\left(D_i(j\omega)N(j\omega)D_i^{-1}(j\omega), K_i(j\omega)\right)\right) . && \text{(C.4)}
\end{aligned}$$

Furthermore, assuming perfect realizations of the $\gamma_i(s)$ scalings, we have that

$$|\gamma_i(j\omega)| = (1-\alpha_i)|\gamma_{i-1}(j\omega)| + \alpha_i \frac{\bar{\mu}_{\tilde{\mathbf{\Delta}}}\left(F_\ell\left(N(j\omega), K_i(j\omega)\right)\right)}{\bar{\mu}_{\tilde{\mathbf{\Delta}}_c}\left(F_\ell\left(N(j\omega), K_i(j\omega)\right)\right)} \tag{C.5}$$

$$= |\gamma_{i-1}(j\omega)|\left(1+\alpha_i\left(\frac{\bar{\mu}_{\tilde{\mathbf{\Delta}}}\left(F_\ell\left(N(j\omega), K_i(j\omega)\right)\right)}{\bar{\mu}_{\tilde{\mathbf{\Delta}}_c}\left(F_\ell\left(N(j\omega), K_i(j\omega)\right)\right)}\frac{1}{|\gamma_{i-1}(j\omega)|} - 1\right)\right) \tag{C.6}$$

$$= |\gamma_{i-1}(j\omega)|\left(1+\alpha_i\beta_i(\omega)\right) . \tag{C.7}$$

Consequently:

$$\begin{aligned}
\|F_\ell(N_{\mathrm{D}\Gamma_i}(s), K_i(s))\|_{\mathcal{H}_\infty} &= \sup_{\omega}\left(1+\alpha_i\beta_i(\omega)\right)|\gamma_{i-1}(j\omega)| \cdot \\
&\quad \bar{\sigma}\left(F_\ell\left(D_i(j\omega)N(j\omega)D_i^{-1}(j\omega), K_i(j\omega)\right)\right) . \qquad \text{(C.8)}
\end{aligned}$$

In order to achieve a monotonically non-increasing algorithm it is thus required that

$$\begin{aligned}
&\sup_{\omega}\left(1+\alpha_i\beta_i(\omega)\right)|\gamma_{i-1}(j\omega)|\bar{\sigma}\left(F_\ell\left(D_i(j\omega)N(j\omega)D_i^{-1}(j\omega), K_i(j\omega)\right)\right) \leq \\
&\qquad\qquad \left\|F_\ell(N_{\mathrm{D}\Gamma_{i-1}}(s), K_{i-1}(s))\right\|_{\mathcal{H}_\infty} . \qquad \text{(C.9)}
\end{aligned}$$

We now have the inequalities

$$\begin{aligned}
&|\gamma_{i-1}(j\omega)|\bar{\sigma}\left(F_\ell\left(D_i(j\omega)N(j\omega)D_i^{-1}(j\omega), K_i(j\omega)\right)\right) \\
&\leq |\gamma_{i-1}(j\omega)|\bar{\sigma}\left(F_\ell\left(D_{i-1}(j\omega)N(j\omega)D_{i-1}^{-1}(j\omega), K_i(j\omega)\right)\right) && \text{(C.10)} \\
&\leq \sup_\omega |\gamma_{i-1}(j\omega)|\bar{\sigma}\left(F_\ell\left(D_{i-1}(j\omega)N(j\omega)D_{i-1}^{-1}(j\omega), K_i(j\omega)\right)\right) && \text{(C.11)} \\
&\leq \sup_\omega |\gamma_{i-1}(j\omega)|\bar{\sigma}\left(F_\ell\left(D_{i-1}(j\omega)N(j\omega)D_{i-1}^{-1}(j\omega), K_{i-1}(j\omega)\right)\right) && \text{(C.12)} \\
&= \left\|F_\ell(N_{\mathrm{D}\Gamma_{i-1}}(s), K_{i-1}(s))\right\|_{\mathcal{H}_\infty}. && \text{(C.13)}
\end{aligned}$$

The first inequality follows from (C.2) with perfect realizations of the scalings $D_i(s)$. The last inequality follows from (C.1). Since $\alpha_i \in [0;1]$ and $\beta_i(\omega) \geq -1$ it then becomes clear from (C.10)-(C.13) and (C.8) that if $\beta_i(\omega) \leq 0,\ \forall \omega \geq 0$ then

$$\|F_\ell(N_{\mathrm{D}\Gamma_i}(s), K_i(s))\|_{\mathcal{H}_\infty} \leq \left\|F_\ell(N_{\mathrm{D}\Gamma_{i-1}}(s), K_{i-1}(s))\right\|_{\mathcal{H}_\infty} \quad \text{(C.14)}$$

and the algorithm will be monotonically non-increasing for all values of α_i. We may hence choose $\alpha_i = 1$. If $\beta_i(\omega) < 0,\ \forall \omega \geq 0$ it is thus guaranteed that the ∞-norm $\|F_\ell(N_{\mathrm{D}\Gamma_i}(s), K_i(s))\|_{\mathcal{H}_\infty}$ will be reduced during the i'th step of the iteration.

If $\beta_i(\omega) > 0$ for any frequency ω we must choose α so that (C.9) is fulfilled. Since the frequency at which the supremum is reached depends on α_i we must solve the inequality for all frequencies ω. Let $\bar{\alpha}_i$ denote the solution to (C.9) with inequality replaced with equality. It is easily verified that

$$\bar{\alpha}_i(\omega) = \left(\frac{\left\|F_\ell(N_{\mathrm{D}\Gamma_{i-1}}(s), K_{i-1}(s))\right\|_{\mathcal{H}_\infty}}{\bar{\sigma}\left(F_\ell(D_i(j\omega)N(j\omega)D_i^{-1}(j\omega), K_i(j\omega))\right)|\gamma_{i-1}(j\omega)|} - 1\right)\frac{1}{\beta_i(\omega)}. \quad \text{(C.15)}$$

Thus if we choose α_i such that $\alpha_i \leq \min_\omega \bar{\alpha}_i(\omega)$ then (C.9) will be fulfilled and the algorithm will be monotonically non-increasing if $\alpha_i \geq 0$. From (C.10) we have that

$$\left\|F_\ell(N_{\mathrm{D}\Gamma_{i-1}}(s), K_{i-1}(s))\right\|_{\mathcal{H}_\infty} \geq \bar{\sigma}\left(F_\ell(D_i(j\omega)N(j\omega)D_i^{-1}(j\omega), K_i(j\omega))\right)|\gamma_{i-1}(j\omega)| \quad \text{(C.16)}$$

where the equality holds only if $D_i(j\omega) = D_{i-1}(j\omega)$ and $K_i(j\omega) = K_{i-1}(j\omega)$ at the particular frequency at which the maximum is reached. Since $\beta_i(\omega) > 0$ it is then clear from (C.15) and (C.16) that $\alpha \geq 0$. If $\min_\omega \bar{\alpha}_i(\omega) > 0$ we may choose $\alpha_i > 0$ and it will be guaranteed that the ∞-norm $\|F_\ell(N_{\mathrm{D}\Gamma_i}(s), K_i(s))\|_{\mathcal{H}_\infty}$ is reduced during the i'th step of the iteration.

APPENDIX D
RIGID BODY MODEL OF ASTOVL AIRCRAFT

In the following descriptions and numerical values are given for the rigid body aircraft model introduced in Chapter 7, see Figure 7.1.

D.1 Flight Control Computer Hardware $G_c(s)$

The flight control computer and sample and hold delays are each represented by a first order Padé approximation in each input channel:

$$G_{\text{fcc}}(s) = \frac{-0.00725s + 1}{0.00725s + 1} \tag{D.1}$$

$$G_{\text{s/h}}(s) = \frac{-0.00208s + 1}{0.00417s + 1} \, . \tag{D.2}$$

Furthermore, the input signals are converted into demands on the rear nozzle angle θ_{R}, forward thrust T_{F} and rear thrust T_{R} through a 3×3 transformation matrix T given by

$$T = \begin{bmatrix} -2.8610 \cdot 10^{-5} & 5.6070 \cdot 10^{-4} & -1.7920 \cdot 10^{-2} \\ 1.5120 \cdot 10^{-1} & -4.8362 \cdot 10^{-1} & -7.3830 \cdot 10^{-2} \\ -1.6560 \cdot 10^{-1} & -5.2706 \cdot 10^{-1} & 1.3958 \cdot 10^{-1} \end{bmatrix} . \tag{D.3}$$

$G_{\text{c}}(s)$ is then given by

$$G_{\text{c}}(s) = G_{\text{fcc}}(s) G_{\text{s/h}}(s) I_3 \cdot T \, . \tag{D.4}$$

D.2 Engine and Actuation Model $G_{\text{E}}(s)$

A block diagram representation of the engine and actuation model is given in Figure D.1. The thrust generated by the engine compressor is split into two parts, one part is ejected through the rear nozzle at a varying angle θ_{R}, to the horizontal, to produce a thrust T_{F}. The remaining thrust is directed to the forward nozzles where it can be augmented by plenum chamber burning before being ejected at a fixed angle θ_{F}, to produce the thrust T_{F}. The controller generates demands for rear nozzle actuator angle, and the magnitudes of the thrust ejected through the forward and rear nozzles.

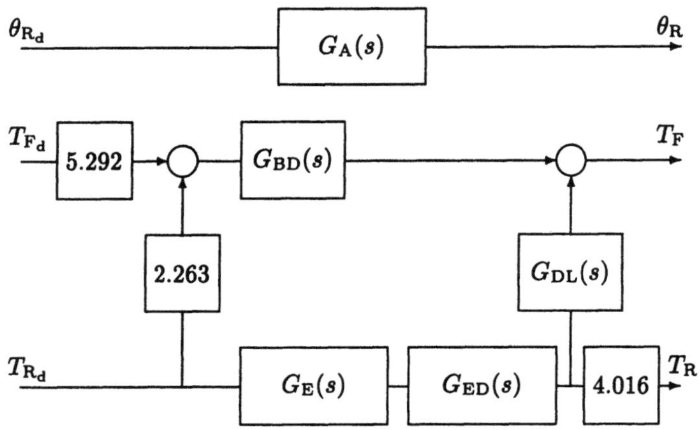

Fig. D.1. *Engine and actuation system.*

The dynamics of the rear nozzle actuator are represented by a second order delay given by,

$$G_A = \frac{w_{n_A}^2}{s^2 + 2\zeta_A w_{n_A} s + w_{n_A}^2} \tag{D.5}$$

where $\zeta_A = 0.7$ and $w_{n_A} = 20.75$.

The plenum chamber burner delay is represented by a first order delay as,

$$G_{BD} = \frac{a_1}{1 + \tau_{BD} s} \tag{D.6}$$

where $a_1 = 0.1885$ and $\tau_{BD} = 0.06$.

The duct lag is also represented as a first order delay as,

$$G_{DL} = \frac{a_2}{1 + \tau_{DL} s} \tag{D.7}$$

where $a_2 = 1.713$ and $\tau_{DL} = 0.085$.

The engine time delay is approximated by a 2nd order Padé approximation as,

$$G_{ED} = \frac{1 - \frac{\tau_{ED}}{2} s + \frac{\tau_{ED}}{96} s^2}{1 + \frac{\tau_{ED}}{2} s + \frac{\tau_{ED}}{96} s^2} \tag{D.8}$$

where $\tau_{ED} = 0.125$.

The engine dynamics are represented by a second order transfer function as,

$$G_E = \frac{a_3 w_{n_E}^2}{s^2 + 2\zeta_E w_{n_E} s + w_{n_E}^2} \tag{D.9}$$

where $a_3 = 0.442$, $\zeta_E = 0.5$ and $w_{n_E} = 8.0$.

D.3 Force Transformation Matrix $\mathbf{F_{mat}}$

The force transformation matrix resolves the forward and rear thrusts and the rear nozzle angle into an axial force, a normal force and a pitching moment. The relationship is:

$$\begin{bmatrix} X_{\mathrm{F}} \\ Z_{\mathrm{F}} \\ M \end{bmatrix} = F_{\mathrm{mat}} \begin{bmatrix} \theta_{\mathrm{R}} \\ T_{\mathrm{F}} \\ T_{\mathrm{R}} \end{bmatrix} \tag{D.10}$$

where F_{mat} is given by

$$F_{\mathrm{mat}} = \frac{1}{\frac{1}{2}\rho V^2 S_{\mathrm{ref}}} \frac{57.3}{2} \cdot$$
$$\begin{bmatrix} -T_{\mathrm{R}} \sin\theta_{\mathrm{R}} & \cos\theta_{\mathrm{F}} & \cos\theta_{\mathrm{R}} \\ T_{\mathrm{R}} \cos\theta_{\mathrm{R}} & -sin\theta_{\mathrm{F}} & -sin\theta_{\mathrm{R}} \\ \frac{T_{\mathrm{R}}(-z_{\mathrm{R}} \sin\theta_{\mathrm{R}} - x_{\mathrm{R}} \cos\theta_{\mathrm{R}})}{l_{\mathrm{T}}} & \frac{x_{\mathrm{F}} \sin\theta_{\mathrm{F}} + z_{\mathrm{F}} \cos\theta_{\mathrm{F}}}{l_{\mathrm{T}}} & \frac{z_{\mathrm{R}} \cos\theta_{\mathrm{R}} - x_{\mathrm{R}} \sin\theta_{\mathrm{R}}}{l_{\mathrm{T}}} \end{bmatrix} \tag{D.11}$$

where the parameters x_{F}, z_{F}, x_{R} and z_{R} denotes the position of the nozzles with respect to the center of gravity, S_{ref} is the wing area, ρ the air density, V the true air speed and finally l_{T} the tail arm.

Substitution of the numerical values for this flight condition and the scalings mentioned above yield the following matrix:

$$F_{\mathrm{mat}} = 28.6499 \begin{bmatrix} -6.621 \cdot 10^{-1} & 1.570 \cdot 10^{-3} & 1.680 \cdot 10^{-3} \\ -9.352 \cdot 10^{-2} & -1.190 \cdot 10^{-2} & -1.188 \cdot 10^{-2} \\ -9.305 \cdot 10^{-2} & 7.410 \cdot 10^{-3} & -6.290 \cdot 10^{-3} \end{bmatrix} . \tag{D.12}$$

D.4 Rigid Aircraft Frame $\mathbf{G_A(s)}$

The linearized equations of motion for the longitudinal dynamics of the rigid body aircraft flying at Mach 0.151, 100 ft and 6° angle of attack are given by the following state-space equations :

$$\frac{d}{dt} \begin{bmatrix} u \\ w \\ \theta \\ q \end{bmatrix} = A \begin{bmatrix} u \\ w \\ \theta \\ q \end{bmatrix} + B \begin{bmatrix} X_{\mathrm{force}} \\ Z_{\mathrm{force}} \\ M \end{bmatrix} \tag{D.13}$$

$$\begin{bmatrix} q \\ \theta \\ \dot{h} \\ a_x \end{bmatrix} = C \begin{bmatrix} u \\ w \\ \theta \\ q \end{bmatrix} + D \begin{bmatrix} X_{\mathrm{force}} \\ Z_{\mathrm{force}} \\ M \end{bmatrix} \tag{D.14}$$

where A, B, C and D are given by,

$$A = \begin{bmatrix} -0.0017 & 0.0413 & -5.3257 & -9.7565 \\ -0.0721 & -0.3393 & 49.5146 & -1.0097 \\ -0.0008 & 0.0138 & -0.2032 & 0.0009 \\ 0 & 0 & 1.0000 & 0 \end{bmatrix} \tag{D.15}$$

$$B = \begin{bmatrix} 0.2086 & -0.0005 & -0.0271 \\ -0.0005 & 0.2046 & 0.0139 \\ -0.0047 & 0.0023 & 0.1226 \\ 0 & 0 & 0 \end{bmatrix} \tag{D.16}$$

$$C = \begin{bmatrix} 0 & 0 & 57.2958 & 0 \\ 0 & 0 & 0 & 57.2958 \\ 0.1045 & -0.9945 & 0.1375 & 51.3791 \\ -0.0002 & 0.0045 & 0 & 0 \end{bmatrix} \tag{D.17}$$

$$D = \begin{bmatrix} 0 & 0 & 0 \\ 0 & 0 & 0 \\ 0 & 0 & 0 \\ 0.0212 & 0 & 0 \end{bmatrix} . \tag{D.18}$$

D.5 Sensor Transfer Matrix $G_S(s)$

The delays in each sensor channel are represented by :

$$G_s(s) = \frac{1 - 0.005346s + 0.0001903s^2}{1 + 0.03082s + 0.0004942s^2} . \tag{D.19}$$

In addition, anti-aliasing filters are included in the pitch rate and pitch attitude channels, given by :

$$G_{aa}(s) = \frac{1}{1 + 0.00398s + 0.0000158s^2} . \tag{D.20}$$

Thus $G_S(s)$ is given by

$$G_S(s) = \begin{bmatrix} G_s(s)G_{aa}(s) & 0 & 0 & 0 \\ 0 & G_s(s)G_{aa}(s) & 0 & 0 \\ 0 & 0 & G_s(s) & 0 \\ 0 & 0 & 0 & G_s(s) \end{bmatrix} . \tag{D.21}$$

APPENDIX E

COMPUTING THE PARAMETER ESTIMATE $\hat{\theta}_N$

In this appendix, a brief description of a general search algorithm for finding the parameter estimate $\hat{\theta}_N$ will be given. The algorithm is easy to implement in eg MATLAB. The search for $\hat{\theta}_N$ is considerably simplified for the special case where the predictor can be written as a linear expression in θ:

$$\hat{y}(k|\theta) = \phi^T(k)\theta \tag{E.1}$$

where $\phi(k)$ is known as the state vector. An example of a model with linear predictor is the well-known ARX model structure

$$A(q)y(k) = B(q)u(k) + e(k) . \tag{E.2}$$

Notice that the ARX model corresponds to the general model (9.1) on page 129 with $F(q) = C(q) = D(q) = 1$. The ARX model predictor is thus given by

$$\hat{y}(k|\theta) = B(q)u(k) + (1 - A(q))y(k) = \phi^T(k)\theta \tag{E.3}$$

for

$$\phi^T(k) = [-y(k-1), \cdots, -y(k-n_a), u(k-1), \cdots, u(k-n_b)] \tag{E.4}$$

$$\theta^T = [a_1, \cdots, a_{n_a}, b_1, \cdots, b_{n_b}] . \tag{E.5}$$

A minimum for the criterion (9.10) can now be computed from

$$V'_N(\hat{\theta}_N, Z^N) = 0 \tag{E.6}$$

$$\Leftrightarrow \quad -\frac{1}{N}\sum_{k=1}^{N} \psi(k, \hat{\theta}_N)\epsilon(k, \hat{\theta}_N) = 0 \tag{E.7}$$

where the model gradient $\psi(k, \hat{\theta}_N)$ is given by

$$\psi(k, \hat{\theta}_N) = \frac{\partial \hat{y}(k|\hat{\theta}_N)}{\partial \theta} = \phi(k) . \tag{E.8}$$

Note that the model gradient for the ARX model structure is independent of the parameter vector θ. We then have:

$$-\frac{1}{N}\sum_{k=1}^{N}\phi(k)(y(k)-\phi^T(k)\hat{\theta}_N)=0 \tag{E.9}$$

$$\Leftrightarrow \quad \sum_{k=1}^{N}\phi(k)\phi^T(k)\hat{\theta}_N=\sum_{k=1}^{N}\phi(k)y(k) \tag{E.10}$$

$$\Leftrightarrow \quad \hat{\theta}_N=\left[\sum_{k=1}^{N}\phi(k)\phi^T(k)\right]^{-1}\sum_{k=1}^{N}\phi(k)y(k)\ . \tag{E.11}$$

Consequently, the expression for $\hat{\theta}_N$ is analytical provided $\sum_{k=1}^{N}\phi(k)\phi^T(k)$ is invertible. Since

$$V_N''(\theta,Z^N)=\frac{1}{N}\sum_{k=1}^{N}\phi(k)\phi^T(k) \tag{E.12}$$

this is ensured if $V_N''(\hat{\theta}_N,Z^N)$ is positive definite.

In the general case, however, the predictor will not be linear in θ and we can not derive analytical expressions for the parameter estimate $\hat{\theta}_N$. Instead, we must use numerical search algorithms based on the gradient (9.14) and Hessian matrix (9.16) of the performance function, see page 130. A popular search scheme is known as *Marquardts algorithm.* The i'th step in Marquardts algorithm is

$$\theta_i=\theta_{i-1}-\left(V_N''(\theta_{i-1},Z^N)+\rho_i I\right)^{-1}V_N'(\theta_{i-1},Z^N) \tag{E.13}$$

where I is a $n_a+n_b+n_f+n_c+n_d$ unity matrix and where ρ_i should be chosen as small as possible (to increase convergence speed) but so that $V_N(\theta_i,Z^N) < V_N(\theta_{i-1},Z^N)$. Usually the approximation

$$V_N''(\theta,Z^N)\approx\frac{1}{N}\sum_{k=1}^{N}\phi(k,\theta)\phi^T(k,\theta) \tag{E.14}$$

is used for the Hessian. The corresponding search scheme for (E.14) is sometimes denoted *Gauss-Newtons algorithm*, see [KR82].

E.1 Computation of $\phi(k,\theta)$ and $\psi(k,\theta)$

The remaining question is now how to efficiently compute the state vector $\phi(k,\theta)$ and the model gradient $\psi(k,\theta)$ for $1\le t\le N$. For this purpose it proves convenient to write the predictor in the ***pseudo-linear*** form

$$\hat{y}(k|\theta)=\phi^T(k,\theta)\theta\ . \tag{E.15}$$

For the general predictor (9.8), $\phi^T(k,\theta)$ can be found as follows. Rearrange (9.8) on page 129 to get

$$C(q)F(q)\hat{y}(k|\theta) = F(q)\left[C(q) - D(q)A(q)\right] y(k) + D(q)B(q)u(k) \quad \text{(E.16)}$$

$$\Leftrightarrow \quad -C(q)F(q)\left[y(k) - \hat{y}(k|\theta)\right] = -F(q)D(q)A(q)y(k) + D(q)B(q)u(k) \quad \text{(E.17)}$$

$$\Leftrightarrow \quad \epsilon(k,\theta) = y(k) - \hat{y}(k|\theta) = \frac{D(q)}{C(q)}\left[A(q)y(k) - \frac{B(q)}{F(q)}u(k)\right] . \quad \text{(E.18)}$$

Now introduce

$$w(k,\theta) = \frac{B(q)}{F(q)}u(k) , \qquad v(k,\theta) = A(q)y(k) - w(k,\theta) \quad \text{(E.19)}$$

to get

$$\epsilon(k,\theta) = \frac{D(q)}{C(q)}v(k,\theta) . \quad \text{(E.20)}$$

We now have:

$$\begin{aligned} w(k,\theta) &= b_1u(k-1) + \cdots + b_{n_\mathrm{b}}u(k-n_\mathrm{b}) - f_1w(k-1,\theta) - \cdots - \\ &\quad - f_{n_\mathrm{f}}w(k-n_\mathrm{f},\theta) && \text{(E.21)} \\ v(k,\theta) &= y(k) + a_1y(k-1) + \cdots + a_{n_\mathrm{a}}y(k-n_\mathrm{a}) - b_1u(k-1) - \cdots - \\ &\quad - b_{n_\mathrm{b}}u(k-n_\mathrm{b}) + f_1w(k-1,\theta) + \cdots + f_{n_\mathrm{f}}w(k-n_\mathrm{f},\theta) && \text{(E.22)} \\ \epsilon(k,\theta) &= y(k) + a_1y(k-1) + \cdots + a_{n_\mathrm{a}}y(k-n_\mathrm{a}) - b_1u(k-1) - \cdots - \\ &\quad - b_{n_\mathrm{b}}u(k-n_\mathrm{b}) + f_1w(k-1,\theta) + \cdots + f_{n_\mathrm{f}}w(k-n_\mathrm{f},\theta) + \\ &\quad + d_1v(k-1,\theta) + \cdots + d_{n_\mathrm{d}}v(k-n_\mathrm{d},\theta) - c_1\epsilon(k-1,\theta) - \cdots - \\ &\quad - c_{n_\mathrm{c}}\epsilon(k-n_\mathrm{c},\theta) && \text{(E.23)} \end{aligned}$$

and may write the predictor as

$$\begin{aligned} \hat{y}(k|\theta) &= y(k) - \epsilon(k,\theta) = -\left(\epsilon(k,\theta) - y(k)\right) && \text{(E.24)} \\ &= -a_1y(k-1) - \cdots - a_{n_\mathrm{a}}y(k-n_\mathrm{a}) + b_1u(k-1) + \cdots + \\ &\quad + b_{n_\mathrm{b}}u(k-n_\mathrm{b}) - f_1w(k-1,\theta) - \cdots - f_{n_\mathrm{f}}w(k-n_\mathrm{f},\theta) + \\ &\quad + c_1\epsilon(k-1,\theta) + \cdots + c_{n_\mathrm{c}}\epsilon(k-n_\mathrm{c},\theta) - d_1v(k-1,\theta) - \cdots - \\ &\quad - d_{n_\mathrm{d}}v(k-n_\mathrm{d},\theta) && \text{(E.25)} \\ &= \phi^T(k,\theta)\theta && \text{(E.26)} \end{aligned}$$

with

$$\begin{aligned} \phi^T(k,\theta) = [&-y(k-1), \cdots, -y(k-n_\mathrm{a}), u(k-1), \cdots, u(k-n_\mathrm{b}), \\ &-w(k-1,\theta), \cdots, -w(k-n_\mathrm{f},\theta), \epsilon(k-1,\theta), \cdots, \epsilon(k-n_\mathrm{c},\theta), \\ &-v(k-1,\theta), \cdots, -v(k-n_\mathrm{d},\theta)] \qquad \text{(E.27)} \end{aligned}$$

and θ given by (9.9) on page 130. From (E.26) a convenient expression for the *predicted output vector*

$$\hat{Y}(\theta, Z^N) = [\hat{y}(k_s|\theta), \hat{y}(k_s+1|\theta), \cdots, \hat{y}(N|\theta)]^T \tag{E.28}$$

can be derived. k_s is given by

$$k_s = \max\{n_a, n_b, n_f, n_c, n_d\} + 1 \,. \tag{E.29}$$

We thus do not start the prediction until the state vector $\phi(k,\theta)$ is filled with measurements. Now let $\Phi(\theta, Z^N)$ be given by

$$\Phi(\theta, Z^N) =$$

$$\begin{bmatrix}
-y(k_s\text{-}1) & -y(k_s) & -y(k_s+1) & \cdots & -y(N\text{-}1) \\
\vdots & \vdots & \vdots & \vdots & \vdots \\
-y(k_s\text{-}n_a) & -y(k_s\text{-}n_a+1) & -y(k_s\text{-}n_a+2) & \cdots & -y(N\text{-}n_a) \\
u(k_s\text{-}1) & u(k_s) & u(k_s+1) & \cdots & u(N\text{-}1) \\
\vdots & \vdots & \vdots & \vdots & \vdots \\
u(k_s\text{-}n_b) & u(k_s\text{-}n_b+1) & u(k_s\text{-}n_b+2) & \cdots & u(N\text{-}n_b) \\
-w(k_s\text{-}1,\theta) & -w(k_s,\theta) & -w(k_s+1,\theta) & \cdots & -w(N\text{-}1,\theta) \\
\vdots & \vdots & \vdots & \vdots & \vdots \\
-w(k_s\text{-}n_f,\theta) & -w(k_s\text{-}n_f+1,\theta) & -w(k_s\text{-}n_f+2,\theta) & \cdots & -w(N\text{-}n_f,\theta) \\
\epsilon(k_s\text{-}1,\theta) & \epsilon(k_s,\theta) & \epsilon(k_s+1,\theta) & \cdots & \epsilon(N\text{-}1,\theta) \\
\vdots & \vdots & \vdots & \vdots & \vdots \\
\epsilon(k_s\text{-}n_c,\theta) & \epsilon(k_s\text{-}n_c+1,\theta) & \epsilon(k_s\text{-}n_c+2,\theta) & \cdots & \epsilon(N\text{-}n_c,\theta) \\
-v(k_s\text{-}1,\theta) & -v(k_s,\theta) & -v(k_s+1,\theta) & \cdots & -v(N\text{-}1,\theta) \\
\vdots & \vdots & \vdots & \vdots & \vdots \\
-v(k_s\text{-}n_d,\theta) & -v(k_s\text{-}n_d+1,\theta) & -v(k_s\text{-}n_d+2,\theta) & \cdots & -v(N\text{-}n_d,\theta)
\end{bmatrix}$$

Then it is easy to see that

$$\hat{Y}(\theta, Z^N) = \Phi^T(\theta, Z^N)\theta \tag{E.30}$$

$$E(\theta, Z^N) = Y - \hat{Y}(\theta, Z^N) \tag{E.31}$$

$$V_N(\theta, Z^N) = \frac{1}{N-k_s}\frac{1}{2}E^T(\theta, Z^N)E(\theta, Z^N) \tag{E.32}$$

where the prediction error vector $E(\theta, Z^N)$ is given by

$$E(\theta, Z^N) = [\epsilon(k_s,\theta), \epsilon(k_s+1,,\theta), \cdots, \epsilon(N,\theta)]^T \,. \tag{E.33}$$

The above expression for $\hat{Y}(\theta, Z^N)$, $E(\theta, Z^N)$ and $V_N(\theta, Z^N)$ can be easily and efficiently implemented in MATLAB, see Appendix F.

Let us now return to the computation of the model gradient $\psi(k,\theta)$. From Equation (E.26) we have

$$\psi(k,\theta) = \frac{\partial \hat{y}(k|\theta)}{\partial \theta} = \phi(k,\theta) + \frac{\partial \phi^T(k,\theta)}{\partial \theta}\theta \tag{E.34}$$

$$= \left[\frac{\partial \hat{y}(k|\theta)}{\partial a_1}, \cdots, \frac{\partial \hat{y}(k|\theta)}{\partial a_{n_a}}, \frac{\partial \hat{y}(k|\theta)}{\partial b_1}, \cdots, \frac{\partial \hat{y}(k|\theta)}{\partial b_{n_b}}, \frac{\partial \hat{y}(k|\theta)}{\partial f_1}, \cdots, \right.$$
$$\left. \frac{\partial \hat{y}(k|\theta)}{\partial f_{n_f}}, \frac{\partial \hat{y}(k|\theta)}{\partial c_1}, \cdots, \frac{\partial \hat{y}(k|\theta)}{\partial c_{n_c}}, \frac{\partial \hat{y}(k|\theta)}{\partial d_1}, \cdots, \frac{\partial \hat{y}(k|\theta)}{\partial d_{n_d}}\right]^T . \tag{E.35}$$

Let us derive an expression for the derivative $\psi_i(k,\theta) = \partial \hat{y}(k|\theta)/\partial a_i$ where $1 \le i \le n_a$:

$$\psi_i(k,\theta) = \frac{\partial \hat{y}(k|\theta)}{\partial a_i} = \phi_i(k,\theta) + \frac{\partial \phi^T(k,\theta)}{\partial a_i}\theta \tag{E.36}$$

$$= -y(k-i) + \frac{\partial \phi^T(k,\theta)}{\partial a_i}\theta \tag{E.37}$$

where

$$\frac{\partial \phi^T(k,\theta)}{\partial a_i} = \left[-\frac{\partial y(k-1)}{\partial a_i}, \cdots, -\frac{\partial y(k-n_a)}{\partial a_i}, \frac{\partial u(k-1)}{\partial a_i}, \cdots, \frac{\partial u(k-n_b)}{\partial a_i}, \right.$$
$$-\frac{\partial w(k-1,\theta)}{\partial a_i}, \cdots, -\frac{\partial w(k-n_f,\theta)}{\partial a_i}, \frac{\partial \epsilon(k-1,\theta)}{\partial a_i}, \cdots, \frac{\partial \epsilon(k-n_c,\theta)}{\partial a_i},$$
$$\left. -\frac{\partial v(k-1,\theta)}{\partial a_i}, \cdots, -\frac{\partial v(k-n_d,\theta)}{\partial a_i}\right] \tag{E.38}$$

$$= \left[0, \cdots, 0, -\frac{\partial \hat{y}(k-1|\theta)}{\partial a_i}, \cdots, -\frac{\partial \hat{y}(k-n_c|\theta)}{\partial a_i}, \right.$$
$$\left. -y(k-1-i), \cdots, -y(k-n_d-i)\right] \tag{E.39}$$

since $\epsilon(k,\theta) = y(k) - \hat{y}(k|\theta)$ and

$$v(k) = y(k) + a_1 y(k-1) + \cdots + a_{n_a} y(k-n_a) - w(k,\theta) . \tag{E.40}$$

Notice that $w(k,\theta)$ is independent of a_i. We then have

$$\frac{\partial \hat{y}(k|\theta)}{\partial a_i} = -y(k-i) - c_1 \frac{\partial \hat{y}(k-1|\theta)}{\partial a_i} - \cdots - c_{n_c} \frac{\partial \hat{y}(k-n_c|\theta)}{\partial a_i} -$$
$$d_1 y(k-1-i) - \cdots - d_{n_d} y(k-n_d-i) . \tag{E.41}$$

Thus

$$\left[1 + c_1 q^{-1} + \cdots + c_{n_c} q^{-n_c}\right] \frac{\partial \hat{y}(k|\theta)}{\partial a_i} =$$
$$\left[1 + d_1 q^{-1} + \cdots + d_{n_d} q^{-n_d}\right] y(k-i) \tag{E.42}$$

and

$$\psi_i(k,\theta) = \frac{\partial \hat{y}(k|\theta)}{\partial a_i} = -\frac{D(q)}{C(q)} y(k-i) = \frac{D(q)}{C(q)} \phi_i(k,\theta) \,. \tag{E.43}$$

Define

$$\psi_{\rm A}(k,\theta) = [\psi_1(k,\theta), \cdots, \psi_{n_{\rm a}}(k,\theta)]^T \tag{E.44}$$

$$\phi_{\rm A}(k,\theta) = [\phi_1(k,\theta), \cdots, \phi_{n_{\rm a}}(k,\theta)]^T \tag{E.45}$$

such that

$$\psi_{\rm A}(k,\theta) = \frac{D(q)}{C(q)} \phi_{\rm A}(k,\theta) \,. \tag{E.46}$$

The first $n_{\rm a}$ elements of the model gradient $\psi(k,\theta)$ is consequently found by filtering of the first $n_{\rm a}$ elements of the state vector $\phi(k,\theta)$ with the filter $D(q)/C(q)$. Unlike for the ARMAX model structure

$$A(q)y(k) = B(q)u(k) + C(q)e(k) \tag{E.47}$$

where it can be shown, see eg [Knu93], that

$$\psi(k,\theta) = \frac{1}{C(q)} \phi(k,\theta) \tag{E.48}$$

the filters for the general model structure are not equal for the different "parts" of $\phi(k,\theta)$. Using similar arguments as above it is straightforward to show that

$$\psi_{n_{\rm a}+i}(k,\theta) = \frac{\partial \hat{y}(k|\theta)}{\partial b_i} = \frac{D(q)}{F(q)C(q)} u(k-i) = \frac{D(q)}{F(q)C(q)} \phi_{n_{\rm a}+i}(k,\theta) \tag{E.49}$$

$$\psi_{n_{\rm a}+n_{\rm b}+i}(k,\theta) = \frac{\partial \hat{y}(k|\theta)}{\partial f_i} = -\frac{D(q)}{F(q)C(q)} w(k-i,\theta) = \frac{D(q)}{F(q)C(q)} \phi_{n_{\rm a}+n_{\rm b}+i}(k,\theta) \tag{E.50}$$

$$\psi_{n_{\rm a}+n_{\rm b}+n_{\rm c}+i}(k,\theta) = \frac{\partial \hat{y}(k|\theta)}{\partial c_i} = \frac{1}{C(q)} \epsilon(k-i,\theta) = \frac{1}{C(q)} \phi_{n_{\rm a}+n_{\rm b}+n_{\rm c}+i}(k,\theta) \tag{E.51}$$

$$\psi_{n_{\rm a}+n_{\rm b}+n_{\rm f}+n_{\rm c}+i}(k,\theta) = \frac{\partial \hat{y}(k|\theta)}{\partial d_i} = -\frac{1}{C(q)} v(k-i,\theta) = \frac{1}{C(q)} \phi_{n_{\rm a}+n_{\rm b}+n_{\rm f}+n_{\rm c}+i}(k,\theta) \,. \tag{E.52}$$

Thus with

$$\phi_{\rm BF}(k,\theta) = [\phi_{n_{\rm a}+1}, \cdots, \phi_{n_{\rm a}+n_{\rm b}+n_{\rm f}}(k,\theta)]^T \tag{E.53}$$

$$\phi_{\rm CD}(k,\theta) = [\phi_{n_{\rm a}+n_{\rm b}+n_{\rm f}+1}, \cdots, \phi_{n_{\rm a}+n_{\rm b}+n_{\rm f}+n_{\rm c}+n_{\rm d}}(k,\theta)]^T \tag{E.54}$$

the model gradient can be written

$$\psi(k,\theta) = \begin{bmatrix} \dfrac{D(q)}{C(q)}\phi_{\mathrm{A}}(k,\theta) \\ \\ \dfrac{D(q)}{F(q)C(q)}\phi_{\mathrm{BF}}(k,\theta) \\ \\ \dfrac{1}{C(q)}\phi_{\mathrm{CD}}(k,\theta) \end{bmatrix} . \tag{E.55}$$

Notice that for the special case of the ARMAX model structure ($D(q) = F(q) = 1$) the general expression for $\psi(k,\theta)$ (E.55) reduce to (E.48).

In order to find a convenient expression for the gradient $V_N'(\theta, Z^N)$ and the Hessian $V_N''(\theta, Z^N)$ introduce

$$y_{\mathrm{f}}(k) = -\frac{D(q)}{C(q)}y(k) \tag{E.56}$$

$$u_{\mathrm{f}}(k) = \frac{D(q)}{F(q)C(q)}u(k) \tag{E.57}$$

$$w_{\mathrm{f}}(k,\theta) = -\frac{D(q)}{F(q)C(q)}w(k,\theta) \tag{E.58}$$

$$\epsilon_{\mathrm{f}}(k,\theta) = \frac{1}{C(q)}\epsilon(k,\theta) \tag{E.59}$$

$$v_{\mathrm{f}}(k,\theta) = -\frac{1}{C(q)}v(k,\theta) \tag{E.60}$$

and

$$\Psi(\theta, Z^N) =$$

$$\begin{bmatrix}
-y_f(k_s-1) & -y_f(k_s) & -y_f(k_s+1) & \cdots & -y_f(N-1) \\
\vdots & \vdots & \vdots & \vdots & \vdots \\
-y_f(k_s-n_a) & -y_f(k_s-n_a+1) & -y_f(k_s-n_a+2) & \cdots & -y_f(N-n_a) \\
u_f(k_s-1) & u_f(k_s) & u_f(k_s+1) & \cdots & u_f(N-1) \\
\vdots & \vdots & \vdots & \vdots & \vdots \\
u_f(k_s-n_b) & u_f(k_s-n_b+1) & u_f(k_s-n_b+2) & \cdots & u_f(N-n_b) \\
-w_f(k_s-1,\theta) & -w_f(k_s,\theta) & -w_f(k_s+1,\theta) & \cdots & -w_f(N-1,\theta) \\
\vdots & \vdots & \vdots & \vdots & \vdots \\
-w_f(k_s-n_f,\theta) & -w_f(k_s-n_f+1,\theta) & -w_f(k_s-n_f+2,\theta) & \cdots & -w_f(N-n_f,\theta) \\
\epsilon_f(k_s-1,\theta) & \epsilon_f(k_s,\theta) & \epsilon_f(k_s+1,\theta) & \cdots & \epsilon_f(N-1,\theta) \\
\vdots & \vdots & \vdots & \vdots & \vdots \\
\epsilon_f(k_s-n_c,\theta) & \epsilon_f(k_s-n_c+1,\theta) & \epsilon_f(k_s-n_c+2,\theta) & \cdots & \epsilon_f(N-n_c,\theta) \\
-v_f(k_s-1,\theta) & -v_f(k_s,\theta) & -v_f(k_s+1,\theta) & \cdots & -v_f(N-1,\theta) \\
\vdots & \vdots & \vdots & \vdots & \vdots \\
-v_f(k_s-n_d,\theta) & -v_f(k_s-n_d+1,\theta) & -v_f(k_s-n_d+2,\theta) & \cdots & -v_f(N-n_d,\theta)
\end{bmatrix}$$

Then it is easily checked that

$$V_N'(\theta, Z^N) = -\frac{1}{N - k_s}\Psi(\theta, Z^N)E(\theta, Z^N) \tag{E.61}$$

$$V_N''(\theta, Z^N) \approx \frac{1}{N - k_s}\sum_{k=k_s}^{N} \psi(k,\theta)\psi^T(k,\theta) \tag{E.62}$$

$$= \frac{1}{N - k_s}\Psi(\theta, Z^N)\Psi^T(\theta, Z^N) \tag{E.63}$$

The above expressions for $V_N'(\theta, Z^N)$ and $V_N''(\theta, Z^N)$ can be easily and efficiently implemented in MATLAB, see Appendix F.

Having determined computable expressions for the criterion function $V_N(\theta, Z^N)$ and its first and second derivatives we may construct simple and effective search algorithms based on Marquardts iteration scheme. If we want to get rid of initial condition effects using delayed start of the criterion function, simply replace k_s with $N_1 \geq k_s$ in (E.32), (E.61) and (E.63) and let

$$E(\theta, Z^N) = [\epsilon(N_1, \theta), \epsilon(N_1 + 1, \theta), \cdots, \epsilon(N, \theta)]^T\ . \tag{E.64}$$

APPENDIX F

A MATLAB FUNCTION FOR COMPUTING $\Phi(\theta)$ and $\Psi(\theta)$

The following is an example of how the expressions for $\Phi(\theta)$ and $\Psi(\theta)$ derived in Appendix E may be efficiently implemented into MATLAB.

```
function [FI,PSI] = mk_signals(theta,nn,U,Y)
% [FI,PSI] = mk_signals(theta,nn,U,Y,) creates
% the state vector matrix FI and the model gradient
% matrix PSI from the parameter vector theta,
% nn=[na,nb,nf,nc,nd], input vector U and output
% vector Y.

% S. Toffner-Clausen, Last revised:  95.02.21
% Copyright (c) by the authors
% All Rights Reserved.

N = length(U);
ks = max(nn)+1;
na=nn(1);nb=nn(2);nf=nn(3);nc=nn(4);nd=nn(5);
ns = sum(nn);
Ns = N-ks+1;
PSI = zeros(ns,Ns);                        % allocating space for PSI
FI = zeros(ns,Ns);                         % allocating space for FI
jj = ks:N;
A = [1 theta(1:na)'];                      % A polynomial
B = theta(na+1:na+nb)';                    % B polynomial
F = [1 theta(na+nb+1:na+nb+nf)'];          % F polynomial
C = [1 theta(na+nb+nf+1:na+nb+nf+nc)'];    % C polynomial
D = [1 theta(na+nb+nf+nc+1:ns)'];          % D polynomial
w = filter([0 B],F,U');                    % use filter.m for fast
v = filter(A,1,Y')-w;                      % computation of signals
e = filter(D,C,v);
yf = filter(-D,C,Y');
uf = filter(D,conv(F,C),U');
wf = filter(-D,conv(F,C),w);
ef = filter(1,C,e);
vf = filter(-1,C,v);
```

```
for i = 1:na                                    % computing FI and PSI
  FI(i,:) = -Y(jj-i)';
  PSI(i,:) = yf(jj-i);
end;
for i = 1:nb
  FI(na+i,:) = U(jj-i)';
  PSI(na+i,:) = uf(jj-i);
end;
for i = 1:nf
  FI(na+nb+i,:) = -w(jj-i);
  PSI(na+nb+i,:) = wf(jj-i);
end;
for i = 1:nc
  FI(na+nb+nf+i,:) = e(jj-i);
  PSI(na+nb+nf+i,:) = ef(jj-i);
end;
for i = 1:nd
  FI(na+nb+nf+nc+i,:) = -v(jj-i);
  PSI(na+nb+nf+nc+i,:) = vf(jj-i);
end;
```

APPENDIX G

COMPUTING THE θ ESTIMATE THROUGH QR FACTORIZATION

Recall the performance function for the least squares parameter estimate given as Equation (11.29) on page 172:

$$V_N(\theta, Z^N) = (Y - \Phi\theta)^T (Y - \Phi\theta) \tag{G.1}$$

$$= \|Y - \Phi\theta\|_2^2 \tag{G.2}$$

where $\|\cdot\|_2$ is the usual Euclidean norm. Given an $(N \times N)$ orthonormal matrix T[1] the performance function $V_N(\theta, Z^N)$ is clearly not affected by the *orthonormal transformation*:

$$V_N(\theta, Z^N) = \|T(Y - \Phi\theta)\|_2^2 \ . \tag{G.3}$$

Now choose T such that:

$$T\Phi = \begin{bmatrix} Q \\ 0 \end{bmatrix} \tag{G.4}$$

where Q is an $(n \times n)$ upper triangular matrix, where n is the order of the parametric model. Then rewrite Equation (G.4) as:

$$T^T T\Phi = T^T \begin{bmatrix} Q \\ 0 \end{bmatrix} \tag{G.5}$$

$$\Leftrightarrow \quad \Phi = T^T \begin{bmatrix} Q \\ 0 \end{bmatrix} . \tag{G.6}$$

Equation (G.6) is known as the *QR Factorization* of Φ. QR factorizations is easily performed in e.g. MATLAB.

Having determined a T and Q according to (G.6) introduce:

$$TY = \begin{bmatrix} \vartheta \\ \varrho \end{bmatrix} \tag{G.7}$$

where ϑ is a n vector and ϱ is a $N - n$ vector. Inserting (G.4) and (G.7) into (G.3) it is easily seen that:

[1] An orthonormal matrix T satisfies $T^T T = I$.

$$V_N(\theta, Z^N) = \left\| \begin{bmatrix} \vartheta \\ \varrho \end{bmatrix} - \begin{bmatrix} Q \\ 0 \end{bmatrix} \theta \right\|_2^2 \tag{G.8}$$

$$= \left\| \begin{bmatrix} \vartheta - Q\theta \\ \varrho \end{bmatrix} \right\|_2^2 \tag{G.9}$$

$$= \|\vartheta - Q\theta\|_2^2 + \|\varrho\|_2^2 \, . \tag{G.10}$$

The performance criterion given by (G.10) is clearly minimized for:

$$Q\hat{\theta}_N = \vartheta \tag{G.11}$$

$$\Leftrightarrow \qquad \hat{\theta}_N = Q^{-1}\vartheta \, . \tag{G.12}$$

The minimum of $V_N(\theta, Z^N)$ is then:

$$V_N(\hat{\theta}_N, Z^N) = \|\varrho\|_2^2 \, . \tag{G.13}$$

Notice that from (G.4) it follows that:

$$Q^T Q = \Phi^T T^T T \Phi = \Phi^T \Phi = \mathcal{R}(N) \tag{G.14}$$

where $\mathcal{R}(N)$ is the coefficient matrix for the non-transformed estimation problem, see Equation (11.37) on page 173. It can now be shown that the ratio between the smallest and largest eigenvalue of Q is the square root of that of $\mathcal{R}(N)$, see [Lju87, pp 276]. Consequently the inversion of Q is better conditioned that the conversion of $\Phi^T\Phi$ and the parameter estimate given by Equation (G.12) has superior numerical properties than the estimate given by (11.37) on page 173.

G.1 Transforming the Residuals

When applying maximum likelihood estimation on the parameter vector ζ the residuals ϵ shall be transformed into a non-singular distribution:

$$\varpi = R^T \epsilon \tag{G.15}$$

where ϖ has non singular distribution. The choice of R affects the condition number of the covariance matrix Σ for ϖ. Since the numerical solution to the maximum likelihood estimate involves inversion of Σ this should be well conditioned. Inspired by the above QR factorization we suggest that R is chosen as:

$$R^T = \kappa \tilde{T} \tag{G.16}$$

where κ is a scalar scaling factor and $\tilde{T}$ is the last $(N - p)$ columns of T found by the QR factorization above.

This particular transformation yielded very nice numerical properties of the maximum likelihood estimate. κ scales the numerical value of the loglikelihood function.

APPENDIX H

FIRST AND SECOND ORDER DERIVATIVES OF $\ell(\varpi/U,\zeta)$

It will now be demonstrated how one may derive explicit expressions for the partial derivatives of the loglikelihood function $\ell(\varpi|U,\zeta)$ with respect to ζ. Knowledge of these derivatives enables us to construct the *Fisher Information Matrix* given by:

$$M \triangleq E\left\{\left[\frac{\partial \ell}{\partial \zeta}\right]\left[\frac{\partial \ell}{\partial \zeta}\right]^T\right\} = -E\left\{\frac{\partial^2 \ell}{\partial \zeta^2}\right\} \tag{H.1}$$

$$\Rightarrow \qquad [M_\zeta]_{ij} = E\left\{\frac{\partial \ell}{\partial \zeta_i}\cdot\frac{\partial \ell}{\partial \zeta_j}\right\}. \tag{H.2}$$

Remember that the Hessian matrix H is given as:

$$H = \frac{\partial^2 \ell}{\partial \zeta^2} \tag{H.3}$$

so that:

$$M = -E\{H\}. \tag{H.4}$$

Thus, Fishers Information Matrix is the estimated Hessian.

H.1 Partial First Order Derivatives of $\ell(\varpi|U,\zeta)$

Recall Equation (11.111) on page 181:

$$\ell(\varpi|U,\zeta) = -\frac{1}{2}\ln(\det\Sigma) - \frac{1}{2}\varpi^T\Sigma^{-1}\varpi + k \tag{H.5}$$

where Σ is given by:

$$\Sigma = R^T X C_\eta X^T R + R^T C_\nu R. \tag{H.6}$$

In order to evaluate the partial derivatives we need the following two Lemma's:

Lemma H.1. *Consider an invertible square matrix Σ parameterized by a set of scalars $\{\zeta_1, \cdots, \zeta_n\}$. The partial derivative of $\ln\det(\Sigma)$ with respect to one of the parameterizing constants ζ_i may be written:*

$$\frac{\partial \ln\det(\Sigma)}{\partial \zeta_i} = tr\left\{\Sigma^{-1}\frac{\partial \Sigma}{\partial \zeta_i}\right\}. \tag{H.7}$$

Lemma H.2. *Consider a matrix Σ as above. The partial derivative of Σ^{-1} with respect to ζ_i may be written:*

$$\frac{\partial \Sigma^{-1}}{\partial \zeta_i} = -\Sigma^{-1}\frac{\partial \Sigma}{\partial \zeta_i}\Sigma^{-1}. \tag{H.8}$$

The proofs of these Lemma's are straightforward and may be found in [GGN92]. Applying the Lemma's to Equation (H.5) one obtain:

$$\begin{aligned}
\frac{\partial \ell(\varpi|U,\zeta)}{\partial \zeta_i} &= -\frac{1}{2}\frac{\partial}{\partial \zeta_i}\{\ln\det\Sigma\} - \frac{1}{2}\varpi^T\frac{\partial}{\partial \zeta_i}\{\Sigma^{-1}\}\varpi + 0 && \text{(H.9)}\\
&= -\frac{1}{2}\text{tr}\left\{\Sigma^{-1}\frac{\partial \Sigma}{\partial \zeta_i}\right\} + \frac{1}{2}\varpi^T\Sigma^{-1}\frac{\partial \Sigma}{\partial \zeta_i}\Sigma^{-1}\varpi && \text{(H.10)}\\
&= -\frac{1}{2}\text{tr}\left\{\Sigma^{-1}\left(R^TX\frac{\partial C_\eta}{\partial \zeta_i}X^TR + R^T\frac{\partial C_\nu}{\partial \zeta_i}R\right)\right\}\\
&\quad + \frac{1}{2}\varpi^T\Sigma^{-1}\left(R^TX\frac{\partial C_\eta}{\partial \zeta_i}X^TR + R^T\frac{\partial C_\nu}{\partial \zeta_i}R\right)\Sigma^{-1}\varpi && \text{(H.11)}
\end{aligned}$$

When the parameterizations of the covariance matrices $C_\eta(\beta)$ and $C_\nu(\gamma)$ have been chosen the corresponding partial derivatives with respect to each element in ζ may then be determined, and from (H.11) the partial derivatives of $\ell(\varpi|U,\zeta)$ can be found.

H.2 Partial Second Order Derivatives of $\ell(\varpi|U,\zeta)$

Applying Equation (H.2) with (H.11) will yield the desired expression for the *Fisher Information Matrix.* In [GGN92] it is shown that the Fisher Matrix can be written:

$$\begin{aligned}
[M_\zeta]_{ij} &= E\left\{\frac{\partial \ell}{\partial \zeta_i}\cdot\frac{\partial \ell}{\partial \zeta_j}\right\} && \text{(H.12)}\\
&= \frac{1}{2}\text{tr}\left\{\Sigma^{-1}\frac{\partial \Sigma}{\partial \zeta_i}\Sigma^{-1}\frac{\partial \Sigma}{\partial \zeta_j}\right\} && \text{(H.13)}\\
&= \frac{1}{2}\text{tr}\left\{\Sigma^{-1}\left(R^TX\frac{\partial C_\eta}{\partial \zeta_i}X^TR + R^T\frac{\partial C_\nu}{\partial \zeta_i}R\right)\cdot\right.\\
&\qquad\left.\Sigma^{-1}\left(R^TX\frac{\partial C_\eta}{\partial \zeta_j}X^TR + R^T\frac{\partial C_\nu}{\partial \zeta_j}R\right)\right\}. && \text{(H.14)}
\end{aligned}$$

We may then use (H.14) to obtain an estimate of the Hessian matrix. With the first and second order partial derivatives we may then construct powerful search algorithms for the maximum of the loglikelihood function.

The partial derivatives of the noise covariance for the investigated parameterizations is presented in Appendix I. The results for the undermodeling is given in Appendix J.

APPENDIX I

PARTIAL DERIVATIVES OF THE NOISE COVARIANCE

The partial derivatives of the noise covariance matrix for the different assumptions on the covariance structure will be derived. The partial derivatives is used to compute the gradient and Fisher for the loglikelihood function of the transformed residuals ϖ when searching for the maximum of the loglikelihood function.

Two different parameterization of the noise covariance has been investigated:

$$C_{\nu_1} = \sigma_e^2 \cdot I_N \tag{I.1}$$

$$C_{\nu_2}(i,j) = \begin{cases} \sigma_e^2 \left(1 + \frac{(c-a)^2}{1-a^2}\right), & i = j \\ \frac{\sigma_e^2 (-a)^M \left(1 + c^2 - a * c - c/a\right)}{1-a^2}, & |j-i| = M \end{cases} . \tag{I.2}$$

For C_{ν_1} the obvious solution is:

$$\frac{\partial C_{\nu_1}}{\partial \sigma_e^2} = I_N . \tag{I.3}$$

For C_{ν_2} a few hand calculations yield the following three derivatives:

$$\frac{\partial C_{\nu_2}(i,j)}{\partial a} = \begin{cases} \sigma_e^2 \left(\frac{2a(c-a)^2 - 2(c-a)(1-a^2)}{(1-a^2)^2}\right), & i = j \\ \left(\frac{\sigma_e^2(-a)^M \left[\left(1+c^2-a*c-c/a\right)\left(M+(2-M)a^2\right)+\left(a-a^3\right)\left(c/a^2-c\right)\right]}{a(1-a^2)^2}\right), & |j-i| = M \end{cases} \tag{I.4}$$

$$\frac{\partial C_{\nu_2}(i,j)}{\partial c} = \begin{cases} \sigma_e^2 \left(\frac{2(c-a)}{1-a^2}\right), & i = j \\ \sigma_e^2 \left(\frac{(-a)^M \left(2c-a-a^{-1}\right)}{1-a^2}\right), & |j-i| = M \end{cases} \tag{I.5}$$

$$\frac{\partial C_{\nu_2}(i,j)}{\partial \sigma_e^2} = \begin{cases} \left(1 + \frac{(c-a)^2}{1-a^2}\right), & i = j \\ \left(\frac{(-a)^M (1+c^2-ac-c/a)}{1-a^2}\right), & |j-i| = M \end{cases} . \tag{I.6}$$

APPENDIX J
PARTIAL DERIVATIVES OF THE UNDERMODELING COVARIANCE

The partial derivatives of the undermodeling impulse response covariance matrix for the different assumptions on the covariance structure will be derived. The partial derivatives is used to compute the gradient and Fisher for the loglikelihood function of the transformed residuals ϖ when searching for the maximum of the loglikelihood function.

Three different parameterization of the undermodeling covariance has been investigated:

$$C_{\eta_1} = \alpha \cdot I_L \tag{J.1}$$

$$C_{\eta_2} = \operatorname*{diag}_{1\leq k\leq L} \{\alpha\lambda^k\} \tag{J.2}$$

$$C_{\eta_3} = \left[\alpha\lambda^0 \cdots \alpha\lambda^{L-1}\right]^T \left[\alpha\lambda^0 \cdots \alpha\lambda^{L-1}\right] . \tag{J.3}$$

The partial derivatives are easily found as:

$$\frac{\partial C_{\eta_1}}{\partial \alpha} = I_L \tag{J.4}$$

$$\frac{\partial C_{\eta_2}}{\partial \alpha} = \operatorname*{diag}_{1\leq k\leq L} \lambda^k \tag{J.5}$$

$$\frac{\partial C_{\eta_2}}{\partial \lambda} = \operatorname*{diag}_{1\leq k\leq L} \alpha k\lambda^{k-1} \tag{J.6}$$

$$\frac{\partial C_{\eta_3}}{\partial \alpha} = 2\alpha \left[\lambda^0 \cdots \lambda^{L-1}\right]^T \left[\lambda^0 \cdots \lambda^{L-1}\right] \tag{J.7}$$

$$\frac{\partial C_{\eta_3}}{\partial \lambda} = \alpha^2 \begin{bmatrix} 0 & 1 & \cdots & (L-1)\lambda^{L-2} \\ 1 & 2\lambda & \cdots & L\lambda^{L-1} \\ \vdots & \vdots & \ddots & \vdots \\ (L-1)\lambda^{L-2} & L\lambda^{L-1} & \cdots & 2(L-1)\lambda^{2L-3} \end{bmatrix} . \tag{J.8}$$

APPENDIX K
ARMA(1) NOISE COVARIANCE MATRIX

Assume that the process noise $\nu(k)$ is given by the ARMA(1) description:

$$\nu(k) = H(q) = \frac{1 + cq^{-1}}{1 + aq^{-1}} e(k) \tag{K.1}$$

where $e(k)$ is white noise with variance σ_e^2. If $h_\nu(k)$ is the impulse response of $H(q)$ we may write $\nu(k)$ as:

$$\nu(k) = \sum_{\kappa=0}^{\infty} h_\nu(\kappa) e(k - \kappa) . \tag{K.2}$$

The (i, j)'th element of the covariance C_ν is then defined by:

$$C_\nu(i,j) \triangleq E\{\nu(i)\nu(j)\} \tag{K.3}$$

$$= E\left\{\sum_{\kappa=0}^{\infty} h_\nu(\kappa) e(i-\kappa) \sum_{\lambda=0}^{\infty} h_\nu(\lambda) e(j-\lambda)\right\} \tag{K.4}$$

$$= E\left\{\sum_{\kappa=0}^{\infty}\sum_{\lambda=0}^{\infty} h_\nu(\kappa) h_\nu(\lambda) e(i-\kappa) e(j-\lambda)\right\} . \tag{K.5}$$

Now since:

$$E\{e(i-\kappa)e(j-\lambda)\} = 0 \qquad (i-\kappa) \neq (j-\lambda) \tag{K.6}$$

$$E\{e(i-\kappa)e(j-\lambda)\} = \sigma_e^2 \quad (i-k) = (j-l) \tag{K.7}$$

we may write the covariance as:

$$C_\nu(i,j) = \sigma_e^2 \sum_{\kappa=0}^{\infty} h_\nu(\kappa) h_\nu(\kappa + j - i) . \tag{K.8}$$

Given the ARMA(1) noise description (K.1) we may write the difference equation for $\nu(k)$ as:

$$\nu(k) = -a\nu(k-1) + e(k) + ce(k-1) . \tag{K.9}$$

The impulse response $h_\nu(k)$ is then obtained by setting $e(k) = \delta_0(k)$:

$$h_\nu(0) = 1 \tag{K.10}$$
$$h_\nu(1) = -a + c \tag{K.11}$$
$$h_\nu(2) = -a(-a + c) \tag{K.12}$$
$$h_\nu(3) = (-a)^2(-a + c) \tag{K.13}$$
$$h_\nu(n) = (-a)^{n-1}(-a + c) \tag{K.14}$$

or:

$$h_\nu(k) = \begin{cases} 1\,, & k = 0 \\ (-a)^{k-1}(-a + c)\,, & k > 0 \end{cases} \quad . \tag{K.15}$$

We may now compute the (i, j)'th element of C_ν as follows. For $j - i = 0$, we have

$$C_\nu(i, j) = \sigma_e^2 \left(1 + \sum_{\kappa=1}^{\infty} \left((-a)^{\kappa-1}(-a + c)\right)^2\right) \tag{K.16}$$
$$= \sigma_e^2 \left(1 + (-a + c)^2 \sum_{\kappa=1}^{\infty} \left((-a)^2\right)^{\kappa-1}\right) \tag{K.17}$$
$$= \sigma_e^2 \left(1 + (-a + c)^2 \frac{1}{1 - a^2}\right) . \tag{K.18}$$

For $j - i = 1$, the result becomes:

$$C_\nu(i, j) = \sigma_e^2 \left((-a + c) + \sum_{\kappa=1}^{\infty} (-a)^{\kappa-1}(-a + c)(-a)^{\kappa}(-a + c)\right) \tag{K.19}$$
$$= \sigma_e^2 \left((-a + c) + (-a + c)^2(-a)^1 \sum_{\kappa=1}^{\infty} \left((-a)^2\right)^{\kappa-1}\right) \tag{K.20}$$
$$= \sigma_e^2 \left((-a + c) + (-a + c)^2(-a)^1 \frac{1}{1 - a^2}\right) . \tag{K.21}$$

Finally, we may generalize to $j - i = M$

$$C_\nu(i, j) = \sigma_e^2 \Bigg((-a)^{M-1}(-a + c) + \sum_{\kappa=1}^{\infty} (-a)^{\kappa-1}(-a + c)(-a)^{\kappa-1+M}(-a + c)\Bigg) \tag{K.22}$$
$$= \sigma_e^2 \left((-a)^{M-1}(-a + c) + (-a + c)^2(-a)^M \sum_{\kappa=1}^{\infty} \left((-a)^2\right)^{\kappa-1}\right) \tag{K.23}$$
$$= \sigma_e^2 \left((-a)^{M-1}(-a + c) + (-a + c)^2(-a)^M \frac{1}{1 - a^2}\right) \tag{K.24}$$

for $M > 0$. However, since C_ν is symmetric we may generalize the above expressions. A little rearranging then gives:

$$C_\nu(i,j) = \begin{cases} \sigma_e^2 \left(1 + \dfrac{(c-a)^2}{1-a^2}\right), & j = i \\ \sigma_e^2 \left((-a)^M \dfrac{1 + c^2 - ac - c/a}{1-a^2}\right), & |j-i| = M \end{cases} \quad \text{(K.25)}$$

where $|\cdot|$ denotes absolute value.

APPENDIX L
EXTRACTING PRINCIPAL AXIS FROM FORM MATRIX

Here it will shown how the principal axis and the angle of an ellipse with form matrix P can be extracted from the form matrix. Let the ellipse in Figure L.1 with major and minor principal axis a and b respectively and angle A be given by the ellipse equation

$$x^T P^{-1} x = 1 . \tag{L.1}$$

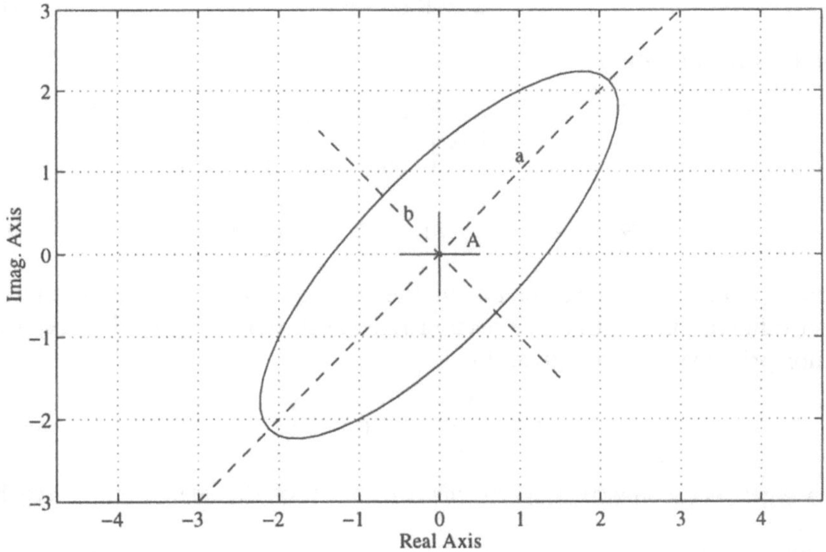

Fig. L.1. *Ellipse with major and minor principal axis a and b respectively and angle A.*

In order to find the major and minor principal axis and the angle we perform an eigenvalue decomposition of the form matrix P^{-1}:

$$P^{-1} = VDV^{-1} \tag{L.2}$$

where $D = \text{diag}\{\lambda_1, \lambda_2\}$ is a diagonal vector of eigenvalues and V is a 2-by-2 matrix whose columns are the corresponding eigenvectors. We may then write

$$
\begin{aligned}
& x^T V D V^{-1} x = 1 && \text{(L.3)}\\
\Rightarrow \quad & x^T V D^{\frac{1}{2}} D^{\frac{1}{2}} V^{-1} x = 1 && \text{(L.4)}\\
\Rightarrow \quad & y^T y = 1 && \text{(L.5)}
\end{aligned}
$$

with

$$
\begin{aligned}
& y = D^{\frac{1}{2}} V^{-1} x && \text{(L.6)}\\
\Rightarrow \quad & x = \left(D^{\frac{1}{2}} V^{-1}\right)^{-1} y && \text{(L.7)}\\
\Leftrightarrow \quad & x = V D^{-\frac{1}{2}} y \,. && \text{(L.8)}
\end{aligned}
$$

Thus the ellipse given by (L.2) can be seen as the circle (L.5) transformed by

$$x = V D^{-\frac{1}{2}} y \,. \qquad \text{(L.9)}$$

For $V = I_2$ we obtain

$$
\begin{aligned}
& \begin{bmatrix} x_1 \\ x_2 \end{bmatrix} = \begin{bmatrix} \lambda_1^{-\frac{1}{2}} & 0 \\ 0 & \lambda_2^{-\frac{1}{2}} \end{bmatrix} \begin{bmatrix} y_1 \\ y_2 \end{bmatrix} && \text{(L.10)}\\
\Leftrightarrow \quad & x_1 = \frac{1}{\sqrt{\lambda_1}} y_1 \,, \quad x_2 = \frac{1}{\sqrt{\lambda_2}} y_2 && \text{(L.11)}
\end{aligned}
$$

corresponding to a pure scaling in the $y_1 - y_2$ plane. Assuming that the eigenvalue decomposition has been ordered such that $\lambda_1 \le \lambda_2$ the major and minor principal axis are given by

$$a = \frac{1}{\sqrt{\lambda_1}} \,, \qquad b = \frac{1}{\sqrt{\lambda_2}} \qquad \text{(L.12)}$$

The eigenvector matrix V performs a rotation of the scaled circle, see Figure L.2.

Then it is easy to check that β in Figure L.2 may be written as the following transformation of α

$$\beta = R\alpha \,, \qquad R = \begin{bmatrix} \cos(A) & -\sin(A) \\ \sin(A) & cos(A) \end{bmatrix} . \qquad \text{(L.13)}$$

Thus the angle A of the ellipse is given by the solution to the equation

$$V = \begin{bmatrix} \cos(A) & -\sin(A) \\ \sin(A) & cos(A) \end{bmatrix} . \qquad \text{(L.14)}$$

The solution in degrees can then be determined as follows

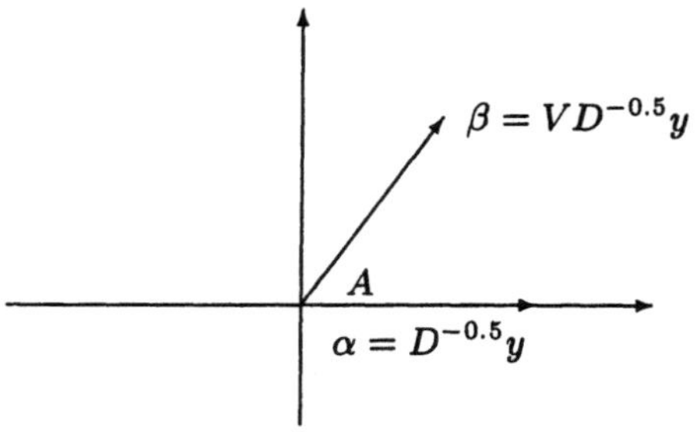

Fig. L.2. *Rotation of ellipse.*

First quadrant:

$$\left.\begin{array}{l}\cos(A) \geq 0 \\ \sin(A) \geq 0\end{array}\right\} \Rightarrow \left.\begin{array}{l}V_{11} \geq 0 \\ V_{21} \geq 0\end{array}\right\} \Rightarrow A = \arccos(V_{11})\frac{180}{\pi} . \tag{L.15}$$

Second quadrant:

$$\left.\begin{array}{l}\cos(A) < 0 \\ \sin(A) \geq 0\end{array}\right\} \Rightarrow \left.\begin{array}{l}V_{11} < 0 \\ V_{21} \geq 0\end{array}\right\} \Rightarrow A = \arccos(V_{11})\frac{180}{\pi} . \tag{L.16}$$

Third quadrant:

$$\left.\begin{array}{l}\cos(A) < 0 \\ \sin(A) < 0\end{array}\right\} \Rightarrow \left.\begin{array}{l}V_{11} < 0 \\ V_{21} < 0\end{array}\right\} \Rightarrow A = 360° - \arccos(V_{11})\frac{180}{\pi} . \tag{L.17}$$

Fourth quadrant:

$$\left.\begin{array}{l}\cos(A) \geq 0 \\ \sin(A) < 0\end{array}\right\} \Rightarrow \left.\begin{array}{l}V_{11} \geq 0 \\ V_{21} < 0\end{array}\right\} \Rightarrow A = 360° - \arccos(V_{11})\frac{180}{\pi} . \tag{L.18}$$

APPENDIX M
DETERMINING OPEN LOOP UNCERTAINTY ELLIPSES

The open loop Nyquist $G(e^{j\omega T_s})K(e^{j\omega T_s})$ with estimated error bounds can be quite useful in assessing the robustness of the closed loop system. Clearly, provided the plant is stable a necessary and sufficient condition for closed loop stability is that none of the open loop uncertainty ellipses include the Nyquist point $(-1,0)$. In this appendix, we will show how the form matrices for the open loop uncertainty ellipses are determined from the plant uncertainty ellipses and the controller.

Clearly, the open loop uncertainty ellipse at frequency ω is centered at the nominal open loop Nyquist $G(e^{j\omega T_s})K(e^{j\omega T_s})$. The ellipse is amplified by $|K(e^{j\omega T_s})|$ and rotated $\arg(K(e^{j\omega T_s}))$ degrees in comparison with the plant uncertainty ellipse at that frequency. We wish to determine the transformation matrix R which must be multiplied onto the form matrix $P_{\tilde{g}}$ for the plant uncertainty ellipse to obtain the form matrix for the open loop ellipse P_{ol}. Let the plant uncertainty ellipse at frequency ω be given by

$$z = x^T P_{\tilde{g}}^{-1} x \, . \tag{M.1}$$

Then let us start by considering the two vectors y and v given in Figure M.1.

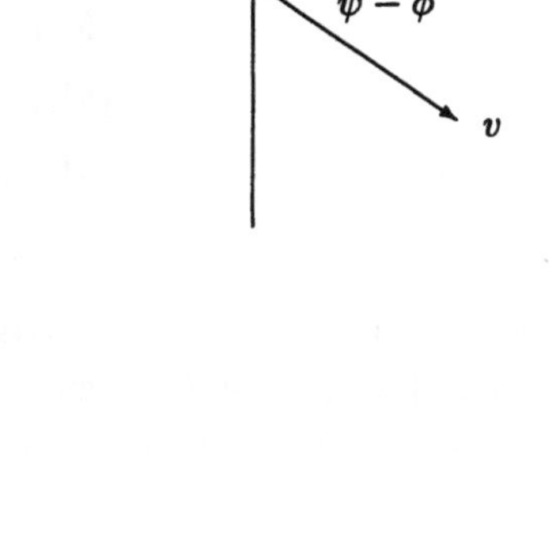

Fig. M.1. *Rotating the vector y ϕ degrees.*

Let $|y| = |v|$ and let y and v be given by

$$y = \begin{bmatrix} y_1 \\ y_2 \end{bmatrix} = \begin{bmatrix} |y|\cos(\psi) \\ |y|\sin(\psi) \end{bmatrix} \tag{M.2}$$

$$v = \begin{bmatrix} v_1 \\ v_2 \end{bmatrix} = \begin{bmatrix} |v|\cos(\psi - \phi) \\ |v|\sin(\psi - \phi) \end{bmatrix} . \tag{M.3}$$

Thus v is the vector y rotated ϕ deg. Let us determine the transformation matrix R which performs this rotation. Thus:

$$v = Ry \tag{M.4}$$

$$\Rightarrow \quad \begin{bmatrix} |v|\cos(\psi - \phi) \\ |v|\sin(\psi - \phi) \end{bmatrix} = R \begin{bmatrix} |y|\cos(\psi) \\ |y|\sin(\psi) \end{bmatrix} \tag{M.5}$$

$$\Rightarrow \quad \begin{bmatrix} \cos(\psi - \phi) \\ \sin(\psi - \phi) \end{bmatrix} = \begin{bmatrix} R_{11} & R_{12} \\ R_{21} & R_{22} \end{bmatrix} \begin{bmatrix} \cos(\psi) \\ \sin(\psi) \end{bmatrix} \tag{M.6}$$

$$\Rightarrow \quad \begin{cases} \cos(\psi - \phi) = R_{11}\cos(\psi) + R_{12}\sin(\psi) \\ \\ \sin(\psi - \phi) = R_{21}\cos(\psi) + R_{22}\sin(\psi) \end{cases} . \tag{M.7}$$

The equations (M.7) are fulfilled for the transformation matrix R given by

$$R = \begin{bmatrix} \cos(-\phi) & -\sin(-\phi) \\ \sin(-\phi) & \cos(-\phi) \end{bmatrix} . \tag{M.8}$$

Note that ϕ in Figure M.1 was a negative rotation. If we use standard sign convention for ϕ, the transformation matrix R becomes

$$R = \begin{bmatrix} \cos(\phi) & -\sin(\phi) \\ \sin(\phi) & \cos(\phi) \end{bmatrix} . \tag{M.9}$$

If $|v| = k|y|$, the result is

$$v = kRy \tag{M.10}$$

with R given by (M.8).

Then let us consider the ellipse equation for $z = 1$:

$$x^T P_{\bar{g}}^{-1} x = 1 \tag{M.11}$$

$$\Rightarrow \quad x^T V D V^{-1} x = 1 \tag{M.12}$$

$$\Rightarrow \quad x^T V D^{\frac{1}{2}} D^{\frac{1}{2}} V^{-1} x = 1 \tag{M.13}$$

$$\Rightarrow \quad \bar{z}^T \bar{z} = 1 . \tag{M.14}$$

Here $P_{\bar{g}} = VDV^{-1}$ is the eigenvalue decomposition of the form matrix and $\bar{z}$ is obviously given by $\bar{z} = D^{\frac{1}{2}} V^{-1} x$. (M.14) is the equation for a circle. Consequently, a point $\bar{z}'$ on a circle can be transformed to a point x' on the ellipse with form matrix $P_{\bar{g}} = VDV^{-1}$ via

$$x' = \left(D^{\frac{1}{2}}V^{-1}\right)^{-1}\bar{z}' = VD^{-\frac{1}{2}}\bar{z}' \,. \tag{M.15}$$

Then a point y' on the open loop ellipse amplified by k and rotated by ϕ is given by the transformation (M.10):

$$y' = kRx' = kRVD^{-\frac{1}{2}}\bar{z}' \tag{M.16}$$

$$\Rightarrow \quad \bar{z}' = \left(kRVD^{-\frac{1}{2}}\right)^{-1}y' = \frac{1}{k}D^{\frac{1}{2}}V^{-1}R^{-1}y' \tag{M.17}$$

and

$$\bar{z}^T\bar{z} = 1 \tag{M.18}$$

$$\Rightarrow \quad \frac{1}{k^2}y^T\left(D^{\frac{1}{2}}V^{-1}R^{-1}\right)^T\left(D^{\frac{1}{2}}V^{-1}R^{-1}\right)y = 1 \tag{M.19}$$

$$\Rightarrow \quad \frac{1}{k^2}y^TR^{-T}VD^{\frac{1}{2}}D^{\frac{1}{2}}V^{-1}R^{-1}y = 1 \tag{M.20}$$

$$\Rightarrow \quad \frac{1}{k^2}y^TR^{-T}P_{\bar{g}}^{-1}R^{-1}y = 1 \tag{M.21}$$

$$\Rightarrow \quad y^TP_{\text{ol}}^{-1}y = 1 \tag{M.22}$$

where P_{ol} is given by

$$P_{\text{ol}} = k^2RP_{\bar{g}}R^T \tag{M.23}$$

which is the desired transformation for $k = |K(e^{j\omega T_s})|$ and $\phi = \arg(K(e^{j\omega T_s}))$.

INDEX

MIX
Papier aus verantwortungsvollen Quellen
Paper from responsible sources
FSC® C105338

If you have any concerns about our products,
you can contact us on
ProductSafety@springernature.com

In case Publisher is established outside the EU,
the EU authorized representative is:
Springer Nature Customer Service Center GmbH
Europaplatz 3, 69115 Heidelberg, Germany

Printed by Libri Plureos GmbH
in Hamburg, Germany